Post-Harvest Pathology of Fruits and Vegetables

Edited by

Colin Dennis

Campden Food Preservation Research
Association, Chipping Campden,
Gloucestershire, U.K.

1983

ACADEMIC PRESS

A Subsidiary of Harcourt Brace Jovanovich, Publishers

London   New York

Paris   San Diego   San Francisco   São Paulo

Sydney   Tokyo   Toronto

ACADEMIC PRESS INC. (LONDON) LTD.
24–28 Oval Road
London NW1 7DX

U.S. Edition published by
ACADEMIC PRESS INC
111 Fifth Avenue
New York, New York 10003

British Library Cataloguing in Publication Data
Post-harvest pathology of fruits and vegetables.
    1. Fruit—Pathology   2. Vegetables—Pathology
    I. Dennis, Colin
    634   SB360

    ISBN 0–12–210680–6 ✓
    LCCCN 83–70586

Text set in 10/12 pt Linotron 202 Times, printed and bound
in Great Britain at The Pitman Press, Bath

Post-Harvest Pathology of Fruits and Vegetables

Contributors

C. Dennis   Campden Food Preservation Research Association, Chipping Campden, Gloucestershire GL55 6LD, U.K.

K. L. Edney   East Malling Research Station, East Malling, Maidstone, Kent ME19 6BJ, U.K.

B. Garrod   School of Biological Sciences, University of East Anglia, Norwich NR4 7TJ, U.K.

J. D. Geeson   Agricultural Research Council, Food Research Institute, Colney Lane, Norwich NR4 7UA, U.K.

B. G. Lewis   School of Biological Sciences, University of East Anglia, Norwich NR4 7TJ, U.K.

C. Logan   Department of Agriculture, Plant Pathology Research Division, New Forge Lane, Belfast BT9 5PX, Northern Ireland.

B. M. Lund   Agricultural Research Council, Food Research Institute, Colney Lane, Norwich NR4 7UA, U.K.

R. B. Maude   National Vegetable Research Station, Wellesbourne, Warwickshire CV35 9EF, U.K.

T. R. Swinburne   East Malling Research Station, Maidstone, Kent ME19 6BJ, U.K.

Preface

Although there are a number of published texts on plant pathology, including those on crop plants, the topic of post-harvest pathology has tended to be neglected in such texts. The increased research activity on diseases during storage and marketing in the last 10 years or so has resulted in much published information in a wide range of journals and conference proceedings. The present book brings together all of the up-to-date information on a range of fruits and vegetables and provides an authoritative, specialized text on post-harvest pathology. All of the chapters contain an extensive list of references, providing the reader with a source of information for further study.

Seven of the nine chapters deal with specific crops or groups of crops and describe the major diseases in terms of symptoms, biology, epidemiology, effect of pre- and post-harvest factors on development and where possible include sections on physiological and biochemical aspects in relation to degradation of tissues and mechanisms of resistance. These chapters, in common with the final chapter on "Bacterial Spoilage", also include sections on control measures currently practised in the light of present information. In this respect, the importance of knowledge of pre-harvest factors and spread of diseases in the field is clearly recognized and emphasized where appropriate.

One of the characteristic features that many post-harvest diseases have in common is that at some stage between arrival of inoculum and the development of progressive diseases, the growth of the pathogen is arrested. Such "quiescent infections" are the subject of the opening chapter, which critically reviews the factors responsible for the dormant phase and highlights the growing evidence that the resistance of immature fruit is attributable to post-infectional changes.

The second chapter on "Soft Fruit" (strawberries, raspberries, blackberries, loganberries, gooseberries) highlights the importance of *Botrytis cinerea* together with the Zygomycetes, *Mucor piriformis*, *Rhizopus sexualis* and *R. stolonifer*, the relative incidence being affected by a number of pre- and post-harvest factors. Similarly, in Chapter 3, which deals with "Top Fruit" (apples and pears), factors affecting development of diseases caused by *Gloeosporium* spp., *Monilia fructigena*, *Phytophthora syringae*,

Botrytis cinerea and *Penicillium expansum* are discussed. Both of these chapters contain sections on physiology which complement several of the points raised in Chapter 1.

Chapter 4 on "Onions" gives special emphasis to the recent increase in knowledge of neck-rot disease (caused by *Botrytis allii*) and consequent improvement in control. Additional sections in this chapter deal with diseases caused by *Aspergillus*, *Penicillium* and bacteria as well as some minor fungal diseases. The following chapter on "Carrots" includes sections on *Botrytis cinerea*, *Mycocentrospora acerina*, *Sclerotinia sclerotiorum*, *Rhizoctonia carotae* and *Stemphylium radicinum* but gives particular emphasis to the first two of these fungi, especially in relation to the chemical basis for resistance and the role of wound healing.

Chapter 6 describes the diseases of "Brassicas" (cabbages, cauliflower, broccoli, Brussels sprouts) and in particular those of winter white cabbage (*Botrytis cinerea*, *Mycosphaerella brassicola*, *Alternaria* spp., *Phytophthora porri*), which reflects our greater knowledge of this crop relative to other brassicas. The next chapter is on "Salad Crops" and discusses diseases of tomatoes, cucumbers, peppers, lettuce and celery. *Botrytis cinerea* (grey mould) is a major cause of post-harvest disease of these crops and in the case of the fruits can cause serious losses, especially if they are subjected to chilling injury. Other sections cover *Alternaria* spp. (tomatoes, peppers), *Didymella bryoniae* (cucumber), *Mycocentrospora acerina* (celery) and *Sclerotinia* spp. (celery, lettuce).

Chapter 8 on "Potatoes" covers all of the major fungal diseases caused by *Phytophthora infestans*, *P. erythroseptica*, *Fusarium* spp., *Phoma* spp., *Pythium ultimum*, *Oospora lactis* and *Polyscytalum pustulans*, but indicates the current importance of gangrene (*Phoma exigua* var. *foveata*) as the dominant storage pathogen in the U.K.

The lack of information on bacterial diseases in the above chapters is due to the fact that the final chapter is solely concerned with such diseases. In particular, the importance of soft-rotting bacteria (erwinias, pseudomonads, bacilli, clostridia) in causing post-harvest spoilage of vegetables is discussed. The properties and epidemiology of each group of bacteria are described together with general accounts of the mechanisms and factors affecting soft-rotting and the control measures currently practised.

March 1983   Colin Dennis

Contents

4 Onions

R. B. MAUDE

5 Carrots

B. G. LEWIS AND B. GARROD

6 Brassicas

J. D. GEESON

1

Quiescent Infections in Post-Harvest Diseases

T. R. SWINBURNE

I. Introduction

Post-harvest diseases of fruit and vegetables represent one of the most severe sources of loss of production. Such losses can be remarkably high (Derbyshire and Shipway, 1978), and their economic cost is proportionally greater than for losses in the field because costs of harvesting, transport and storage must be added to those of production. Growers and merchants would wish to discard produce that will not reach the consumer before decay, but unfortunately the infections that initiate this decay are not usually visible, or predictable, at the time of harvest and overt symptoms become apparent only after ripening, a stage usually equated with edibility in fruit crops. The delay between infection and symptom production in these diseases can be lengthy, for example, over 5 months in anthracnose disease of banana (Simmonds, 1941). The physiological factors underlying these phenomena have interested many pathologists and some of the hypotheses that have been advanced will be explored in this review. Such information is of more than academic interest (Simmonds, 1941) and may provide clues to the methods that should be adopted in attempting to control wastage (Wardlaw *et al.*, 1939).

A. Quiescent or Latent Infections?

One of the most characteristic features that many post-harvest diseases have in common is that at some stage between the arrival of the inoculum and the development of progressive disease, the growth of the pathogen is arrested. Such arrested infections are usually described as "latent" (Verhoeff, 1974). The meaning of the terms "latent infection" and "latency" were abundantly clear to post-harvest pathologists. Thus

Verhoeff's (1974) definition of latency was "a quiescent or dormant parasitic relationship which after a time changes to an active one". However, these terms are now frequently used by others, notably epidemiologists, with quite different meanings. Gäumann's (1950) definition of latency included any "symptomless precursor of the disease proper", and in epidemiology the period of time between infection and overt symptom production is taken to be the "latent period" (Butt and Royale, 1980). Medical scientists, some epidemiologists (Vanderplank, 1963) and virologists (Federation of British Plant Pathologists, 1973) use the term "latency" for the period between infection and infectiousness. Hayward (1974) held that any symptomless growth of bacterial pathogens or saprophytes constituted a latent infection and this was recorded in the same volume as Verhoeff's (1974) more restricted definition.

To avoid further confusion it seems expedient to leave the term "latent infection" for these wider uses and adopt "quiescent infection" for those instances where growth of a pathogen is inhibited for some time. The term "quiescent infection" was originally used by Jenkins and Reinganum (1965) to describe so-called latent infections that could be seen with the naked eye, a somewhat artificial distinction. The term "infection" has also been defined in so many ways that ambiguity has become inevitable (Butt and Royale, 1980). Hirst and Schein (1965) included initiation of germination, germ-tube elongation, appressorium formation, penetration and subsequent colonization as stages in the overall process of infection. A period of quiescence can be enforced in the pathogen at any of these stages, including the initiation of germination, and thus the term "quiescent infection" is taken to include even some examples where the inoculum remains ungerminated, which has previously been referred to as latent contamination (Byrde and Willetts, 1977). In adopting such a definition there is an implication that failure to germinate or to develop beyond any subsequent stage is due to adverse physiological conditions temporarily imposed by the host on the pathogen and not simply to the direct effects of an unfavourable climatic regime.

Thus a quiescent infection in the context of post-harvest diseases involves the inhibition of development of the pathogen through physiological conditions imposed by the host until some stage of maturation has been accomplished.

II. Host Factors Involved in Quiescent Infections

A consideration of the variety of stages at which pathogens become arrested in quiescent infections of fruit crops (Table 1.1) indicates that it is

Table 1.1 *Examples of quiescent infections in fruit crops and the stages at which infection is arrested*

Crop	Pathogen	Stage of Development	Reference
Apple	*Gloeosporium perennans*	Appressoria	Edney (1958)
Apple	*Diaporthe perniciosa*	Colonization	Nawawi and Swinburne (1970)
Apple	*Nectria galligena*	Colonization	Swinburne (1971)
Apricot	*Monilinia fructicola*	Pre-germination	Jerome (1958)
		Stomatal penetration	Wade (1956)
Avocado	*Colletotrichum gloeosporioides*	Appressoria	Binyamini and Schiffmann-Nadel (1972)
Avocado	*Botryosphaeria ribis*	Stomatal penetration	Horne and Palmer (1935)
Banana	*Colletotrichum musae*	Colonization	Simmonds (1941)
		Appressoria	Muirhead and Deverall (1981)
Capsicum	*Colletotrichum capsici*	Colonization	Adikaram (1981)
Capsicum	*Glomerella cingulata*	Appressoria	Adikaram (1981)
Citrus	*Colletotrichum gloeosporioides*	Appressoria	Brown (1975)
Mango	*Colletotrichum gloeosporioides*	Colonization	Baker *et al.* (1940)
Pawpaw	*Colletotrichum gloeosporioides*	Colonization	Baker *et al.* (1940)
Raspberry	*Botrytis cinerea*	Colonization	Powelson (1960)
Strawberry	*Botrytis cinerea*	Colonization	Jarvis (1962)
Tomato	*Colletotrichum phomoides*	Appressoria	Fulton (1948)
		Colonization	Fulton (1948)

extremely unlikely that any one set of factors could provide an explanation to cover all contingencies. Indeed, it is probable that in each host–parasite interaction several factors may be operating simultaneously or sequentially. An example of this, anthracnose of banana, will be discussed below. For convenience, some of the physiological factors underlying the temporary resistance involved in quiescent infections will be considered according to the stages at which the pathogen is arrested.

A. Inhibition of Germination of Fungal Propagules

Many of the pathogens responsible for post-harvest diseases are fungi whose propagules arrive at the surface of the fruit during the growing period. With many, if not most, of these pathogens, the propagules are thin-walled asexual spores which do not usually undergo a fixed constitutive dormancy period. The presence of ungerminated spores on the surfaces of harvested fruit has been recognized as the source of infection for a number of diseases, including brown rot of apricot incited by *Monilinia fructicola* (Jerome, 1958) and apple rots incited by *Penicillium expansum* (Kidd and Beaumont, 1924). Both of these fungi are able to rot their respective hosts before harvest if a suitable entry point, such as a wound, is available (Swinburne, 1970; Byrde and Willetts, 1977). If propagules fail to germinate simply for lack of water, then such inocula constitute "latent contamination", *sensu* Byrde and Willetts (1977). Phillips and Harvey (1975) recovered as many as 170 000 conidia of *Monilinia* from surface washings of individual fruit before harvest. However, Jenkins and Reinganum (1965) have questioned the ability of conidia of *Monilinia* to survive ungerminated on the surface for long periods, especially in the presence of surface moisture. Jerome's (1958) experiments suggest that this is possible, but as the spores germinate freely in water their failure to do so during a period of wetness requires explanation. The possibility that inhibitory substances are present at the surface of unripe fruit does not seem to have been explored directly. Wade and Cruickshank (1973) showed that the addition of fruit extracts to the inoculum droplet resulted in the formation of progressive lesions and presumed this to be due to host inhibitors of fungal pectic enzymes being overcome.

B. Stimulation of Germination and Appressorium Formation

Although evidence of inhibition of germination of fruit surfaces is scant, there are examples where stimulation of germination has been observed (Grover, 1971). Conidia of *Colletotrichum piperatum* germinated much more freely on leachates of red fruit of *Capsicum frutescens* than on green fruit. Even greater differences were found between red and green fruit with respect to appressoria formation. This phenomenon was attributed to the presence of sucrose in leachates of the ripe fruit. This is surprising as sucrose and other simple sugars are usually associated with inhibition of appressorium formation by encouraging vegetative growth of the germ tubes (Mercer *et al.*, 1971). Attempts to repeat Grover's experiments with *C. capsici* and *Glomerella cingulata* gave an entirely opposite result: germination and particularly appressorium formation were greater on unripe than on ripe fruit (Adikaram, 1981).

Conidia of *C. musae* germinate more readily on the surface of banana fruits at any stage of ripeness than in water, and appressoria production is particularly stimulated on the surface of unripe fruit (Swinburne, 1976). Anthranilic acid detected in fruit leachates accounted for almost all of the stimulation observed. It was shown later (Harper and Swinburne, 1979) that conidia rapidly degrade anthranilic acid to 2, 3-dihydroxybenzoic acid (DHBA) and it was inferred that the latter was the stimulant. 2, 3-Dihydroxybenzoic acid is involved in the active transport of iron in a number of micro-organisms and it was shown (Harper *et al.*, 1980) that other synthetic iron-chelating agents also non-specifically stimulated germination and appressorium formation. As conidia produced on media deficient in iron germinated freely without added chelating agents, it was concluded (Harper *et al.*, 1980) that iron was associated with a site of inhibition, possibly in the conidial wall.

Under conditions of iron deficiency, bacteria produce chelating agents with a high affinity for this metal (Neilands, 1974); such compounds are termed siderophores. A siderophore produced by a *Pseudomonas* sp. was found to be highly stimulatory to germination and appressorium formation (McCracken and Swinburne, 1979) and most of the bacteria isolated from banana fruit surfaces produced such siderophores (McCracken and Swinburne, 1980).

Similar stimulation of pycnidiospores of *Diaporthe perniciosa* was observed with leachates of apple fruit (Brown and Swinburne, 1978); the stimulants were identified as *p*-coumaryl quinic, chlorogenic and caffeic acids. The involvement of phenolic acids in the stimulation of infection was perhaps surprising, as such compounds are often associated with resistance (Kosuge, 1969) and chlorogenic acid was directly implicated in resistance of immature apple fruit to *Gloeosporium perennans* (Hulme and Edney, 1960).

These examples suggest that the chemical environment at the surface of immature fruit may be particularly conducive to the rapid germination of fungal propagules and the formation of infection structures, perhaps more so than with mature fruit where the presence of sugars in the leachate delay the formation of appressoria (Swinburne, 1976).

C. Germination of Appressoria

Many of the pathogens associated with quiescent infections take advantage of wounds or natural openings to gain access to the fruit, e.g. *Nectria galligena* (Swinburne, 1971) and *Botrytis cinerea* (Powelson, 1960). Such organisms do not usually penetrate intact fruit and the process of coloniz-ation immediately follows germination. The genera *Gloeosporium* and *Colletotrichum* which figure prominently in any list of post-harvest diseases, particularly of tropical fruit, form characteristic, sometimes deeply pigmented, appressoria (Emmett and Parbery, 1975). In their pri-mary function as structures to aid penetration, appressoria become closely attached to the outer layers of the cuticle, sometimes by means of a mucilaginous secretion (Fig. 1.1). The action of penetration is both mech-anical and enzymic (Emmett and Parbery, 1975), and cutinase activity may aid the partial burying of appressoria in the cuticular layers (Fig. 1.2) (Binyamini and Schiffmann-Nadel, 1972).

There are conflicting reports on the role of appressoria in the dormant phase of quiescent infections. Several workers have contended that the production of penetration tubes and limited colonization of the cuticle and outer layers of the fruit follow shortly after the appressoria are formed in many of these post-harvest diseases. Thus both Simmonds (1941) and Chak-ravarty (1957) found sub-cuticular hyphae of *C. musae* on immature banana fruits shortly after inoculation. Similar observations were made with *C. gloeosporioides* on mango and pawpaw (Baker *et al.*, 1940; Simmonds, 1941). In addition to such direct observations, early penetration has been inferred from experiments with surface sterilants. In one of the earliest demonstrations of quiescent infection, Shear and Wood (1913) showed that immersion of citrus fruits in mercuric chloride some time after inoculation with *C. gloeosporioides* failed to control subsequent colonization. Simmonds (1941) compared the effects of alcohol, formaldehyde and mer-curic chloride treatments on conidia and appressoria of *C. musae* and found that the latter were relatively resistant to such toxicants. He concluded, however, that the failure of surface sterilization with these materials to prevent anthracnose of banana was indicative of early penetration. Muir-head and Deverall (1981) have made a careful histological study of the infection of bananas by *C. musae*, which has dispelled some of the earlier confusion. They were able to distinguish between the development of the dark, thick-walled appressoria and those which remain thin-walled and hyaline (termed proto-appressoria by Emmett and Parbery, 1975). None of the dark appressoria produced infection threads until ripening had com-menced. The hyaline appressoria formed penetration hyphae on unripe fruit, which resulted in necrosis of the underlying epidermal cells, remin-

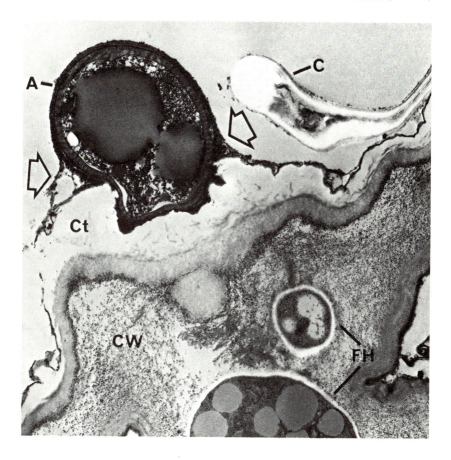

Fig. 1.1 Transmission electron micrograph of *Colletotrichum musae* on ripe banana fruit showing a thick-walled, pigmented appressorium (A) partially embedded in the host cuticle (Ct) and adhering to the surface with mucilage-like deposits (arrowed). The remnants of the conidium (C) can be seen at the surface and hyphae (FH) in the host cell wall (CW). Magnification *c.* 8000. Specimen prepared by T. R. Swinburne, sectioned and photographed by T. W. Fraser, Queen's University of Belfast.

iscent of a hypersensitive reaction. These hyphae failed to develop further and progressive anthracnose lesions followed colonization of ripened fruit by hyphae from the dark appressoria. The resistance of unripe fruit to early infections will be discussed in a subsequent section, but evidently the quiescent infection survives as ungerminated appressoria. Recent experiments with iron-chelating agents support these conclusions (Swinburne and Brown, 1982). As well as increasing the number of appressoria, DHBA dramatically increases the proportion of dark compared with hyaline types. Addition of DHBA to the inoculum drop on unripe fruit reduces or eliminates early necrosis as very few penetrations occur at this stage. When allowed to ripen, fruit inoculated with conidia in DHBA develop anthracnose lesions much earlier than those given comparable inoculations made in water (Brown and Swinburne, 1981). Evidently, as so few of the conidia in DHBA gave rise to early penetration, the resistance mechanisms of the fruit were not triggered.

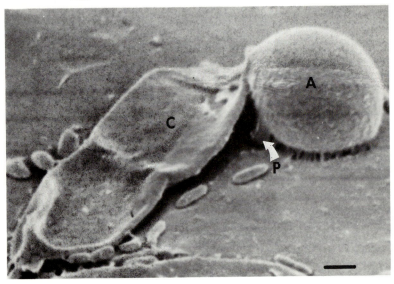

Fig. 1.2 Scanning electron micrograph of *Glomerella cingulata* on the surface of a *Capsicum annuum* fruit. The conidium (C) has produced an appressorium (A) which has been pushed away from the surface of the cuticle by the penetration tube (P) revealing the zone of adherence. Magnification *c.* 8000. Specimen prepared and photographed by N. Adikaram, Queen's University of Belfast.

Infections of hard avocado fruit by *C. gloeosporioides* remain dormant as appressoria and sub-cuticular hyphae could not be detected until the fruit softened with ripening (Binyamini and Schiffmann-Nadel, 1972). Tangerines washed before degreening with ethylene showed much less infection with *C. gloeosporioides* than those washed after degreening (Brown, 1975). It was concluded that factors associated with ripening (following degreening) stimulated appressorial germination and penetration which placed the infection beyond reach of the washing process.

Dormant appressoria are obviously resistant to externally applied toxicants and this coupled with their submersion in the cuticular waxes

(Binyamini and Schiffmann-Nadel, 1972) or location in lenticels (Edney, 1956) made control somewhat difficult until the introduction of such systemic fungicides as the benzimidazoles for post-harvest dips (Eckert, 1977).

There is little or no information available on the mechanisms governing appressorial dormancy. The application of sugars and other nutrients usually results in regrowth from appressoria (e.g. Muirhead and Deverall, 1981). The levels of soluble carbohydrates usually increase on ripening and these may leak through the cuticle to the pathogen. Ethylene may also stimulate appressorial germination but, as Brown (1975) pointed out, it is not possible to distinguish direct effects of such treatments on the pathogen from indirect effects on the host, although *in vitro* experiments may be helpful in this respect. Further information on factors influencing germination of appressoria may be very useful in developing control measures, especially where insensitivity to fungicides has become common.

D. Colonization

Most attention on the mechanisms governing quiescent infection has been focused on events following penetration, which marked the onset of a "latent infection" in earlier definitions (Verhoeff, 1974). For wound pathogens involved in quiescent infections these considerations are obviously important (e.g. *N. galligena*) and even some that produce appressoria become troublesome diseases when fruit is damaged by wounding and early contact is established between the internal tissues of the host and the pathogen (e.g. anthracnose of banana; Simmonds, 1963).

The resistance of immature fruit tissue to colonization by fungal pathogens might reasonably be expected to be related to the resistance mechanisms of the plant as a whole, except that in post-harvest diseases the pathogen survives this initial response. Thus those mechanisms which have been postulated for the resistance of plants to necrotrophic fungi in general find their counterpart in the literature of post-harvest diseases. Simmonds (1963) eloquently summarized these into four categories which Verhoeff (1974) condensed to three in the most recent review of the subject. These were:

(1) Toxic compounds that inhibit the pathogen are present in unripe but not ripe fruit.
(2) Unripe fruit does not provide a suitable substrate to fulfil the nutritional and energy requirements of the pathogen.
(3) The enzyme "potential" of the fungus is inadequate to colonize unripe fruit.

To these must be added a fourth hypothesis: the production of phytoalexins in unripe fruit following infection (Swinburne, 1978), for which Simmonds (1963) found no evidence.

1. *Preformed Fungitoxic Compounds*

The concept that preformed inhibitors present in unripe fruit were responsible for retarding the growth of pathogens has attracted many workers (Simmonds, 1963; Verhoeff, 1974). Following the discovery of the fungitoxicity of phenolic compounds and their derivatives, the tannins (Cook and Taubenhaus, 1911) have received most attention. Such compounds represent likely candidates for this role as they are frequently found in much higher concentrations in immature than in ripe fruit (van Buren, 1970). The weakness of the argument stems from the difficulties of relating toxicity to their *in vivo* concentration. For example, chlorogenic acid present in the peel of unripe apple fruit has been implicated in their resistance to *Gloeosporium perennans* (Hulme and Edney, 1960), but the concentrations required for inhibition exceed those normally found (Schultz, 1978) and changes in concentration do not coincide with changes in susceptibility (Noble, 1981). Similarly, the alleged involvement of phenolic or tannin-like inhibitors of *C. musae* in banana peel (Chakravarty, 1957) could not be substantiated in Simmonds' (1963) experiments. The latter author did extract fungitoxic principle from the peel of unripe banana fruit with hot but not cold water. However, it is not possible to relate this finding to the mechanism governing inhibition, as this may be an artefact of the extraction itself. The observation that 3, 4-dihydroxybenzaldehyde was present in fungitoxic concentrations in unripe banana fruit (Mulvena *et al.*, 1969) has not been substantiated by other workers. Infusion of apple fruit with solutions. containing tannin failed to improve resistance to rotting by *Botryosphaeria ribis* (Sitterly and Shay, 1960). Tannins are also involved in the inactivation of extracellular hydrolytic enzymes; this will be considered in a later section.

The involvement of preformed inhibitors in quiescent infections of tomato by *B. cinerea* is convincing (Verhoeff and Liem, 1975). The glyco-alkaloid tomatine, present in high concentrations in the peel of green tomatoes, is inhibitory to mycelial growth of the pathogen at those concentrations. However, although the fungus can be re-isolated from fruit inoculated when green (Verhoeff, 1970), no further development of the lesions occurs after ripening, even though the tomatine cannot be detected. Thus *B. cinerea* in tomato represents an exceptional form of quiescent infection. *Colletotrichum phomoides* establishes a more conventional form of quiescent infection than *B. cinerea* on the same host (Allison, 1952). Following the formation of appressoria, penetration of the green fruit occurs, but further development of the invading hyphae is prevented, allegedly by fungitoxic concentrations of solanine.

2. Nutritional Requirement of the Pathogen

During ripening, fruit undergo a wide range of profound biochemical changes (Hulme, 1970). The hypothesis that these changes result in the production of nutrients vital to the requirements of the pathogen has been advanced by many authors.

One of the most obvious of the changes that occur during fruit ripening is the conversion of insoluble carbohydrate to soluble sugars. In a number of fungal diseases of plants the general resistance to colonization has been related to the sugar content of the tissues (Horsfall and Dimond, 1957). Diseases in which susceptibility was favoured by high sugar levels were referred to as high-sugar diseases and the converse as low-sugar diseases. If increasing sugar levels accounted for the onset of susceptibility to rotting in fruit, then such post-harvest diseases would be high-sugar diseases. The levels of sugar in plant tissue can be manipulated with 2, 4-dichlorophen-oxyacetic acid, 2, 4-dinitrophenol and maleic hydrazide (Horsfall and Dimond, 1957) and these compounds have been used in experiments with fruit (Sitterly and Shay, 1960; Simmonds, 1963). Each of them accelerated the onset of rotting of apple or banana fruits, but the same compounds also hastened the onset of the climacteric. Consequently, the effects of such treatments cannot be ascribed to sugar levels alone. Infusion of apple fruit with sucrose or fructose through the petiole accelerated the onset of suscep-tibility of apple fruit to *Botryosphaeria ribis* but, as Sitterly and Shay (1960) pointed out, this could be attributed to a reduction in the toxicity of anti-fungal compounds in the presence of sugars.

The weakness of the nutrition hypothesis as a single explanation for quiescent infection is exposed in experiments that have shown that the fungal pathogens responsible for rotting of ripe fruit can be cultured satis-factorily on the nutrients extracted from unripe fruit (Simmonds, 1963; Bompeix, 1966).

3. The "Enzyme Potential" of the Pathogen

One of the most common symptoms in post-harvest disease is the disor-ganization of host tissue brought about by enzymes, secreted by the path-ogen, which attack the host walls. Of these enzymes, most attention has been given to those degrading pectic substances (Bateman and Basham, 1976), but other enzymes are also involved (Wallace *et al.*, 1962a). Accord-ing to Simmonds' (1963) hypothesis, quiescent infections may be attribut-able to the failure of the pathogen to produce adequate levels of such enzymes until ripening leads to suitable changes in the cell wall structure.

Theoretically, an inadequate enzyme potential could arise in any of three ways:

(1) Such enzymes are not all constitutive (Bateman and Basham, 1976); thus it is possible that immature fruits lack the necessary inducing substrates.
(2) If the enzymes are produced, access to the labile sites within the wall may be blocked by cross-linking.
(3) Enzymes may be inactivated by inhibitors present in higher quantities in immature fruit.

Polygalacturonase was produced by *Glomerella cingulata*, *Botryosphaeria ribis* and *Physalospora obtusa* most rapidly on a medium containing the water-insoluble residues of immature (resistant) apple fruit than with corresponding extracts of ripe (susceptible) fruit (Wallace *et al.*, 1962a). Similar results were obtained with *C. capsici* and *G. cingulata* grown on water-insoluble components of acetone powders prepared from unripe or ripe fruit of *Capsicum annuum* and *Capsicum frutescens* (Adikaram, 1981). The addition of pectic enzymes to the inoculum of *C. musae* accelerated anthracnose lesion development on banana (Simmonds, 1963; Shillingford and Sinclair, 1980), but Simmonds' careful experiments indicated that this was probably attributable to the nutritional value of the enzyme preparations to the pathogen rather than to their activity. The fact that increasing sugar levels normally repress pectic enzyme production by micro-organisms does not seem to have been considered in relation to the onset of rotting in ripening fruit. However, there is evidence to suggest that inducing substrates are more readily available in the cell wall material of the ripe than of the unripe fruit.

It is not clear whether the changes in cell wall structure that make inducers more readily available are the same or merely coincidental with the changes that increase the accessibility of the bonds to enzymic degradation. The pectic substances in immature apple fruits were found to be more extensively cross-linked than those in mature fruits, and the onset of susceptibility has been associated with the exchange of polyvalent for monovalent cations (Wallace *et al.*, 1962b). Treatment of apples with calcium delays rotting in fruit by *Gloeosporium perennans* (Sharples and Little, 1970) and losses were shown to occur earlier in orchards producing fruit low in this element (Edney and Perring, 1973). The low levels of pectin methylesterase production by *Glomerella cingulata* were thought to be a limiting factor in the invasion of immature apples by this organism (Wallace *et al.*, 1962a) and it was suggested that demethylation during normal ripening exposed the pectin chains to the cleavage enzymes that were produced.

As cell walls are apparently composed of carbohydrate–cation–protein complexes (Keegstra *et al.*, 1973), proteases may be equally as important as pectic enzymes in rotting but have received much less attention (Kúc and

Williams, 1962; Porter, 1966). When *C. capsici* and *G. cingulata* were cultured on water-insoluble residues of cell walls of unripe *Capsicum* fruit, much higher protease activities were detected than in corresponding cultures from ripe fruit (Adikaram, 1981). Although this suggests that protease production may be a factor determining the ability to attack unripe fruit, Swinburne (1975a) found that those fungi which rotted unripe apple fruit lacked a protease but those involved in quiescent infections produced such enzymes both *in vitro* and *in vivo*. The involvement of proteases in the rotting process and as a factor in quiescent infections warrants further study.

One of the difficulties encountered in work on cell wall degrading enzymes stems from their inactivation in the host tissues, which necessitates making assumptions about their activities from the observed degradation of wall tissue (Cole and Wood, 1961). Polyphenols and tannins are responsible for much of this inactivation (Byrde *et al., 1960). In one of the earliest papers on quiescent infections, Cook and Taubenhaus (1911) implicated tannins in the resistance of immature banana fruits and support for this has been obtained more recently (Green and Morales, 1967). In dismissing this hypothesis, Simmonds (1963) pointed out that such substrates were localized within the latex vessels and therefore not generally available. Although Byrde (1957) was able to demonstrate increased susceptibility of some cider varieties of apple to rotting by Sclerotinia fructigena when polyphenoloxidase activity was blocked by inhibitors,* the involvement of phenolic compounds or their oxidation products in quiescent infections is by no means certain.

Another class of inhibitors of cell wall degrading enzymes has been demonstrated recently, namely proteins (Albersheim and Anderson, 1971), and their presence in plum and peach fruits (Fielding, 1981) and in *Capsicum* fruit (Brown and Adikaram, 1983) has been confirmed. Such inhibitors show little specificity in their action against pectic enzymes produced by some pathogens and non-pathogens (Anderson and Albersheim, 1972; Fielding, 1981). Pectinase activity produced either *in vitro* or *in vivo* by *B. cinerea* was much less inhibited by cell proteins from *Capsicum* fruit than were corresponding enzymes from *G. cingulata* (Brown and Adikaram, 1983). *Botrytis cinerea* can rot an immature *Capsicum* fruit but *G. cingulata* can only attack the ripened fruit. These differential reponses suggest that such inhibitors might play some role in quiescent infections and warrant further investigation.

4. *Phytoalexin Production*

The accumulation of anti-fungal compounds or phytoalexins in host tissues following infection or challenge by fungi, bacteria or viruses has been extensively documented (Bailey and Mansfield, 1982). Whereas the causal link between phytoalexins and resistance of plant tissues generally is the subject of a continuing debate, their involvement in quiescent infections is less contentious (Wheeler, 1975). Phytoalexins have been demonstrated with a number of fruit crops (Swinburne, 1978).

Nectria galligena invades wounds and lenticels of apple fruits of the cultivar Bramley's Seedling before harvest (Swinburne, 1975b), but fruit rotting does not become severe until after harvest. Limited colonization takes place following the initial invasion and the presence of phytoalexin activity in the necrotic tissue formed is readily demonstrated (Swinburne, 1971) and was found to be related to the synthesis of benzoic acid (Brown and Swinburne, 1971). Benzoic acid was found to be toxic only as the undissociated molecule, and the slight increases in pH of cell sap observed in ripening lead to a marked loss of fungitoxic activity (Brown and Swinburne, 1973a). This, in conjunction with increasing sugar levels as postulated by Sitterly and Shay (1960), enables the pathogen to degrade benzoic acid, ultimately to CO_2 (Brown and Swinburne, 1973b), and resume active growth. The elicitor of benzoic acid synthesis was found to be protease produced by the pathogen (Swinburne, 1975a). As a number of proteases from several sources evoked the same response, the elicitor is clearly non-specific. Two other apple pathogens involved in quiescent infections, *Diaporthe perniciosa* and *Gloeosporium perennans*, were also found to secrete proteases, *in vivo*, whereas with *Penicillium expansum*, *B. cinerea*, *Phytophthora cactorum*, *S. fructigena* and *Aspergillus niger* none was found in rotted tissues and these organisms can rot immature fruit without provoking benzoic acid accumulation (Swinburne, 1975a). It was suggested that this mechanism may be the basis of quiescent infections generally in this host. The results obtained for *N. galligena* and *G. perennans* in the cultivar Bramley's Seedling have been confirmed and extended to the dessert cultivar, Cox's Orange Pippin (Noble, 1981).

The resistance of unripe banana to anthracnose incited by *C. musae*, as mentioned above, has been ascribed to several factors. Simmonds (1963) was unable to demonstrate the production of phytoalexins in response to infection but these have been found recently using alternative methods of extraction (Brown and Swinburne, 1980). When conidia were placed on the surface of unripe fruit, necrotic spots developed beneath the inoculum droplets. Solvent extracts of this tissue when chromatographed on thin layer chromatography plates were shown to contain five fungitoxic compounds that were not present in healthy tissue. As the fruit ripened these compounds diminished and at the time when the lesions were expanding none could be detected, but until these compounds have been identified it will not be possible to determine the physiological basis for their toxicity nor the methods of degradation. Experiments with elicitors have indicated that the pathogen is responsible for this degradation. A glucan-like fraction of the walls of hyphae and conidia of *C. musae* elicited both necrosis and the accumulation of the two major phytoalexins found with live inoculum. However, these compounds remained, undiminished, throughout ripening in the absence of the fungus.

Phytoalexins have also been implicated in another anthracnose disease, that of peppers, incited by *Glomerella cingulata* and *C. capsici* (Adikaram, 1981; Adikaram *et al.*, 1982a). Stoessl *et al.* (1972) have shown that when inocula of non-pathogens of *Capsicum* were inserted into the cavity of the ripening fruit, phytoalexins, notably capsidiol, accumulated in the infection drop. When unripe *Capsicum* fruits were inoculated with *G. cingulata*, a phytoalexin was also readily demonstrated in tissue extracts, but this proved not to be capsidiol but another, possibly related compound, tentatively named capsicannol (Adikaram *et al.*, 1982c), which, unlike capsidiol, is only slightly water soluble. In ripening fruit both capsicannol and capsidiol accumulated, but at the stage when lesion expansion was beginning both compounds were absent. It is known that capsidiol can be oxidized to capsenone and further metabolized to less toxic compounds by a number of fungi (Stoessl *et al.*, 1977). The fate of capsicannol in ripening fruit is not yet understood but, unlike the phytoalexins in banana anthracnose, it diminishes in concentration in the absence of the pathogen, albeit more slowly, than when live inoculum is present.

An interesting feature of the investigations made with *G. cingulata* and *Capsicum* fruit became apparent when inocula depleted in iron were used (Adikaram *et al.*, 1982b). Lesions developed on unripe fruit with such inocula or with iron-replete inocula in the presence of low levels of iron-chelating agent (EDTA, 10^{-3} M). Only small quantities of capsicannol accumulated following such inoculations and evidently these were too small to check the development of the fungus.

The accumulation of a phytoalexin-like compound, 3-methyl-6-methoxy-8-hydroxy-3, 4-dihydroisocoumarin, has also been implicated in the resistance of carrots to rotting by *B. cinerea* (Goodliffe and Heale, 1977), and further consideration to phytoalexins in carrots is given by Lewis and Garrod in Chapter 5. In further observations on the resistance of tomato fruit to the same pathogen, Glazener (1982) found that phenolic compounds accumulated after infection, and that this was accompanied by an increase in phenylalanine ammonia lyase activity. These phenolic compounds were not directly toxic to the pathogen but were incorporated in lignin-like materials which may serve to restrict mycelium to a few epidermal cells.

Thus there is growing evidence that in many quiescent infections the resistance of the immature fruit is attributable to post-infectional changes.

III. Summary and Conclusions

The variety of the hypotheses that have been put forward over the years to account for quiescent infections in physiological terms suggests that it would be risky to promulgate a unifying theory. The diversity of hosts and pathogens concerned in such infections also mitigates against such a theory having general applicability. Nevertheless, it is possible that there are elements in common to those diseases which have been explored to date and that it would be profitable to seek these same elements in other post-harvest diseases.

The most satisfactory explanations of the temporary resistance expressed by some immature fruits to colonization by fungal pathogens involve post-infectional changes rather than pre-infection characteristics. These post-infectional changes may be lignin synthesis and associated cell wall alterations or, more likely, the synthesis of fungitoxic substances, phytoalexins.

To suggest that phytoalexin synthesis may account for the observed resistance does not entirely dismiss the importance of factors present before infection. The acidity of the cell sap or the levels of soluble sugars present has been shown to be an important factor in the toxicity of one phytoalexin, namely benzoic acid, in apple fruit. So although the nutrition hypothesis cannot be substantiated by experiments in which the fungus is grown on media prepared from immature fruit, such compounds may act indirectly. Similarly, claims that changes in resistance can be correlated with changes in concentrations of preformed inhibitors cannot be substantiated by reference to their *in vitro* toxicity. However, the concentrations of these compounds in fruit at sequential stages of development may reflect a synthetic capability for aromatic compounds generally, including phytoalexins.

There is evidence to suggest that changes in host cell walls during maturation increase their ability to induce the production of wall-degrading enzymes by the pathogen. That this is not a determinant of the onset of susceptibility is shown by experiments with *Capsicum* fruits infected by *G. cingulata*. Inoculation of immature fruit with iron-deficient conidia leads to immediate lesion development without phytoalexin accumulation, at a stage when the cell walls are relatively inefficient inducers of pectic enzymes. Enzyme induction is likely to be important in the speed of lesion expansion rather than in lesion development itself.

Survival of the pathogen during the period that the host remains resistant is achieved by one of two strategies. A wound pathogen, such as *N. galligena* in apple, can evidently survive exposure to the phytoalexin and resumes growth when changes in ancillary factors, such as acidity, reduce the inhibitory effect. In the second strategy the pathogen avoids exposure to phytoalexins by remaining dormant as appressoria. Those hyphae of *C. musae* which penetrate unripe banana fruit die, and only those hyphae which emerge from appressoria after fruit ripening has begun can colonize the tissues. Factors governing the dormancy of appressoria do not necessarily relate to phytoalexin accumulation and may involve the release of nutritional substances from the senescing fruit. It may be possible to break the dormancy of appressoria and control the disease if the requisite substances could be applied exogenously without accelerating fruit ripening.

References

Adikaram, N. K. B. (1981). A study of the latent infection of fruit of *Capsicum* spp. by *Colletotrichum capsici* and *Glomerella cingulata*. Ph.D. Thesis, Queen's University of Belfast.

Adikaram, N. K. B., Brown, A. E. and Swinburne, T R. (1982a). Phytoalexin involvement in the latent infection of *Capsicum annuum* L. fruit by *Glomerella cingulata* (Stonem.). *Physiol. Pl. Path.* **21**, 161–170.

Adikaram, N. K. B., Brown, A. E. and Swinburne, T. R. (1982b). Rotting of immature *Capsicum frutescens* L. fruit by iron depleted *Glomerella cingulata* (Stonem.). *Physiol. Pl. Path.* **21**, 171–177.

Adikaram, N. K. B., Grimshaw, J. T., Grimshaw, J., Blake, P. and Austin, D. (1982c). Personal communication.

Albersheim, P. and Anderson, A. (1971). Host–pathogen interactions. III. Proteins from plant cell walls inhibit polygalacturonases secreted by plant pathogens. *Proc. natn. Acad. Sci. U.S.A.* **68**, 1815–1819.

Allison, P. V. (1952). Relation of solanine content of tomato fruits to colonization by *Colletotrichum phomoides*. *Phytopathology* **42**, 1.

Anderson, A. J. and Albersheim, P. (1972) Host–pathogen interactions. V. Comparison of the abilities of proteins isolated from three varieties of *Phaseolus vulgaris* to inhibit the endopolygalacturonases secreted by three races of *Colletotrichum lindemuthianum*. *Physiol. Pl. Path.* **2**, 339–346.

Bailey, J. A. and Mansfield, J. W. (1982). "Phytoalexins." Blackie, London.

Baker, R. E. D., Crowdy, S. H. and McKee, R. K. (1940). A review of latent infections caused by *Colletotrichum gloeosporioides* and allied fungi. *Trop. Agric.* **17**, 128–132.

Bateman, D. F. and Basham, H. G. (1976). Degradation of plant cell walls and membranes by microbial enzymes. *In* "Encyclopedia of Plant Physiology" (R. Heitefuss and P. H. Williams, Eds), 316–355. Springer-Verlag, Berlin.

Binyamini, N. and Schiffmann-Nadel, M. (1972). Latent infection in avocado fruit due to *Colletotrichum gloeosporioides*. *Phytopathology* **62**, 592–594.

Bompeix, G. (1966). Contribution à l'étude de la maladie des taches lenticellaires des pommes "Golden Delicious" en France. *Mem. Fac. Sci. Rennes. Univ., Brest*, 121 pp.

Brown, A. E. and Adikaram, N. K. B. (1983). The differential inhibition of pectic enzymes from *Glomerella cingulata* and *Botrytis cinerea* by cell wall protein from *Capsicum annuum* fruit. *Phytopath. Z.*, **106**, 27–38.

Brown, A. E. and Swinburne, T. R. (1971). Benzoic acid: an anti-fungal compound formed in Bramley's Seedling apple fruits following infection by *Nectria galligena* Bres. *Physiol. Pl. Path.* **1**, 469–475.

Brown, A. E. and Swinburne, T. R. (1973a). Factors affecting the accumulation of benzoic acid in Bramley's Seedling apples infected with *Nectria galligena*. *Physiol. Pl. Path.* **3**, 91–99.

Brown, A. E. and Swinburne, T. R. (1973b). Degradation of benzoic acid by *Nectria galligena* Bres. *in vitro* and *in vivo*. *Physiol. Pl. Path.* **3**, 453–459.

Brown, A. E. and Swinburne, T. R. (1978). Stimulants of germination and appressorial formation by *Diaporthe perniciosa* in apple leachate. *Trans. Br. mycol. Soc.* **71**, 405–411.

Brown, A. E. and Swinburne, T. R. (1980). The resistance of immature banana fruits to anthracnose (*Colletotrichum musae* (Berk. & Curt.) Arx.). *Phytopath. Z.* **99**, 70–80.

Brown, A. E. and Swinburne, T. R. (1981). Influence of iron and iron chelators on formation of progressive lesions by *Colletotrichum musae* on banana fruits. *Trans. Br. mycol. Soc.* **77**, 119–124.

Brown, G. E. (1975). Factors affecting post-harvest development of *Colletotrichum gloeosporioides* in citrus fruit. *Phytopathology* **65**, 404–409.

Butt, D. J. and Royale, D. J. (1980). The importance of terms and definitions for a conceptually unified epidemiology. *In* "Comparative Epidemiology" (J. Palti and J. Kranz, Eds), 29–45. PUDOC, Wageningen.

Byrde, R. J. W. (1957). The varietal resistance of fruit to brown rot. II. The nature of resistance in some varieties of cider apple. *J. hort. Sci.* **32**, 227–238.

Byrde, R. J. W. and Willetts, H. J. (1977). "The Brown Rot Fungi of Fruit, Their Biology and Control." Pergamon Press, Oxford.

Byrde, R. J. W., Fielding, A. H. and Williams, A. H. (1960). The role of oxidased polyphenols in the varietal resistance of apples to brown rot. *In* "Phenolics in Plants in Health and Disease" (J. B. Pridham, Ed.), 95–99. Pergamon Press, Oxford.

Chakravarty, T. (1957). Anthracnose of banana (*Gloeosporium musarum* Cke. et Massee), with special reference to latent infection in storage. *Trans. Br. mycol. Soc.* **40**, 337–345.

Cole, M. and Wood, R. K. S. (1961). Types of rot, rate of rotting, and analysis of pectic substances in apples rotted by fungi. *Ann. Bot.* **25**, 417–434.

Cook, M. T. and Taubenhaus, J. J. (1911). The relation of parasitic fungi to the contents of the cells of the host. I. Toxicity of tannin. *Delaware agric. Exp. Stn Bull.* **91**, 77 pp.

Derbyshire, D. M. and Shipway, M. R. (1978). Control of post-harvest deterioration in vegetables in the U.K. *Outl. Agric.* **9**, 246–252.

Eckert, J. W. (1977). Control of post-harvest diseases. *In* "Antifungal Compounds" (M. R. Siegel and H. D. Sisler, Eds), Vol. 1, 269–352. Marcel Dekker. New York.

Edney, K. L. (1956). The rotting of apples by *Gloeosporium perennans* Zeller and Childs. *Ann. appl. Biol.* **44**, 113–128.

Edney, K. L. (1958). Observations on the infection of Cox's Orange Pippin apples by *Gloeosporium perennans* Zeller & Childs. *Ann. appl. Biol.* **46**, 622–629.

Edney, K. L. and Perring, M. A. (1973). Study of farms with high incidence of *Gloeosporium* infection. *Rep. E. Malling Res. Stn 1972*, 157.

Emmett, R. W. and Parbery, D. G. (1975). Appressoria. *A. Rev. Phytopath.* **13,** 147–167.

Federation of British Plant Pathologists (1973). "A Guide to the Use of Terms in Plant Pathology", Phytopathological Papers, No. 17. Commonwealth Mycological Institute, Kew.

Fielding, A. H. (1981). Natural inhibitors of fungal polygalacturonases in infected fruit tissues. *J. gen. Microbiol.* **123,** 337–381.

Fulton, J. P. (1948). Infection of tomato fruits by *Colletotrichum phomoides*. *Phytopathology* **38,** 235–246.

Gäumann, E. (1950). "Principles of Plant Infection." Crosby, Lockwood, London.

Glazener, J. A. (1982). Accumulation of phenolic compounds in cells and formation of lignin-like polymers in cell walls of young tomato fruits after inoculation with *Botrytis cinerea*. *Physiol. Pl. Path.* **20,** 11–25.

Goodliffe, J. P. and Heale, J. B. (1977). Factors affecting the resistance of stored carrot roots to *Botrytis cinerea*. *Ann. appl. Biol.* **85,** 163.

Green, G. L. and Morales, C. (1967). Tannins as the cause of latency in anthracnose infections of tropical fruits. *Turrialba* **17,** 447–449.

Grover, R. K. (1971). Participation of host exudate chemicals in appressorium formation of *Colletotrichum piperatum*. *In* "Ecology of Leaf Surface Microorganisms" (T. F. Preece and C. H. Dickinson, Eds), 509–518. Academic Press, London.

Harper, D. B. and Swinburne, T. R. (1979). 2, 3-Dihydroxybenzoic acid and related compounds as stimulants of germination of conidia of *Colletotrichum musae* (Berk. & Curt) Arx. *Physiol. Pl. Path.* **14,** 363–370.

Harper, D. B., Swinburne, T. R., Moore, S. K., Brown, A. E. and Graham, H. (1980). A role for iron in germination of conidia of *Colletotrichum musae. J. gen. Microbiol.* **121,** 169–174.

Hayward, A. C. (1974). Latent infections by bacteria. *A. Rev. Phytopath.* **12,** 87–97.

Hirst, J. M. and Schein, R. D. (1965). Terminology of infection processes. *Phytopathology* **55,** 1157.

Horne, W. T. and Palmer, D. F. (1935). The control of *Dothiorella* rot of avocado fruits. *Univ. Calif. Coll. Agric., agric. Exp. Stn Bull.* 594.

Horsfall, J. G. and Dimond, A. E. (1957). Interactions of tissue sugar, growth substances, and disease susceptibility. *Z. PflKrankh. PflPath. PflSchutz* **64,** 415–419.

Hulme, A. C. (1970). "The Biochemistry of Fruits and their Products", Vol. 1. Academic Press, London.

Hulme, A. C. and Edney, K. L. (1960). Phenolic substances in the peel of Cox's Orange Pippin apples with reference to infection by *G. perennans. In* "Phenolics in Plants in Health and Disease" (J. B. Pridham, Ed.), 87–94. Pergamon Press, London.

Jarvis, W. R. (1962). The infection of strawberry and raspberry fruits by *Botrytis cinerea* Fr. *Ann. appl. Biol.* **50,** 569–575.

Jenkins, P. T. and Reinganum, C. (1965). Occurrence of a quiescent infection of stone fruits caused by *Sclerotinia fructicola* (Wint.) Rehm. *Aust. J. agric. Res.* **16,** 131–140.

Jerome, S. M. R. (1958). Brown rot of stone fruits: latent contamination in relation to spread of the disease. *J. Aust. Inst. agric. Sci.* **24,** 132–140.

Keegstra, K., Talmadge, K. W., Bauer, W. D. and Albersheim, P. (1973). The structure of plant cell walls. III. A model of the walls of suspension-cultured sycamore cells based on the interconnections of the macromolecular components. *Pl. Physiol.* **51**, 188–197.

Kidd, M. N. and Beaumont, B. A. (1924). Apple rot fungi in storage. *Trans. Br. mycol. Soc.* **10**, 98–118.

Kosuge, T. (1969). The role of phenolics in host response to infection. *A. Rev. Phytopath.* **7**, 195–222.

Kúc, J. and Williams, E. B. (1962). Production of proteolytic enzymes by four pathogens of apple fruit. *Phytopathology* **52**, 739.

McCracken, A. R. and Swinburne, T. R. (1979). Siderophores produced by saprophytic bacteria as stimulants of germination of conidia of *Colletotrichum musae. Physiol. Pl. Path.* **15**, 331–340.

McCracken, A. R. and Swinburne, T. R. (1980). Effect of bacteria isolated from surfaces of banana fruits on germination of *Colletotrichum musae* conidia. *Trans. Br. mycol. Soc.* **74**, 212–213.

Mercer, P. C., Wood, R. K. S. and Greenwood, A. D. (1971). Initial infection of *Phaseolus vulgaris* by *Colletotrichum lindemuthianum. In* "Ecology of Leaf Surface Micro-organisms" (T. F. Preece and C. H. Dickinson, Eds), 381–389. Academic Press, London.

Muirhead, I. F. and Deverall, B. J. (1981). Role of appressoria in latent infection of banana fruits by *Colletotrichum musae. Physiol. Pl. Path.* **19**, 77–84.

Mulvena, D., Webb, E. C. and Zarner, B. (1969). 3, 4-Dihyroxybenzaldehyde, a fungistatic substance from green Cavendish bananas. *Phytochemistry* **8**, 393–395.

Nawawi, A. and Swinburne, T. R. (1970). Observations on the infection and rotting of apples var. Bramley's Seedling by *Diaporthe perniciosa. Ann. appl. Biol.* **66**, 245–255.

Neilands, J. B. (1974). "Microbial Iron Metabolism: A Comprehensive Treatise." Academic Press, New York.

Noble, J. P. (1981). The biochemical basis of latency in bitter-rot of apple. Ph.D. Thesis, University of Birmingham.

Phillips, D. J. and Harvey, J. M. (1975). Selective medium for detection of inoculum of *Monilinia* spp. on stone fruits. *Phytopathology* **65**, 1233–1236.

Porter, F. M. (1966). Protease activity in diseased fruits. *Phytopathology* **56**, 1424–1425.

Powelson, R. L. (1960). Initiation of strawberry fruit rot caused by *Botrytis cinerea. Phytopathology* **50**, 491–494.

Schultz, F. A. (1978). Some physiological and biochemical aspects of the action mechanisms of fungal parasites during fruit storage. *Fruits* **33**, 15–21.

Sharples, R. O. and Little, R. C. (1970). Experiments on the use of calcium sprays for bitter pit control in apples. *J. hort. Sci.* **45**, 49–56.

Shear, C. I. and Wood, A. K. (1913). Studies of fungus parasites belonging to the genus *Glomerella. U.S. Dept Agric. Bureau Pl. Ind. Bull.* 252.

Shillingford, C. A. and Sinclair, J. B. (1980). Macerating enzyme production by *Colletotrichum musae* and *Fusarium semitectum*, incitants of banana fruit decay. *Phytopath. Z.* **97**, 127–135.

Simmonds, J. H. (1941). Latent infection in tropical fruits discussed in relation to the part played by species of *Gloeosporium* and *Colletotrichum. Proc. R. Soc. Qd* **52**, 92–120.

Simmonds, J. H. (1963). Studies in the latent phase of *Colletotrichum* species causing ripe rots of tropical fruits. *Qd J. agric. Sci.* **20**, 373–424.

Sitterly, W. R. and Shay, J. R. (1960). Physiological factors affecting the onset of susceptibility of apple fruit to rotting by fungus pathogens. *Phytopathology* **50**, 91–93.

Stoessl, A., Unwin, C. H. and Ward, E. W. B. (1972). Post-infectional inhibitors from plants I. Capsidiol, an antifungal compound from *Capsicum frutescens*. *Phytopath. Z.* **74**, 141–152.

Stoessl, A., Robinson, J. R., Rock, G. L. and Ward, E. W. B. (1977). Metabolism of capsidiol in sweet pepper: some possible implications for phytoalexin studies. *Phytopathology* **67**, 64–66.

Swinburne, T. R. (1970). Fungal rotting of apples. I. A survey of the extent and cause of current fruit losses in Northern Ireland. *Rec. agric. Res. North. Ire.* **18**, 15–19.

Swinburne, T. R. (1971). The infection of apples, cv. Bramley's Seedling, by *Nectria galligena* Bres. *Ann. appl. Biol.* **68**, 253–262.

Swinburne, T. R. (1975a). Microbial proteases as elicitors of benzoic acid accumulation in apples. *Phytopath. Z.* **82**, 152–162.

Swinburne, T. R. (1975b). European canker of apple (*Nectria galligena*). *Rev. Pl. Path.* **54**, 787–799.

Swinburne, T. R. (1976). Stimulants of germination and appressoria formation by *Colletotrichum musae* (Berk. & Curt.) Arx. in banana leachate. *Phytopath. Z.* **87**, 74–90.

Swinburne, T. R. (1978). Post-infection antifungal compounds in quiescent or latent infections. *Ann. appl. Biol.* **89**, 322–324.

Swinburne, T. R. and Brown, A. E. (1982). Appressoria development and quiescent infections of banana fruit by *Colletotrichum musae*. *Trans. Br. mycol. Soc.* **80**, 176–178.

van Buren, J. (1970). Fruit phenolics. *In* "The Biochemistry of Fruits and their Products" (A. C. Hulme, Ed.), Vol. I, 269–304. Academic Press, London.

Vanderplank, J. E. (1963). "Plant Diseases: Epidemics and Control." Academic Press, New York.

Verhoeff, K. (1970). Spotting of tomato fruits caused by *Botrytis cinerea*. *Neth. J. Pl. Path.* **76**, 219–226.

Verhoeff, K. (1974). Latent infections by fungi. *A. Rev. Phytopath.* **12**, 99–110.

Verhoeff, K. and Liem, J. I. (1975). Toxicity of tomatin to *Botrytis cinerea* in relation to latency. *Phytopath. Z.* **82**, 333–338.

Wade, G. C. (1956). Investigations on brown rot of apricot caused by *Sclerotinia fructicola*. I. The occurrence of latent infection in fruit. *Aust. J. agric. Res.* **7**, 504–515.

Wade, G. C. and Cruickshank, R. (1973). The development of susceptibility of stone fruit to brown rot during ripening. *2nd int. Congr. Pl. Path., Minneap.* (Abstract).

Wallace, J., Kúc, J. and Williams, E. B. (1962a). Production of extracellular enzymes by four pathogens of apple fruit. *Phytopathology* **52**, 1004–1009.

Wallace, J., Kúc, J. and Draudt, H. M. (1962b). Biochemical changes in the water-insoluble material of maturing apple fruit and their possible relationship to disease resistance. *Phytopathology* **52**, 1023–1027.

Wardlaw, C. W., Barker, R. E. D. and Crowdy, S. H. (1939). Latent infections in tropical fruits. *Trop. Agric.* **16**, 275–276.

Wheeler, H. (1975). "Plant Pathogenesis" (I. W. Thomas and B. R. Sabey, Eds), Advanced Series in Agricultural Science, Vol. 2. Springer-Verlag, Berlin.

2
Soft Fruits

C. DENNIS

I. Introduction

The term "soft fruits" includes a range of fruit types, berry fruits, currants, drupes and false fruits and achenes which have become associated because of their culinary qualities rather than any botanical relationship or morphological similarity. They are widely used both as fresh fruit and processed, where they may be canned, frozen, pureed or made into jams, jellies or conserves, or their juices extracted. The most important soft fruits in the U.K. are strawberries (*c.* 50 000 tonnes annually), raspberries (*c.* 20 000 tonnes annually) and blackcurrants (*c.* 20 000 tonnes annually) with minor quantities of gooseberries (*c.* 8000 tonnes annually), blackberries and loganberries (*c.* 3000 tonnes annually).

Once harvested, the length of time for which fresh soft fruits can be stored at temperatures above freezing is primarily limited by the onset of fungal spoilage. The diversity in morphological and anatomical structure, as well as differences in cultural habit, results in considerable differences in susceptibility to fungal infection and thus to post-harvest rotting. The relatively hard pericarp of gooseberries is intrinsically more resistant to fungal attack than the softer tissue of the strawberry fruit or the drupes of raspberries. The most important post-harvest pathogen common to all soft fruits is the grey-mould fungus *Botrytis cinerea* (Edney, 1964; Dennis and Mountford, 1975; Davis and Dennis, 1977; Mason and Dennis, 1978). Other fungi such as *Mucor* and *Rhizopus* species can be of considerable importance on strawberries and to a lesser extent on raspberries (Lowings, 1956; Dennis, 1978; Mason and Dennis, 1978), and *Cladosporium* spp. infect the surface of drupes, especially raspberries (Cappellini *et al.*, 1961; Dennis, 1975; Mason and Dennis, 1978).

Due to their economic importance and their susceptibility to fungal infection, I shall consider in detail the post-harvest pathology of strawberries and raspberries. Blackcurrants, although economically important for production of juice, are processed immediately or stored in very high (≃40%) carbon dioxide atmospheres for relatively short periods when fungal infection is not a serious problem. The major agents causing post-harvest diseases of soft fruit are also pre-harvest pathogens, so their biology before harvest will be discussed where this is relevant to their post-harvest development.

II. Strawberries

A. *Botrytis cinerea* (Grey Mould)

Infection of strawberries by *B. cinerea* results in a brown discolouration of the infected tissue which usually remains relatively firm. The surface of the rotted tissue is initially covered by a white to greyish mycelium which eventually becomes covered by a mat of grey conidiophores bearing the conidia. The sclerotia are rarely, if at all, produced on infected fruits.

Botrytis cinerea is the most important cause of fruit rots prior to harvest and was responsible for substantial losses of strawberry crops in some years prior to the introduction of effective fungicides (Anon, 1975). The fungus can exist as a saprophyte in strawberry plantations and has been observed to sporulate on dead leaves and debris (Jarvis, 1962a; Dennis *et al.*, 1979), and therefore it is not surprising that airborne conidia are always present prior to and during the flowering and fruiting season (Jarvis, 1962b; Jordon, 1978). Jordon and Pappas (1977), using "Rotorod" air samplers, showed that most conidia were caught less than 20 cm above the ground, whereas very few were caught at a height of 300 cm. Such data suggest that the inoculum responsible for flower infection originates mainly from the strawberry plant debris, and this is also consistent with the data of Dennis *et al.* (1979), where *B. cinerea* was consistently isolated from samples of such debris. In addition, Harris and Dennis (1980a) showed that the fungus can be consistently isolated from flowers and fruits, although the relative contamination is influenced by weather conditions, which affect both the rate of sporulation on the debris, infected flowers and fruits and the release of conidia from the conidiophores which is brought about by hygroscopic movements (Jarvis, 1962b). Many conidia are released in warm, moist periods, but relatively few during prolonged dry periods.

Jarvis and Borecka (1968) showed that the "open-flower" stage was most susceptible to blossom rot and fruit infections that remain latent until ripening when, if conditions are favourable, the fruits rot. Jordan (1978), however, showed that flowers opening early were more liable to become infected than those opening late.

Powelson (1960) considered infection of strawberries by *B. cinerea* to be characterized in most cases by the origin of the rot at the stem end. The frequent isolation of *B. cinerea* from the stem end of ripe, healthy fruit and the high percentage of these fruits which rot under moist conditions indicated considerable latent infection. In most cases necrotic stamens and sepals were associated with marketable fruit having latent infections and these necrotic parts were infected by *B. cinerea*. Hyphae of *B. cinerea* were observed in infected petals, stamens, calyces and also in the stem end receptacle tissue of marketable fruit. The receptacle was thought to be infected via the stamens or calyces, with invasion being internal through these tissues and not by contact of the infected flower parts with the fruit surface (Powelson, 1960). Harris and Dennis (1980a) showed no evidence of latent mycelium in the central plug of vascular tissue of harvested fruit, suggesting that the mycelium remains confined to the stem end of the receptacle at the point of attachment of the floral organs from which the rot develops. Jarvis (1962a) also stressed the importance of the presence of necrotic tissue in infection by both mycelium and spores of *B. cinerea*.

Thus at harvest, strawberries will be contaminated by conidia of *B. cinerea* on the fruit surface and on the remains of flower parts which may also already be infected by the fungus, as well as latent infections being present at the stem end. Davis and Dennis (1979) showed that during post-harvest storage a small proportion of the infections were initiated from latent infections (*c.* 15%) at the stem end, but that the majority of infections occurred at points on the surface of the fruit. The latter infections mostly occur directly from propagules on the fruit surface and not from conidia on adhering flower parts, which thus contrasts with pre-harvest infections as reported by Jarvis (1962a). It is therefore not surprising that, if the mode of pre- and post-harvest infection differs, there is often little correlation between the pre- and post-harvest susceptibility of strawberry varieties to infection by *B. cinerea* (Daubeny and Pepin, 1977).

Botrytis cinerea is able to grow and infect strawberries at temperatures as low as 0 °C (Dennis and Cohen, 1976) and thus its development is only delayed by rapid cooling and storage of fruit under recommended commercial conditions (*c.* 2°C).

B. *Mucor piriformis* ("Leak" Disease)

Mucor piriformis was first reported as a serious post-harvest pathogen of strawberries by Lowings (1956), who considered it to be a more important cause of infection than *B. cinerea* in the Kent strawberry crop of 1955. Subsequently, this fungus has been reported in a number of geographical areas to be a serious cause of post-harvest rots of different varieties of fruit destined for both fresh and processed use (Edney, 1964; Dennis, 1975; Dennis and Mountford, 1975; Dennis and Davis, 1977a,b; Davis and Dennis, 1979; Dennis *et al.*, 1979). Reports by Dennis and co-workers prior to 1979 refer to this fungus as *M. mucedo*, but as a result of revision of the taxonomy of the genus by Schipper (1975) the isolates were subsequently identified as *M. piriformis* (see Dennis and Harris, 1979).

Infection of strawberries by *M. piriformis* results in a water-soaked appearance of the infected area, which is very soft and very rapidly becomes covered with asexual sporangiophores and sporangia. *Mucor piriformis* is also able to cause pre-harvest infection and, as with *B. cinerea*, environmental conditions during flowering and fruiting greatly influence the degree of contamination of the fruit and subsequent post-harvest infection (Dennis, 1978; Harris and Dennis, 1980a). Unlike *B. cinerea*, *M. piriformis* does not sporulate on debris in plantations and only infects ripe fruit, unless the fruit is damaged, when white or sometimes green fruit can become infected (Harris and Dennis, 1980a). Thus since there is no continued production of inoculum on the debris and as sporulation usually occurs only on ripe fruit, the increase in inoculum in plantations is limited to the latter part of the season. Hence late season fruit shows a greater contamination by *M. piriformis* (Dennis, 1978) and the fungus is generally more important in causing post-harvest rotting of fruit harvested towards the end of the season.

At the onset of fruiting the debris and soil of strawberry plantations are a potential source of inoculum of *M. piriformis* (Dennis *et al.*, 1979). The initial infection can occur when ripe or damaged white fruits come into contact with dormant spores or mycelium in this material. Similarly, the transfer of fungal propagules from the soil and debris to the fruit by rainsplash is extremely important, particularly for *M. piriformis*, a wet-spored fungus (Ingold, 1978) relying on dispersal by rain. Clearly, the time at which rain storms occur in the fruiting season can have a marked effect on the amount of contamination and subsequent spoilage caused by the fungus. When heavy storms occur at the beginning of the fruiting period, *M. piriformis* can cause a high percentage of post-harvest rotting of early season fruit and may even be more important than at the end of the season if the weather becomes dry (K. M. Browne and C. Dennis, unpublished). Once established on ripe fruits, the levels of *M. piriformis* in the plantation may increase very rapidly if conditions are suitable by virtue of its fast growth rate (sporulation in 3–4 days) and the large numbers of asexual spores produced on each berry. During storage and distribution, strawberries become infected from viable propagules (spores and mycelial fragments) on the surface of the fruit and subsequent spread occurs via fruit contact or sporangia from infected berries contacting healthy fruit. Infection occurs at temperatures as low as 0°C (Dennis and Cohen, 1976) and the rapid maceration results in exudation of juice from the infected berries (hence the name "leak" disease), which runs onto healthy fruit and if it is contaminated with spores will provide a means of spreading the fungus.

Mucor piriformis appears to be present in most of the important strawberry-growing areas in the U.K. (East Anglia, Kent, Midlands) as well as Southern Ireland (Dennis, 1978), although it is rarely found to be the cause of post-harvest disease in Western England (Jordon, 1973).

C. *Rhizopus* spp. ("Leak" Disease)

As early as 1919 Stevens reported *Rhizopus stolonifer* (≡ *R. nigricans*; Von Arx, 1970) to be the predominant spoilage fungus of strawberries harvested late in the season in the U.S.A., although Edney (1964) reported the fungus to be of minor importance in infection of strawberries harvested in Kent. Jordan (1973) and Jordan and Richmond (1975) also reported *R. stolonifer* to be the predominant Phycomycete causing post-harvest rotting of fruit harvested at Long Ashton Research Station, whereas this species has rarely been found to cause rots of fruit in East Anglia, Warwickshire, Kent and Southern Ireland (Dennis and Davis, 1977a; Dennis *et al.*, 1979). In these areas the homothallic species *Rhizopus sexualis* is of considerably greater importance and is often the predominant Phycomycete (Dennis *et al.*, 1979).

In common with *M. piriformis*, neither *Rhizopus* species is able to sporulate on the debris in the plantation and therefore renewal of inoculation does not readily occur as it does with *B. cinerea*. Sporulation occurs only on infected fruits and hence the inoculum often increases as the season progresses (Dennis, 1975; Dennis *et al.*, 1979). The debris and soil provide the initial source of inoculum in a similar way to that of *M. piriformis*. In contrast to *M. piriformis*, however, production of asexual spores by *Rhizopus* species is markedly influenced by environmental conditions (Harris and Dennis, 1980a). Under high humidity *R. stolonifer* is predominantly mycelial in habit, whereas *R. sexualis* produces abundant sexual zygospores. The significance of the zygospores is uncertain as they have not been observed to germinate (Ingold, 1978). Both *Rhizopus* species produce abundant asexual sporangia when the relative humidity is below 80%.

Thus *Rhizopus* species only cause serious problems in post-harvest rotting in warm, dry seasons when levels of inoculum in the plantation and contamination of the fruit, especially that harvested late in the season, will be high (Freeman and Pepin, 1977). Under such conditions, pre-harvest rotting of fruit also occurs. During storage and distribution, infection of fruit from spores or mycelium will not occur if the fruit is rapidly cooled to below 5 °C. Below this temperature neither *R. stolonifer* nor *R. sexualis* is able to grow (Dennis and Cohen, 1976). Indeed, if the fruit is held at 0 °C, the asexual spores of *R. sexualis* rapidly lose viability (Dennis and Blijham, 1979) and fruits stored at this temperature do not become infected with this species when moved to a higher temperature (C. Dennis and K. M. Browne, unpublished data).

D. *Phytophthora cactorum* (Leather Rot)

Phytophthora cactorum has not been reported as a cause of post-harvest rotting of strawberries in the U.K., but does cause disease during marketing of fruit in the U.S.A. (Harvey and Pentzer, 1960). Jordan and Pappas (1977) and Hunter *et al.* (1979), however, have recently reported pre-harvest infection of fruit by this fungus. The proportion of fruit infected by *P. cactorum* was significantly greater on plants sprayed with dicarboximide fungicides, which was most probably due to the lack of competition resulting from the selective suppression of *B. cinerea* by the fungicides.

Harvey and Pentzer (1960) consider that the post-harvest appearance of disease is always associated with wet weather in the field but is also favoured by moderately high temperatures. *Phytophthora cactorum* can penetrate the uninjured surface of the fruit and the infected tissue becomes cream-brown in colour while a superficial white mycelial growth develops, especially under conditions of high humidity. The infected tissue does not soften but becomes tough and leathery, and within it the fungus produces numerous oospores.

Post-harvest infection by *P. cactorum* can be reduced to a minimum by rapid cooling of fruit to below 5 °C and maintaining such a temperature during distribution.

III. Raspberries

A. *Botrytis* cinerea (Grey Mould)

The infection of raspberries by this fungus produces a brown discolouration of the fruit tissue which very rapidly becomes covered with a dense mat of grey conidiophores. The activity of the fungus in raspberry plantations, as in the case of strawberries, greatly influences its importance for harvested fruit and much of the life-cycle described for strawberries also occurs in raspberry plantations. On raspberry plants, mummified fruits and receptacles are common. Small, discrete sclerotia occur on dead raspberry leaves, petioles, peduncles and stems, and large, flat, sub-epidermal sclerotia,

1–2 mm in diameter, occur on mature canes. Damaged canes may carry a mycelium from which a crop of conidiophores can arise; such infected canes frequently show a characteristic "watermark". Because of the relatively cleaner conditions in raspberry plantations, there is virtually no infection of fruit from adhering organic debris (Jarvis, 1961a). Petals are shed as in strawberries, but rarely adhere to the fruit for sufficient time to allow infection; the microclimate of the fruit and its surface contours do not favour petal adherence. Studies by Jarvis (1962a) showed that infection by contact with rotting raspberries and especially with rotting receptacles remaining after the berries were picked comprised 35% of infections, whereas 64% developed from latent infections originating from sapro-phytically based mycelium on moribund flower parts. Such infections can occur at any time between flowering and fruiting. The mycelium can then invade the developing fruitlet. The rate of flowers opening, and hence the proportion of flowers senescing, is not constant throughout the flowering period but increases rapidly at the beginning of the flowering period and then declines. Jarvis (1969) reported that approximately 80% of the flowers of the variety Malling Jewel open during a period of about 12–15 days.

Jarvis (1962a) considered that infection by direct germination of conidia on the surface of the fruit only accounted for 1% of pre-harvest grey-mould rots compared with 64% that originated from infected flower parts attached to the receptacle and 35% that were caused by berries being in physical contact with other berries or receptacles. However, Mason and Dennis (1978) presented evidence that post-harvest infection does not follow this pattern. Indeed, observed outbreaks of *B. cinerea* in the field following a delay in harvesting caused by wet weather or because some ripe berries were not removed by a mechanical harvester showed that sporulating mycelium of *B. cinerea* developed all over the berry surface, not just at the proximal end, as would be expected if the mycelium had developed from a latent infection. Not all berries develop a rot caused by *B. cinerea* immediately after picking, and when the fruit is incubated at 21°C a mycelium of *B. cinerea* may not develop for 5 days even if the berries are not rotted earlier by other fungi. Fungicide sprays during the flowering period decreased the development of post-harvest rots on the proximal row of drupelets but not on other parts of the berry, a result which suggests that proximal infection occurred at the time of spraying but infections elsewhere on the berry occurred later, after the last fungicide spray had lost its effectiveness. These observations, viewed collectively, suggest that some *B. cinerea* infections arise from direct germination of spores on the berry surface, particularly under the conditions of high relative humidity that prevail in containers of fruit after harvest when berries are in close contact with each other. The close proximity of fruit will also result in spread of infection by fruit contact. Mason and Dennis (1978) considered that between 19 and 25% of post-harvest rots were caused by direct germination of conidia, which contrasts with the 1% reported by Jarvis (1962a) for pre-harvest rots.

B. Other Spoilage Fungi

Other fungi that have been reported to cause post-harvest spoilage of raspberries are species of *Rhizopus*, *Mucor*, *Cladosporium* and *Penicillium*, and *Colletotrichum gloeosporioides* (Cappellini *et al.*, 1961; Jarvis, 1973; Dennis, 1975; Dennis and Mountford, 1975; Jennings and Carmichael, 1975; Freeman and Pepin, 1976; Mason and Dennis, 1978). Jarvis (1973), Jennings and Carmichael (1975) and Freeman and Pepin (1976) attributed the "leak" disease of raspberries to *Rhizopus* spp. (especially *R. stolonifer*), although Dennis and Mountford (1975), Dennis (1975) and Mason and Dennis (1978) observed *Mucor piriformis*, *M. hiemalis* and *R. sexualis* to cause similar symptoms (exudation of juice).

The incidence of post-harvest rots caused by all of the above fungi varies considerably both from year to year and within seasons, as well as between locations, suggesting that climatic factors may affect the amount of inoculum available or the resistance of the fruit to infection (Dennis, 1975; Dennis and Mountford, 1975; Mason and Dennis, 1978). The relative importance of *Mucor* and *Rhizopus* species will be influenced by temperature and humidity, as for strawberries. The incidence of *Cladosporium*, *Penicillium* and *Colletotrichum gloeosporioides* also tends to be greater after periods of rain and especially if the fruit are mist when harvested. These pathogens all develop both on the surface of the berries as well as in the cavity left by removal of the fruit from the receptable during harvesting. This removal from the receptacle often ruptures the drupelets internally and thus exposes the tissue to infection, especially as juice often exudes into the cavity, creating ideal nutrient and humidity conditions for growth of fungi.

IV. Blackberries and Loganberries

Dennis and Mountford (1975) reported *B. cinerea* and *M. piriformis* to be the most important post-harvest spoilage fungi on blackberries, whereas *Cladosporium* spp., *Penicillium* spp. and *Colletotrichum gloeosporioides* were of minor importance. On loganberries, *B. cinerea* was the predominant spoilage fungus with *M. piriformis* being of minor importance (Davis and Dennis, 1977).

The conditions favouring infection were similar to those reported for raspberries, except that there was a greater incidence of rots caused by *B. cinerea* developing from latent infections as the receptacle remains attached to the berry at harvest.

V. Gooseberries

The main post-harvest disease is caused by *B. cinerea*, which almost always invades the pericarp via the senescent calyx, using the latter as a food base to infect the ripe fruits. Fruits that become contaminated by soil splash after heavy rain are frequently infected by *M. piriformis*. A third fruit disease which has recently been reported in gooseberries is caused by *Alternaria* and *Stemphylium* (Dennis *et al.*, 1976). The disease primarily occurs in the field prior to harvest and often results in premature fall of infected berries from the bushes. The infection occurs via the remains of the calyx and is initially confined to the seeds enclosed within the pericarp, not usually being visible externally. If, however, the fruits are stored for a few days at ambient temperature prior to consumption or processing, the fungus begins to invade the pericarp tissue. The use of benzimidazole fungicides which are ineffective against the two fungi may have resulted in an increased occurrence of the disease (Dennis *et al.*, 1976).

VI. Physiology of Soft Fruit Pathogens

A. *Botrytis cinerea*

The brown discolouration of soft fruit tissue infected with *B. cinerea* is due to oxidation of fruit phenolics by polyphenol oxidase enzymes. Harris and Dennis (1982a) showed that polyphenol oxidase was highest in green fruits infected with *B. cinerea* and decreased as the fruit ripened. They considered that this was due to a combination of fruit and fungal enzymes in immature fruits, whereas in ripe fruits the fungal enzyme accounted for most of the activity. Since polyphenol oxidase was not detected in healthy, unripe strawberries, it was postulated that latent fruit enzyme was activated as a result of infection by the fungus, similar to that reported by Balasubramani *et al.* (1971) for *Botrytis* species on *Vicia faba*. Harris and Dennis (1982a) also considered that the production of oxidized phenolics was responsible for the rapid inactivation of pectolytic enzymes, which accounted for the lack of maceration and firmness of the infected tissue. Very low endopolygalacturonase activity was detected in extracts of infected tissue, even when extracts were prepared in the presence of a phenol-adsorbing compound (polyvinylpolypyrrolidone) and dithiothreitol, which is known to protect – SH groups. Strains of *B. cinerea* have been consistently reported to produce endopolygalacturonase together with the exo-enzyme and pectin esterase in culture (Verhoeff and Warren, 1972; Harris and Dennis, 1979; Chesson, 1980; Cruickshank and Wade, 1980). The endopolygalacturonases, the enzymes considered most important in maceration, have also been shown to be relatively sensitive to inactivation by naturally occurring fruit phenolics in strawberry (Harris and Dennis, 1982a) and this further accounts for the rapid inactivation of these enzymes in such fruit tissue.

B. *Mucor* and *Rhizopus spp.*

Infection of soft fruits (especially strawberries and raspberries) results in the tissue becoming macerated and completely losing coherence. Consequently, much juice exudes from the infected tissue, which is the reason for the term "leak disease" being applied to rots caused by *Rhizopus* but is equally suitable for those caused by *M. piriformis*. The extensive maceration of strawberry tissue is primarily due to the action of endopolygalacturonase in combination with pectin esterase (Archer, 1979). In this case, the endopolygalacturonase is readily extractable from fruit tissue (Harris and Dennis, 1980b) and is more resistant to naturally occurring fruit phenolics than the enzyme from *B. cinerea* (Harris and Dennis, 1982a). None of the *Mucor* or *Rhizopus* species produce polyphenol oxidase which, together with the insensitivity of the endopolygalacturonases to fruit phenolics, accounts for the high pectolytic activity in extracts of ripe fruit. Infection of immature fruits by *Mucor* and *Rhizopus* species results in detectable but low polyphenol oxidase activity, which is considered to be due to activation of a latent fruit enzyme. The consequent production of oxidized phenolics in such fruits does, however, markedly reduce the pectolytic activity in such tissue compared with that in ripe fruit. Similarly, the oxidized phenolics produced in fruit infected with *B. cinerea* inactivate the endopolygalacturonases of *Mucor* and *Rhizopus* in fruit preserved in sulphite (Harris and Dennis, 1982a).

The stability of certain endopolygalacturonase isoenzymes, particularly of *Rhizopus* species, in sulphite liquor has resulted in softening and breakdown of commercial samples of sulphited strawberries in the U.K. (Archer, 1979; Dennis and Harris, 1979; Dennis et al., 1979; Harris and Dennis, 1979), although there are varietal differences in susceptibility to enzymic breakdown (Harris and Dennis, 1981). Fruit of the variety Harvester is markedly more resistant than the variety Cambridge Favourite, and it is considered that this difference is due to differences in phenolic constituents.

The extensive heat-stability of the pectolytic enzymes of *Rhizopus* and *Mucor* species (see Harris and Dennis, 1980b) does not cause problems in softening and breakdown of canned strawberries and raspberries as has been reported for apricots (Harper *et al.*, 1972; Strand *et al.*, 1981). Although the enzymes from strawberry spoilage fungi show similar bimodal heat-stability (with maximum inactivation at 50–60°C and increased residual activity up to 90–100°C) to the apricot spoilage fungi when tested in culture (Harris and Dennis, 1980b), the phenolic constituents in strawberries and raspberries rapidly inactivate the enzymes at temperatures above 50°C (Harris and Dennis, 1982b). There is slight variation in the thermostability of the fungal enzymes in different strawberry varieties, but with present varieties none of the enzymes are sufficiently stable to cause subsequent breakdown of canned fruit.

VII. Varietal Resistance to Post-Harvest Diseases

The need for fruit-rot resistant lines of strawberry and raspberry has been expressed for many years (Stevens and Wilcox, 1918; Anderson, 1946). The development of mechanical harvesting methods and emergence of fungicide-resistant strains of *B. cinerea*, together with the increased importance of fungicide-insensitive fungi such as *Mucor* and *Rhizopus*, have all intensified the need to identify germ plasm sources for fruit-rot resistance.

Several methods for screening the resistance of soft fruit varieties to post-harvest diseases have evolved (see Maas, 1978). Irvine (1959) and Irvine and Fulton (1959) reported the inability of *B. cinerea* to grow in juice of varieties correlated with fruit resistance, but Jarvis (1961b, 1962c) failed to establish such a correlation. Fruit firmness has also been associated with the incidence of rots caused by *B. cinerea*, firm-fruited cultivars having lower incidences of infection than cultivars with softer fruit (Darrow, 1966). Hondelmann and Richter (1973) also correlated susceptibility with the shift of insoluble to soluble pectins as fruits ripen, decreasing the firmness of the fruit. Cultivars with fruits that maintain firmness over a longer period of time remain less susceptible to *B. cinerea* than fruit of cultivars that soften early. Fruit firmness includes skin toughness and flesh firmness: both are heritable characteristics (Jennings and Carmichael, 1975; Scott and Lawrence, 1975), but are subject to modification by temperature, humidity and culture conditions (Scott and Lawrence, 1975).

Post-harvest incubation studies to differentiate clonal variation in fruit rot susceptibility were first utilized in 1969 by Daubeny and Pepin for raspberries and confirmed by Barritt (1971). This method of screening for resistance to *B. cinerea* and *Rhizopus* fruit rots has become an integral part of the raspberry breeding programmes (Daubeny and Pepin, 1974, 1976; Jennings *et al.*, 1976; Keep, 1976). Daubeny and Pepin (1973) determined that their post-harvest screening method for raspberries could also be applied to strawberry fruit. General agreement was found between their post-harvest tests and reported observations of other workers in other regions on relative susceptibility to *B. cinerea* in the field. Their technique also identified clones with lower susceptibility to *Rhizopus* spp., and the ability of this method to differentiate clonal variations in susceptibility to *B. cinerea* and/or *Rhizopus* based on their own pre- and post-harvest disease data was shown by Daubeny and Pepin in 1977. In a study of 116 genetically diverse strawberry clones, they found that field resistance to *B. cinerea* was not necessarily related to post-harvest resistance to *B. cinerea* or *Rhizopus*, nor was there a relationship between post-harvest resistance to *B. cinerea* and post-harvest resistance to *Rhizopus*. Perhaps their most significant finding was that only five clones showed any indication of combined resistance to both fungi.

Maas and Smith (1978) reported the strawberry variety Earliglow to be a source of post-harvest fruit rot resistance but found that the extent of the resistance was markedly affected by the storage temperature. For example, resistance to *B. cinerea* was not obvious if fruit was stored at 0°C prior to storage at 18°C. Such a report emphasizes the need for a standard screening technique to be adopted in order for different laboratories to compare their data on fruit rot resistance.

Recently, Daubeny *et al.* (1980) screened 37 cultivars and selections (clones) of raspberry for post-harvest susceptibility to *Rhizopus* and concluded that fruit resistance to *Rhizopus* cannot be based solely on selection for firmness, as has been suggested for post-harvest resistance to *B. cinerea* (Jennings and Carmichael, 1975). They suggested that two sources of resistance are apparent: intrinsic tissue resistance in clones with relatively soft fruit and resistance associated with fruit firmness. They also suggested that both intrinsic tissue resistance and firmness are probably involved in post-harvest resistance of strawberry fruit to *B. cinerea*, as an association of firmness and resistance does not extend to all clones (Gooding, 1976; Barritt, 1980).

Daubeny *et al.* (1980) stated that it is essential to combine high levels of fruit rot resistance with fruit firmness if new raspberry cultivars are to have extended fresh market shelf-life and also be adapted to mechanical harvesting. Kichina (1976) reported that clones derived from *Rubus cratigifolius* L. have fruit with outstanding shelf-life compared with other clones. Similarly, genotypes such as SHRI 6820/54 have both types of resistance and will be an effective source for the production of resistant cultivars.

Dennis and Davis (1977a) showed that although the fruit from different strawberry varieties from each of two sites was contaminated with a similar level of fungi, the pattern of development of post-harvest fruit rots revealed marked differences between some varieties. *Botrytis cinerea* and *M. piriformis* were the predominant spoilage fungi for all varieties, with *R. sexualis* and *R. stolonifer* being of minor importance. The relative importance of these fungi changed for all varieties as the season progressed, although to a different degree according to the variety, site and season. Although the level of contamination of the fruit by *B. cinerea* generally increased through the season, the incidence of this fungus in causing fruit rots decreased on most cultivars. This does not necessarily indicate an increased resistance to *B. cinerea* later in the season, as the fruit was contaminated to a greater extent by *M. piriformis* and *Rhizopus* species, which cause rapid spoilage of the fruit at this time.

There was a marked difference in the pattern of development of spoilage fungi among three of the main strawberry cultivars grown in the U.K., namely Cambridge Favourite, Cambridge Vigour and Red Gauntlet. On both plantations fruit of Red Gauntlet was the least susceptible to fungal spoilage and that of Cambridge Vigour was the most susceptible, especially to the Phycomycetes. The cultivars Senga Gigana, Chanil and Grandee were also especially susceptible to the Phycomycetes, Grandee differing in that it was more susceptible to *B. cinerea* early in the season. During these studies it was found that those varieties which were especially susceptible to the Phycomycetes were those which tended to suffer from surface bruising, especially under moist conditions.

VIII. Effect of Pre-Harvest Fungicides on Post-Harvest Diseases

It has already been stated that the activity of the spoilage fungi in growing crops influences the extent of post-harvest infection. Since *B. cinerea* is the most important cause of pre-harvest fruit rots of soft fruit, all of the chemical control schemes have been designed to control this fungus. The importance of applying sprays during the early flowering period is essential for effective control of *B. cinerea* (Jarvis and Borecka, 1968; Jarvis, 1969; Jordan, 1978). Recently, Jordan (1978) also claimed that spraying the debris in plots under polythene tunnels prior to flowering resulted in reduced fruit infection by *B. cinerea*, although consistent results have not been obtained in unprotected crops (M. Simkin, personal communication).

A variety of organic fungicides have been used over the last 20 years; initially, thiram and captan were used but resulted in taints in canned fruit and were replaced by the more effective Elvaron (dichlofluanid) in the mid-1960s. Elvaron has given consistent reduction in post-harvest infection by *B. cinerea* (Freeman and Pepin, 1968; Dennis, 1975; Jordan and Richmond,

1975; Dennis and Davis, 1977b; Davis and Dennis, 1979), but results in increased infection by the Phycomycetes (Dennis, 1975; Dennis and Davis, 1977b; Davis and Dennis, 1979). However, the consistent pre- and post-harvest reduction in infection by *B. cinerea* achieved by the use of Elvaron has resulted in its continued use since the 1960s.

In the early 1970s the introduction of the systemic benzimidazole fungicides (e.g. Benlate, Mildothane, Bavistin, Derosal) for the control of *B. cinerea* on soft fruits initially gave excellent control of both pre- and post-harvest disease (Jordan, 1973; Dennis, 1975). However, the subsequent rapid emergence of resistant strains of the fungus (Jordan and Richmond, 1974; Dennis, 1975) due to the very specific site of action of the chemicals (Davidse, 1975; Dekker, 1977), as well as a selective enhancement of post-harvest infection of soft fruit by *Mucor* and *Rhizopus* species (Jordan, 1973; Dennis, 1975), has resulted in very restricted use of these chemicals on soft fruit crops.

Strains of *B. cinerea* resistant to the recently introduced dicarboximide fungicides, Rovral (iprodione) and Ronilan (vinclozolin), have also been found on soft fruit (Davis and Dennis, 1979; Hunter *et al.*, 1979; Pappas *et al.*, 1979), but have not resulted in a loss of control of *B. cinerea* either pre- or post-harvest (Dennis and Davis, 1979; Davis and Dennis, 1981a). This appears to be due primarily to the inability of the resistant strains to sporulate (Davis and Dennis, 1981b) and therefore they are unable to increase rapidly in a plantation. When they are naturally present or inoculated onto debris in a plantation, they fail to spread effectively onto the flowers and fruit, and hence are of little importance in fruit infections. If, however, dicarboximide-resistant strains are inoculated onto strawberry flowers or fruit, they compete with the dicarboximide-sensitive strains and are not effectively controlled by dicarboximide fungicides (Hunter *et al.*, 1979; Davis and Dennis, 1981b).

The dicarboximides, in common with other effective fungicides against *B. cinerea*, usually cause an increased incidence of Phycomycetes on harvested fruit (Davis and Dennis, 1979; Hunter *et al.*, 1979) and, in the case of strawberries grown in polythene tunnels, Rovral and Ronilan have resulted in a greater incidence of leather rot caused by *Phytophthora cactorum* (Jordan and Pappas, 1977; Hunter *et al.*, 1979).

IX. Use of Modified Atmospheres for Controlling Post-Harvest Diseases

The use of low oxygen atmospheres to control the fungal spoilage of soft fruit is not possible commercially, since it is necessary to control the oxygen level very precisely at 0.5% and even this does not always reduce decay caused by *R. stolonifer*; furthermore, at 0.25% and less of oxygen, persistent off-flavours develop (Couey *et al.*, 1966; Couey and Wells, 1970).

The use of elevated concentrations of carbon dioxide for the storage and distribution of soft fruit, however, has been recommended by a number of authors (Smith, 1963; Couey and Wells, 1970; Wells, 1970; Harris and Harvey, 1973; Harvey and Harris, 1976). High concentrations of carbon dioxide are particularly effective in reducing infection by *B. cinerea* when the fruit is stored at temperatures above 10°C and as the length of storage and distribution time increases (Couey and Wells, 1970; Harvey and Harris, 1976). Blackcurrants are exceptional in that they can be stored in 30–40% carbon dioxide without deleterious effects on their quality for processing, and are thus routinely stored in such an atmosphere at 2–4°C prior to processing into juice (Smith, 1963).

Recent work in England (Topping and Cockburn, 1979; Dennis and Browne, 1980) has investigated the use of carbon dioxide as an adjunct to cooling for storage of strawberries.

A simulated commercial system was used in which trays of fruit were enclosed in a polythene cover to give the same ratio of fruit to surface area of polythene. In tightly sealed packs, without added carbon dioxide, respiration of the fruit resulted in an increase of approximately 1% carbon dioxide per day at 2°C. A survey of commercial packs in which the carbon dioxide level had originally been raised to approximately 15% indicated that the carbon dioxide levels were between 1 and 16%, with the majority below 10%. These low values observed in practice indicate poor sealing of commercial packs.

Data from the simulated packs indicated that the effectiveness of carbon dioxide reducing fungal spoilage varied according to season and time of harvest. For example, in 1977, the addition of 15–20% carbon dioxide to the enclosed atmosphere did not reduce the incidence of fungal rots after storage at 2°C for up to 7 days. In 1978, findings were similar for up to 3 days cool storage, but thereafter the carbon dioxide substantially reduced the incidence of rots due to *B. cinerea* and *M. piriformis*. Taste panel assessments indicated that the carbon dioxide concentration must be below 20% to avoid the rapid development of persistent off-flavours in the fruit. This contrasts with reports from the U.S.A. (Couey and Wells, 1970; Harris and Harvey, 1973) indicating a difference in varietal response or an effect of growing conditions (see also Shaw, 1969).

The inhibitory effect of high carbon dioxide atmospheres is considered to be due to the production of high levels of acetaldehyde and ethyl acetate by the fruit in such atmospheres (Shaw, 1969). Prasad and Stadelbacher (1973, 1974) subsequently reported that fumigation of strawberries and raspberries with acetaldehyde vapour controlled infection by *B. cinerea* and *R.*

stolonifer without affecting the quality of the fruit, although such a system appears not to have been pursued commercially.

Recently, laboratory tests reported by El-Goorani and Sommer (1979) showed that supplementing high CO_2 and low O_2 atmospheres with 9% carbon monoxide resulted in an 84% reduction in infection of strawberries by *B. cinerea*, and that the controlled atmospheres alone gave a 45% reduction in infection compared with that in air. The presence of off-flavours was observed but was considered to be associated with O_2 and CO_2 modification rather than with the addition of CO; however, further work is required to assess the feasibility of using CO commercially.

References

Anderson, M. W. (1946). Strawberry fruit rots and their control. *Trans. hort. Soc. South Ill.* **80,** 239–243.

Anon (1975). "Second Report of the Boards of the Joint Consultative Organisation for Research and Development in Agriculture and Food", p. 108, H.M.S.O., London.

Archer, S. A. (1979). Pectolytic enzymes and degradation of pectin associated with breakdown of sulphited strawberries. *J. Sci. Fd Agric.* **30,** 692–703.

Balasubramani, K. A., Deverall, B. J. and Murphy, J. V. (1971). Changes in respiratory rate, polyphenoloxidase and polygalacturonase activity in and around lesions caused by *Botrytis* in leaves of *Vicia faba. Physiol. Pl. Path.* **1,** 105–113.

Barritt, B. M. (1971). Fruit rot susceptibility of red raspberry cultivars. *Pl. Dis. Reptr* **55,** 135–139.

Barritt, B. M. (1980). Resistance of strawberry clones to *Botrytis* fruit rot. *J. Am. Soc. hort. Sci.* **105,** 160–164.

Cappellini, R. A., Stretch, A. W. and Walton, G. S. (1961). Effects of sulphur dioxide on the reduction of post-harvest decay of Latham red raspberries. *Pl. Dis. Reptr* **45,** 301–303.

Chesson, A. (1980). Maceration in relation to the post-harvest handling and processing of plant material. *J. appl. Bact.* **48,** 1–45.

Couey, H. M. and Wells, J. M. (1970). Low-oxygen or high-carbon dioxide atmospheres to control post-harvest decay of strawberries. *Phytopathology* **60,** 47–49.

Couey, H. M., Follstad, M. N. and Uota, M. (1966). Low-oxygen atmospheres for control of post-harvest decay of fresh strawberries. *Phytopathology* **56,** 1339–1341.

Cruickshank, R. M. and Wade, G. C. (1980). Detection of pectic enzymes in pectin–acrylamide gels. *Analyt. Biochem.* **107,** 177–181.

Darrow, G. M. (1966). "The Strawberry." Holt, Rinehart & Winston, New York.

Daubeny, H. A. and Pepin, H. S. (1969). Variations in susceptibility to fruit rot among red raspberry cultivars. *Pl. Dis. Reptr* **53,** 975–977.

Daubeny, H. A. and Pepin, H. S. (1973). Variations in fruit rot susceptibility of strawberry cultivars and selections as indicated by a post-harvest screening technique. *Can. J. Pl. Sci.* **53,** 341–343.

Daubeny, H. A. and Pepin, H. S. (1974). Variations among red raspberry cultivars and selections in susceptibility to the fruit rot causal organisms *Botrytis cinerea* and *Rhizopus* spp. *Can. J. Pl. Sci.* **54**, 511–516.

Daubeny, H. A. and Pepin, H. S. (1976). Recent developments in breeding for fruit rot resistance in red raspberry. *Acta Hort.* **60**, 63–72.

Daubeny, H. A. and Pepin, H. S. (1977). Evaluation of strawberry clones for fruit rot resistance. *J. Am. Soc. hort. Sci.* **102**, 431–435.

Daubeny, H. A., Pepin, H. S. and Barritt, B. M. (1980). Post-harvest *Rhizopus* fruit rot resistance in red raspberry. *Hort. Sci.* **15**, 35–37.

Davidse, L. C. (1975). Mode of action of methyl benzimidazol-2-yl carbonate (MBC) and the mechanism of resistance against this fungicide in *Aspergillus nidulans*. *In* "Microtubules and Microtubule Inhibitors" (M. Borgers and M. de Brabander, Eds), 483–495. North-Holland/American Elsevier, Amsterdam.

Davis, R. P. and Dennis, C. (1977). The fungal flora of loganberries in relation to storage and spoilage. *Ann. appl. Biol.* **85**, 301–304.

Davis, R. P. and Dennis, C. (1979). Use of dicarboximide fungicides on strawberries and potential problems of resistance in *Botrytis cinerea*. *Proc. Br. Crop Prot. Conf., Pests Dis.* **1**, 193–201.

Davis, R. P. and Dennis, C. (1981a). Studies on the survival and infective ability of dicarboximide-resistant strains of *Botrytis cinerea*. *Ann. appl. Biol.* **98**, 395–402.

Davis, R. P. and Dennis, C. (1981b). Properties of dicarboximide-resistant strains of *Botrytis cinerea*. *Pestic. Sci.* **12**, 521–535.

Dekker, J. (1977). Tolerance and mode of action of fungicides. *Proc. Br. Crop Prot. Conf., Pests Dis.* **3**, 689–697.

Dennis, C. (1975). Effect of pre-harvest fungicides on the spoilage of soft fruit after harvest. *Ann. appl. Biol.* **81**, 227–234.

Dennis, C. (1978). Post-harvest spoilage of strawberries. *Agric. Res. Coun. Res. Rev.* **4**, 38–42.

Dennis, C. and Blijham, J. M. (1979). Effect of temperature on viability of sporangiospores of *Rhizopus* and *Mucor* species. *Trans. Br. mycol. Soc.* **74**, 89–94.

Dennis, C. and Browne, K. M. (1980). Storage in modified atmospheres—soft fruit. *Bienn. Rep. Fd. Res. Inst., 1977 & 1978*, 59.

Dennis, C. and Cohen, E. (1976). The effect of temperature on strains of soft fruit spoilage fungi. *Ann. appl. Biol.* **82**, 51–56.

Dennis, C. and Davis, R. P. (1977a). Susceptibility of strawberry varieties to post-harvest fungal spoilage. *J. appl. Bact.* **42**, 197–206.

Dennis, C. and Davis, R. P. (1977b). The selective effect of fungicides on post-harvest spoilage fungi of strawberries. *Proc. Br. Crop Prot. Conf., Pests Dis.* **1**, 203–210.

Dennis, C. and Davis, R. P. (1979). Tolerance of *Botrytis cinerea* to iprodione and vinclozolin. *Pl. Path.* **28**, 131–133.

Dennis, C. and Harris, J. E. (1979). The involvement of fungi in the breakdown of sulphited strawberries. *J. Sci. Fd Agric.* **30**, 687–691.

Dennis, C. and Mountford, J. (1975). The fungal flora of soft fruits in relation to storage and spoilage. *Ann. appl. Biol.* **79**, 141–147.

Dennis, C., Groom, R. W. and Davis, R. P. (1976). *Alternaria* fruit rot of gooseberry. *Pl. Path.* **25**, 57–58.

Dennis, C., Davis, R. P., Harris, J. E., Calcutt, L. W. and Cross, D. (1979). The relative importance of fungi in the breakdown of commercial samples of sulphited strawberries. *J. Sci. Fd Agric.* **30**, 959–973.

Edney, K. L. (1964). Post-harvest rotting of strawberries. *Pl. Path.* **13**, 87–89.

El-Goorani, M. A. and Sommer, N. F. (1979). Suppression of post-harvest pathogenic fungi by carbon monoxide. *Phytopathology* **69**, 834–838.

Freeman, J. A. and Pepin, H. S. (1968). A comparison of two systemic fungicides with non-systemics for control of fruit rot and powdery mildew in strawberries. *Can. Pl. Dis. Surv.* **48**, 120–123.

Freeman, J. A. and Pepin, H. S. (1976). Control of pre- and post-harvest fruit rot of raspberries by field sprays. *Acta Hort.* **60**, 73–80.

Freeman, J. A. and Pepin, H. S. (1977). Control of post-harvest fruit rot of strawberries by field sprays. *Can. J. Pl. Sci.* **57**, 75–80.

Gooding, H. J. (1976). Resistance to mechanical injury and assessment of shelf-life in fruits of strawberry (*Fragaria* × *Ananassa*). *Hort. Res.* **16**, 71–82.

Harper, K. A., Beattie, B. B., Pitt, J. I. and Best, D. J. (1972). Texture change in canned apricots following infection of the fresh fruit with *Rhizopus stolonifer*. *J. Sci Fd Agric.* **23**, 311–320.

Harris, J. E. and Dennis, C. (1979). The stability of pectolytic enzymes in sulphite liquor in relation to breakdown of sulphited strawberries. *J. Sci. Fd Agric.* **30**, 704–710.

Harris, J. E. and Dennis, C. (1980a). Distribution of *Mucor piriformis*, *Rhizopus sexualis* and *R. stolonifer* in relation to their spoilage of strawberries. *Trans. Br. mycol. Soc.* **75**, 445–450.

Harris, J. E. and Dennis, C. (1980b). Heat stability of endopolygalacturonases of Mucoraceous spoilage fungi in relation to canned fruits. *J. Sci. Fd Agric.* **31**, 1164–1172.

Harris, J. E. and Dennis, C. (1981). Susceptibility of strawberry varieties to breakdown in sulphite by endopolygalacturonase from Zygomycete spoilage fungi. *J. Sci. Fd Agric.* **31**, 87–95.

Harris, J. E. and Dennis, C. (1982a). The influence of berries infected with *Botrytis cinerea* on the enzymic breakdown of sulphited strawberries. *Ann. appl. Biol.* **101**, 109–117.

Harris, J. E. and Dennis, C. (1982b). Heat stability of fungal pectolytic enzymes. *J. Sci. Fd Agric.* **33**, 781–791.

Harris, C. M. and Harvey, J. M. (1973). Quality and decay of California strawberries stored in CO_2-enriched atmospheres. *Pl. Dis. Reptr* **57**, 44–46.

Harvey, J. M. and Harris, C. M. (1976). Temperature maintenance in air shipments of strawberries to Far Eastern Markets. *Proc. Int. Inst. Refr. Conf. Melbourne*, No. 25, 7 pp.

Harvey, J. M. and Pentzer, W. I. (1960). "Market Diseases of Grapes and Other Small Fruits", Agriculture Handbook No. 189, 24–25. United States Department of Agriculture, Washington, D.C.

Hondelmann, W. and Richter, E. (1973). Uber die Anfalligbeit von Erdebeerklonen gegen *Botrytis cinerea* Pers. in Abhangigkeit von Pektinquantitat and Qualitat der Fruchte. *Gartenbauwiss* **38**, 311–314.

Hunter, T., Jordan, V. W. L. and Pappas, A. C. (1979). Control of strawberry fruit rots caused by *Botrytis cinerea* and *Phytophthora cactorum*. *Proc. Br. Crop Prot. Conf.*, *Pests Dis.* **1**, 177–183.

Ingold, C. T. (1978). "The Biology of *Mucor* and its Allies," Studies in Biology No. 88. Edward Arnold, London.

Irvine, T. B. (1959). A study of laboratory methods to determine strawberry varietal resistance to grey mould (*Botrytis cinerea*). M.Sc Thesis, Michigan State University, East Lansing.

Irvine, T. B. and Fulton, R. H. (1959). A study of laboratory methods to determine susceptibility of strawberry varieties to grey mould fruit rot, *Botrytis cinerea.* *Phytopathology* **49**, 452 (Abstract).

Jarvis, W. R. (1961a). Problems in the control of raspberry and strawberry grey mould. *Proc. Br. Crop Prot. Conf., Pests Dis.* **1**, 315–319.

Jarvis, W. R. (1961b). Grey mould of soft fruit. *Rep. Scott. hort. Res. Inst. 1960–1961*, 60–63.

Jarvis, W. R. (1962a). The infection of strawberry and raspberry fruits by *Botrytis cinerea* Fr. *Ann. appl. Biol.* **50**, 569–575.

Jarvis, W. R. (1962b). The dispersal of spores of *Botrytis cinerea* Fr. in a raspberry plantation. *Trans. Br. mycol. Soc.* **45**, 549–559.

Jarvis, W. R. (1962c). Grey mould of soft fruit. *Rep. Scott. hort. Res. Inst. 1961–1962*, 72–74.

Jarvis, W. R. (1969). The phenology of flowering in strawberry and raspberry in relation to grey mould control. *Hort. Res.* **9**, 8–17.

Jarvis, W. R. (1973). Factors affecting post-harvest fungi on strawberry and raspberry fruit. *Bull. Scott. hort. Res. Inst. Assn* **7**, 29–31.

Jarvis, W. R. and Borecka, M. (1968). The susceptibility of strawberry flowers to infection by *Botrytis cinerea* Pers. Ex. Fr. *Hort. Res.* **8**, 147–154.

Jennings, D. L. and Carmichael, E. (1975). Resistance to grey mould (*Botrytis cinerea* Fr.) in red raspberry fruits. *Hort. Res.* **14**, 109–115.

Jennings, D. L., Dale, A. and Carmichael, E. (1976). Raspberry and blackberry breeding at the Scottish Horticultural Research Institute. *Acta Hort.* **60**, 129–133.

Jordan, V. W. L. (1973). The effects of prophylactic spray programmes on the control of pre- and post-harvest diseases of strawberry. *Pl. Path.* **22**, 67–70.

Jordan, V. W. L. (1978). Epidemiology and control of fruit rot *Botrytis cinerea* on strawberry. *Pfl Schutz-Nachr. Bayer* **31**, 1–10.

Jordan, V. W. L. and Pappas, A. C. (1977). Inoculum suppression and control of strawberry *Botrytis*. *Proc. Br. Crop Prot. Conf., Pests Dis.* **2**, 341–348.

Jordan, V. W. L. and Richmond, D. V. (1974). The effects of benomyl on sensitive and tolerant isolates of *Botrytis cinerea* infecting strawberries. *Pl. Path.* **23**, 81–83.

Jordan, V. W. L. and Richmond, D. V. (1975). Perennation and control of benomyl-insensitive *Botrytis* affecting strawberries. *Proc. Br. Crop Prot. Conf., Pests Dis.* **1**, 5–13.

Keep, E. (1976). Progress in *Rubus* breeding at East Malling. *Acta Hort.* **60**, 123–128.

Kichina, V. V. (1976). Raspberry breeding for mechanical harvesting in Northern Russia. *Acta Hort.* **60**, 89–94.

Lowings, P. H. (1956). The fungal contamination of Kentish strawberry fruits in 1955. *Appl. Microbiol.* **4**, 84–88.

Maas, J. L. (1978). Screening for resistance to fruit rot in strawberries and red raspberries: a review. *Hort. Sci.* **13**, 423–426.

Maas, J. L. and Smith, W. L. (1978). "Earliglow", a possible source of resistance to *Botrytis* fruit rot in strawberry. *Hort. Sci.* **13**, 275–276.

Mason, D. T. and Dennis, C. (1978). Post-harvest spoilage of Scottish raspberries in relation to pre-harvest fungicide sprays. *Hort. Res.* **18**, 41–53.

Pappas, A. C., Cooke, B. K. and Jordan, V. W. L. (1979). Insensitivity of *Botrytis cinerea* to iprodione, procymidone and vinclozolin and their uptake by the fungus. *Pl. Path.* **28**, 71–76.

Powelson, R. L. (1960). Initiation of strawberry fruit rot caused by *Botrytis cinerea*. *Phytopathology*, **50**, 491–494.

Prasad, K. and Stadelbacher, G. J. (1973). Control of post-harvest decay of fresh raspberries by acetaldehyde vapour. *Pl. Dis. Reptr* **57**, 795–797.

Prasad, K. and Stadelbacher, G. L. (1974). Effect of acetaldehyde vapour on post-harvest decay and market quality of fresh strawberries. *Phytopathology*, **64**, 948–951.

Schipper, M. A. A. (1975). On *Mucor mucedo, Mucor flavus* and related species. *Stud. Mycol.*, No. 10, 1–33.

Scott, D. H. and Lawrence, F. J. (1975). Strawberries. *In* "Advances in Fruit Breeding" (J. Jarvick and J. N. Moore, Eds), 71–97. Purdue University Press, West Lafayette, Indiana.

Shaw, G. W. (1969). The effect of controlled atmosphere storage on the quality and shelf-life of fresh strawberries with special reference to *Botrytis cinerea* and *Rhizopus nigricans*. Ph.D. Thesis, University of Maryland.

Smith, W. H. (1963). The use of carbon dioxide in the transit and storage of fruits and vegetables. *Adv. Fd Res.* **12**, 95–196.

Stevens, N. E. (1919). Keeping quality of strawberries in relation to their temperature when picked. *Phytopathology*, **9**, 170–177.

Stevens, N. E. and Wilcox, R. B. (1918). "Further Studies on the Rots of Strawberry Fruits", Bulletin No. 686. United States Department of Agriculture, Washington, D.C.

Strand, L. L., Ogawa, J. M., Bose, E. and Rumsey, J. W. (1981). Bimodal heat stability curves of fungal pectolytic enzymes and their implication for softening of canned apricots. *J. Fd Sci.* **46**, 498–505.

Topping, A. J. and Cockburn, J. T. (1979). Storage of soft fruit. *Rep. E. Malling Res. Stn 1978*, 159.

Verhoeff, K. and Warren, J. M. (1972). In vitro and in vivo production of cell wall degrading enzymes by *Botrytis cinerea* from tomato. *Neth. J. Pl. Path.* **78**, 179–185.

Von Arx, J. A. (1970). "The Genera of Fungi Sporulating in Pure Culture." Verlag von J. Cramer, Lehre.

Wells, J. M. (1970). Modified atmosphere, chemical and heat treatment to control post-harvest decay of California strawberries. *Pl. Dis. Reptr* **54**, 431–434.

3

Top Fruit

K. L. EDNEY

I. Introduction

As apples and pears ripen they become susceptible to attack by a variety of fungi to which they were resistant during their period of development on the tree. Much of the rotting that develops during storage is derived from spores that collect on the fruit surface during the growing season but which, with a few exceptions, are incapable of causing rotting until after harvest. The strength of the pathogenicity of each species is reflected in the stage of ripeness at which it is able to attack. Post-harvest rotting also develops if fruit is damaged in some way that enables wound parasites to gain entry.

Ripening fruits are not simply culture media for a large range of sapro-phytic fungi. Examination of the identity of the species causing wastage has shown the same fungi to be responsible in many different countries. The earliest surveys of the causes of rotting list 20–30 different fungi but few references to the scale of losses (Schneider-Orelli, 1912; Ames, 1915; Brooks and Cooley, 1917; Marchal and Marchal, 1921; Colhoun, 1938; Kidd and Beaumont, 1925).

Accurate data on the scale of losses is difficult to obtain and, where fungicidal treatment is applied either before or after harvest, will only indicate the incidence of those species which survive such treatment. The most accurate surveys were made in the 1960s. In England, pathologists of the National Agricultural Advisory Service (now known as the Agricultural Development & Advisory Service) studied losses of Cox's Orange Pippin apple during 1961–65 and recorded total losses from all causes of about 6%, with losses from individual stores ranging from 0.3% to 21.0%. Wastage due to rotting caused about half of the total loss and was caused mainly by *Gloeosporium* spp. and *Monilia fructigena* (Preece, 1967). In Northern Ireland, Swinburne (1970) reported average losses of Bramley's Seedling apple between 1962 and 1967 of 15% in refrigerated gas stores and 3% in air (barn) stores. The main species concerned here were *Nectria galligena* and

Penicillium expansum, which accounted for more than half of all the rots examined.

In France, Bondoux (1967a) has given examples of rotting of apples and pears from a number of different areas, *Gloeosporium* spp., *M. fructigena*, *Botrytis cinerea*, *N. galligenà* and *Penicillium* spp. being the main organisms involved and total losses ranging from 7.58 to 22.6%. In New Zealand, Cooper and Padfield (1965) also cite *Gloeosporium* and *Penicillium* spp. as being amongst the main causes of rotting together with *Glomerella*, *Pleospora* and *Alternaria* spp. in a list of 15 different species. The most recent surveys from Poland (Ostrowski, 1971), East Germany (Katschinski, 1974) and Yugoslavia (Babovic *et al.*, 1979) all include *Penicillium*, *Monilia* and *Gloeosporium* amongst the most important causes of rotting of apple and one has to go to a warmer climate, in India, before a different range of causal organisms is encountered (Vyas *et al.*, 1976).

II. Main Features of Principal Diseases

A. *Gloeosporium album* Osterw.

This species is an important parasite of dessert apples, particularly in Europe, but is of minor importance on pears. The conidial stage was recorded initially by Osterwalder (1907) and has grown in importance with the development of fruit storage. The perfect stage, *Pezicula alba* Guthrie (1959) has been found on the wood of apple, but the ascospores are not regarded as an important source of inoculum so far as the fruit rot is concerned. It parasitizes healthy wood with difficulty and is found mainly on dead wood (Ogilvie, 1935). Both stages have also been found on leaves (Olsson, 1967; Tan and Burchill, 1972). The conidia, which are curved, cylindrical with rounded ends, are produced throughout the year (Burchill and Edney, 1963) and are dispersed from acervuli by rain. Fruit may therefore become contaminated at any time during the growing season. The conditions necessary for infection are not fully understood, although it has been shown that the susceptibility of apples is markedly influenced by the amount of rain falling on the fruit prior to inoculation with conidia (Edney *et al.*, 1977). The rots are circular, slow growing, fairly firm, mid-brown in colour and may produce abundant surface mycelium. They appear well before the end of the storage period under all normal storage regimes.

B. *Gloeosporium perennans* Zeller and Childs

As with *Gloeosporium album*, *G. perennans* is a major cause of rotting of apples, but is of much lesser importance on pears. It is widely encountered in Europe, the U.S.A. and elsewhere. It is a more effective parasite of healthy woody tissue and was initially described as a "perennial canker" in America by Zeller and Childs (1925). The perfect stage, *Pezicula malicorticis* (Jacks) Nannf., was first reported in 1913 in America but the ascospores are not considered to be an important source of inoculum so far as the fruit rot is concerned. Two types of conidia are produced in acervuli throughout the year and are dispersed by rain. The macro-conidia are cylindric–fusoid whereas the micro-conidia are spherical. They germinate on the surface of the fruit and may form a latent infection in the lenticels. Subsequent development in store usually takes place earlier than for *G. album* with the production of slow-growing, circular rots, frequently with a yellow centre, which have earned the name "target" or "bull's-eye" rot.

C. *Monilia fructigena* Pers.

This species is a wound parasite and is capable of attacking apple and pear fruits before and after harvest and may attack woody parts to form a spur canker on fruit. It causes lesions that are firm, irregular in shape, spread rapidly and in the orchard produce masses of chains of brown oval conidia from which the term "brown rot" is derived. It is the conidial stage of *Sclerotinia fructigena* Aderh. and Ruhl. Sporulation under storage conditions is very limited but a profuse, white mycelium may be produced under moist conditions.

Development of orchard-derived infections on fruit continues in store and spread by contact between sound and infected fruit is also important and may lead to a tenfold increase in the incidence of rotting (Edney, 1978). The rots remain firm and as they lose water gradually become hard. In the orchard such fruits eventually become mummified and act as a source of infection the following year.

D. *Phytophthora syringae* (Klebahn)

Although first reported as a cause of apple and pear rot in Ireland in 1922 by Lafferty and Pethybridge (1922) and in England in 1929 by Ogilvie (1931), rotting on an epidemic scale was not experienced until the 1970s. Serious outbreaks have been reported from England (Edney, 1978) and Switzerland (Bolay, 1977). The fungus is soil-borne, serious losses of fruit being associated with heavy rainfall which splashes fruit close to the ground with infected soil (Edney, 1978). Harris (1979) has shown that fallen apple and pear leaves contain oospores of *Phytophthora syringae*, in the case of apple, often in large numbers. These are considered to be the major source of inoculum in the orchard. After a resting period the oospores germinate in wet conditions to form sporangia; these in turn release zoospores which are the main agents of infection (Edney and Chambers, 1981a). Infection occurs through lenticels and requires a period of wetness, 24 h at 15°C (Edney, 1978). If conditions are suitable, lesions appear before harvest, the development of visible and other developing infections continues in store and may cause high losses by the spread of rotting by contact with sound fruit (Edney, 1978). The lesions are mainly light greenish-brown in colour, or occasionally darker brown; on pear they are usually dark brown. Rotted tissue is frequently invaded by other soft-rotting fungi.

E. *Botrytis cinerea* Fr.

This fungus is a wound parasite and can cause heavy losses of apples and pears, although it is probably more important on the latter. It sometimes causes a dry eye-rot of apple in the orchard (Wilkinson, 1942), which is non-progressive but develops in store to form a normal rot of the flesh. The lesions are light to mid-brown and irregular in shape. Abundant surface mycelium is produced and sporulation is prolific, producing ovate–ellipsoid conidia which in the mass are grey in colour: hence the name "grey mould". Sclerotia may form on the fruit surface and are a feature of well-advanced lesions. *Botrytis* spreads very readily by contact on apple and this may cause a 16-fold increase in infection during storage (Edney, 1978).

A feature of the disease in recent years has been the development of strains that are resistant to the benzimidazole group of fungicides. As a result the amount of wastage, particularly of pears, has increased after a period when good control had been obtained.

F. *Penicillium expansum* Thom.

A common saprophyte which sporulates profusely, *Penricillium expansum* is the most important species of *Penicillium* attacking apples and pears. Invading mainly through wounds or bruises (Wright and Smith, 1954; Spalding *et al.*, 1969), it is also capable of invading lenticels under favourable conditions (Baker and Heald, 1934). The rot produced is pale in colour and watery in texture. During storage, conidia are produced in coremia which are initially white and become blue-green as the spores mature, giving rise to the description "blue mould". The incidence of this disease is markedly increased if fruit is washed or transported in water in pack-houses (hydro-handling), because this leads to a build-up of inoculum washed from the fruit surface (Burton and Dewey, 1981). Spread by contact between infected and sound fruit also occurs (Edney, 1978).

III. Minor Diseases

The term "minor" is used to describe diseases that are of lesser importance when regarded in the light of the total wastage they cause world-wide. It is acknowledged, however, that in some countries they may always be of considerable importance and that they may also cause losses of epidemic proportions in isolated instances.

A. *Alternaria* spp.

A number of different species of *Alternaria* have been reported on apple and pear (Tweedy and Powell, 1963), *A. tenuis* occurring most frequently. A weak parasite, it produces a dark brown, shallow rot on damaged tissue. On fruit with open calyces it may cause rotting in the core area (Ceponis *et al.*, 1969) and has been reported from Israel (Ben-Arie, 1967) as a secondary infection of bitter pit lesions.

B. *Cladosporium herbarium* Link.

This is chiefly of interest because of its association with scald and other types of physiological disorders of English cultivars of apple. On Laxton Fortune and Ellison's Orange, soft (ribbon) scald which causes shallow areas of moribund tissue may be colonized with the formation of a dark, black-brown rot which is mainly confined to the scalded area and does not penetrate the flesh deeply.

C. *Cylindrocarpon mali (Allesch.)* Wallenw.

As a cause of lenticel rotting, *Cylindrocarpon mali* attacks late in the storage period and, with the exception of Northern Ireland (Swinburne, 1970) and Eire (Kavanagh and Glynn, 1966), is usually of minor importance. It is the conidial stage of *Nectria galligena* which in Europe is better known as a cause of canker of apple and pear wood. Cylindrical, septate conidia are produced in sporodochia throughout the growing season and are dispersed by rain. In the orchard they may infect fruit to cause a dry eye-rot, but normally do not cause rotting until after harvest. In store a circular, mid-brown, softish rot is produced.

D. *Fusarium* spp.

Several species of *Fusarium* have been recorded on apple and pear, those most frequently encountered being *F. avenaceum* (Fr.) Sacc., *F. culmorum* (W. G. Sm) Sacc. and *F. lateritium* Nees (ex Fr.) (Kidd and Beaumont,

1924; Swinburne, 1970). All are very weak parasites causing lenticel rots towards the end of the storage period. Infection is derived from the sickle-shaped septate conidia which are produced in sporodochia on dead wood.

E. *Phomopsis mali* Roberts

The conidial stage of *Diaporthe perniciosa* Marchal, this fungus is consistently included in records of fungi causing rotting of apple and pear, although it is not regarded as a major pathogen. It occasionally causes significant losses of cv. Bramley's Seedling in England and Northern Ireland. The sources of infection are die-backs on apple and pear wood which produce two types of conidia in pycnidia: the so-called "a-spores", which are oval, and the "b-spores", which are filiform. Conidia are dispersed by rain from tendrils which exude from the pycnidia. They penetrate either through the cuticle or the lenticels, the soft-textured rots developing late in storage, and are frequently situated at the stalk end (Ayub and Swinburne, 1970). Stromatic bodies develop during storage, and spores are produced from pycnidial locules within these.

F. *Trichothecium roseum* Link.

Infection by this species is most common under non-refrigerated conditions since it does not grow below 4°C (Borecki and Profic, 1962) and with the widespread use of storage temperatures below this level its importance has diminished. On apple it is frequently associated with lesions caused by the scab fungus, *Venturia inaequalis*. It produces pear-shaped, two-celled, pink conidia, from which the name "pink rot" is derived.

Many more fungi are capable of rotting apples and pears (Mezzetti, 1958; Terui, 1959; Pratella, 1960; Bondoux, 1964; Katschinski, 1974; Babovic *et al.*, 1979), but are of very little commercial importance at present.

IV. Factors Affecting the Incidence of Rotting

The mode of attack and strength of pathogenicity varies with each fungus. Some attack mainly through wounds, whereas in sound fruit the main point of ingress is the lenticel. It is in the latter situation that ripening is particularly closely related to the development of rotting.

For apples, Horne (1934) demonstrated that the time taken for invasion to start via the lenticel varied considerably, but that if the skin was punctured, spread within the cortical tissues showed less variation. He referred to the resistance of cortical tissues to radial advance as "internal" and to the resistance encountered in the lenticel as "external". Although this simple concept has long since been overtaken by more detailed observations, the differentiation between the two aspects of resistance remains valid. A more recent view has been given by Bompeix (1978a) in relation to the susceptibility of apples to *Pezicula alba* and *P. malicorticis*. He divided the total period of growth and storage into three phases. During the first of these, the "refractory" phase, infection is impossible, even if artificially assisted, because the tissue is resistant to maceration enzymes and to the presence of pre- and post-infection inhibitors. In the "interim" phase, resistance can be overcome if tissue is wounded or enzymes or nutrients are administered at the site of penetration by infiltrating the lenticels. In the final, "susceptible" phase, natural invasion is possible, there being no post-infection inhibitors present, and a reduction in the level of pre-infection inhibitors and in resistance to the activity of endo-polygalacturonase (endo-PG) and endo-polymethylgalacturonase (endo-PMG) enzymes. In this phase there is also an increase in the synthesis of pectolytic enzymes and the availability of adequate sources of carbon and nitrogen. These three phases are also reflected in infection by other fungi, although the features of the "susceptible" phase associated with *P. alba* or *P. malicorticis* will not necessarily be the same.

In addition to the effect of host factors, the incidence of rotting is also influenced by the numbers of viable spores present at potential infection sites when fruit ripens sufficiently to permit infection to develop. The amount of inoculum present is closely related to weather conditions during the growing season, particularly where the spores are dispersed by rain, e.g. *Gloeosporium* and *Phytophthora*.

The rate of ripening is governed by the conditions of storage so that the incidence of rotting at any point in time is a reflection of the interaction between all of these factors.

A. The Lenticel

Lenticels develop during the growing season from tears which develop on the skin when the expansion of the cortical tissues proceeds at a rate greater than that of the skin. Hairs and stomata yield first to the strain created and as they disappear a meristem is formed which produces a suberized barrier that either partly or completely seals the openings created. This process has been described by Zschokke (1897), Kidd and Beaumont (1925), Tetley (1930) and Krapf (1961). The degree of closure varies, a completely closed lenticel being one from which no air can be withdrawn if the fruit is submerged in water and subjected to a reduced pressure. Where this occurs, a physical barrier to infection is created. Bompeix (1966) found a positive correlation between the thickness of the cork layer and susceptibility to rotting. The importance of intact lenticels as a barrier to infection was demonstrated by Moreau *et al.* (1965), who submerged Golden Delicious apples in a suspension of spores of *Gloeosporium album* and then applied a reduced pressure of 70 cm mercury for 5 min, finally restoring to normal pressure rapidly. The treatment ruptures the lenticels and the development of infection is accelerated.

Penetration of the lenticel may take place via gaps in the suberized layer or in the angle between the epidermis and the tissue sealing the lenticel pore (Edney, 1958; Bompeix, 1967). The junction between the two types evidently constitutes a point of weakness, possibly due to strains set up by the expansion of the epidermis during growth.

The lenticel cavity at harvest contains a collection of micro-organisms in varying stages of development. These are prevented to some degree from developing to cause a rot by the suberized barrier lining the cavity, but also, as in the "open" lenticel where this is either poorly developed or incomplete, by the resistance of the cortical tissues. The nature of this resistance is complex but has been associated with several factors including the composition of the cell wall, in particular the importance of the calcium content and the presence of inhibitory phenolic substances.

B. Calcium

Studies of the mineral content of apples in many countries have established a negative relationship between the calcium content of fruit and the incidence of the physiological disorder bitter pit (Wilkinson and Fidler, 1973). In the course of this work, note was taken of the influence of calcium content on other physiological disorders and also on rotting, and, as a result, a relationship between the incidence of rotting and calcium content was established. Sharples and Little (1970) showed that spraying apples with calcium salts reduced the incidence of rotting. Fuller (1976) compared samples of Cox's Orange Pippin from four different sources and showed that rotting caused by *Gloeosporium perennans* developing from an applied spore load was highest in samples with a low calcium content at harvest.

The same relationship has been shown by Jackson *et al.* (1977) in a study of the effect of shading during growth on the size and chemical composition of apple fruit. Shading so that the trees received only part of full daylight was found to reduce fruit size through reductions in cell size and the number of cells per fruit. The smaller fruits had higher concentrations of calcium and a lower incidence of rotting. The importance of the high susceptibility to rotting caused by calcium deficiency lies not only in the incidence of rotting during a given period of storage but also in the additional demands made on fungicidal treatments. In a study of persistent infection of stored apples by *Gloeosporium* spp., Edney (1976) demonstrated that on farms where fungicidal treatment was relatively unsuccessful, the susceptibility of the fruit was high and was associated with low calcium and high nitrogen content.

The role of calcium in cell membrane permeability and decomposition has been investigated by Bangerth (1973), who demonstrated with electron micrographs the disintegration of membranes in cells from pitted tissue. In a study of the outer tissues of Cox's Orange Pippin apples, Fuller (1976) stored fruit with low and high calcium content at the time of harvest, in air at 3°C. Examination with the electron microscope revealed that, although after 53 days of storage the ultrastructure of both low and high calcium samples was normal, after 81 days the low calcium samples showed symptoms of membrane breakdown, and this tendency continued with further storage, with more cells exhibiting breakdown of membranes and organelles and more symptoms of cell wall breakdown. The disorganization of cell walls was very severe in long-stored, low calcium fruits and was frequently observed in the central part of the walls, indicating that it may have originated in the middle lamella. Although the biochemistry of cell wall structure and the breakdown due either to senescence or fungal attack has attracted the attention of research workers for many years, many aspects of these problems are still unresolved.

C. Composition of the Cell Wall

The association of calcium content with cell breakdown during ripening does not establish a causal connection between the two factors. The softening of cell walls occurs when divalent metal ions, particularly calcium, are transferred into the cell. Calcium has long been thought to be important as a cross-linkage between polygalacturonide chains in plant cell walls (Doesburg, 1957). However, current opinion suggests that the migration of calcium ions is not responsible for softening. Knee (1973) and Knee *et al.* (1975) have shown that apple fruit cell walls contain a relatively unbranched polyuronide, some of which becomes soluble on ripening and a polyuronide with many side-chains of galactose and arabinose residues which remain insoluble. The unbranched polyuronide is found in the middle lamella and solubilization of this component is associated with cell separation. The methylation of free carboxyl groups during ripening could result in a weakening of the middle lamella (Knee, 1977). Although apple cell walls become more susceptible to breakdown by polygalacturonase during ripening (Knee *et al.*, 1975), autolytic degradation of the host tissue is thought to be a more important cause of susceptibility to fungal attack. However, studies of the circumstances in which rotting develops must relate the loss of resistance to pectolytic enzymes to the actual production of the enzymes by the pathogens concerned; there are very few such studies. Using *Botryosphaeria ribis*, *Glomerella cingulata*, a *Physalospora* sp. and *Physalospora obtusa*, Wallace *et al.* (1962) compared the protein and inorganic components of Golden Delicious apples over the period when they become susceptible to rotting. Analysis showed a loss of protein from the water-insoluble parts of the fruit during maturation until it became susceptible to rotting, the level remaining more or less constant thereafter. The amounts of calcium, copper, iron and magnesium similarly fell, whereas the level of potassium was found to increase with susceptibility. The decrease in polyvalent cations was found to be roughly proportional to the decrease in protein content and a relationship between polyvalent cation and protein solubility was suggested. A pectin–protein–metal complex resistant to fungal hydrolysis and therefore unable to serve as a carbon source was thought to be involved in the resistance of immature apples to the pathogens concerned.

D. Fruit Size and Maturity at Harvest

The influence of fruit size on the calcium content of apples (Jackson *et al.*, 1977; Perring, 1979) is clearly a major factor in the storage potential including susceptibility to rotting. Variation of susceptibility with size of fruit has been demonstrated by Edney (1958), who loaded fruit of different sizes with spores of *G. Sperennans*, and Olsson (1965a), who inoculated fruit and measured the rate of spread of lesions. However, it is unlikely that calcium content is the only factor that varies with size. Martin and Lewis (1952) showed that large fruits from light-cropping trees have larger cells and a higher rate of respiration per unit protein than the smaller cells of fruits from heavy-cropping trees. They suggested that the more rapid senescence and susceptibility to storage disorders of the larger apples might be due to the higher rate of respiration. In a study of the effect of thinning on fruit size, Sharples (1964, 1968a) also found that the larger fruits from thinned trees were more susceptible to rotting caused by *G. perennans*. These differences disappeared if fruits from thinned and unthinned trees were picked at similar points on the climacteric rise in respiration. The onset of the climacteric was advanced by thinning and Sharples considered that the differences observed may largely be a consequence of differences in maturity at harvest.

The significance of this last factor has been investigated by several workers. Edney (1964b) applied spores of *G. perennans* to Cox's Orange

Pippin apples 1 week before the first of three dates of picking. The incidence of rotting in stores increased sharply with the lateness of picking. Wilkinson and Sharples (1967) also found that rotting increased with lateness of pick of Cox's Orange Pippin. In Australia Tindale (1966) found rotting to be greatest on Golden Delicious that had been picked late and in Hungary, Krasztev Kemenes (1977) found that *Penicillium expansum* infection increased with fruit size and lateness of picking. It must be borne in mind that a further consequence of late picking is the longer period to which the fruit is exposed to infection, which in itself would lead to a higher level of rotting. The merit of working with an applied spore load on trees known to carry little or no natural infection is that the results only reflect the influence of maturity.

E. Phenolic Compounds

These compounds have been studied chiefly in relation to latent infections of apple by *Gloeosporium* spp. and this work is described elsewhere in this volume (see Ch. 1). Observations and differences in susceptibility to rotting between the red and yellow-green sides of apples have been made by a number of workers and resistance has been ascribed to the higher levels of phenolic compounds present. Ndubizu (1976) found that extracts of Red Delicious apples contained *p*-coumaryl-quinic and chlorogenic acids which inhibited the growth of *Botrytis cinerea* spores and the mycelial growth of *B. cinerea*, *P. expansum* and *Alternaria* spp., and noted the parallelism between decrease in phenolic compounds as fruit matured and the rise in susceptibility. Bykova *et al.* (1977) and Agarwal and Bisen (1977) have shown that the production of phenolic compounds is associated with the development of rotting and concluded that they may be related to the development of resistance. Fawcett and Spencer (1967) have also shown that a number of anti-fungal substances are present in the juice of Edward VII apples attacked by *Sclerotinia fructigena*, two of which are 4-hydroxy-3-methoxybenzoic acid and 4-hydroxybenzoic acid. These arise from the juice rather than the cell walls. The same authors later demonstrated that *S. fructigena* can convert chlorogenic and quinic acids into compounds with greater anti-fungal activity in culture (Fawcett and Spencer, 1968). Mudretsova-Viss *et al.* (1975) found that resistance to rotting after inoculation with *Alternaria tenuis* and *P. expansum* was greater on the dark red side of apples than on the yellow-red side, but Edney (1964b) was unable to show any such differences with regard to rotting of Cox's Orange Pippin by *G. perennans*. The difficulty in assessing these data lies in the fact that other differences in biochemical change and cell wall composition are associated with the position of the fruit on the tree and its exposure to light. Thus

Atkinson (1974) showed in a comparison of tree spacings that apples grown on widely spaced trees developed more rotting in store and that the fruit, because of its large size, contained a lower concentration of calcium. These results conform with the known effects of different periods of exposure to light, the wider spacings resulting in larger apples. In these circumstances the more coloured fruit was less rather than more resistant to rotting.

Phenolic compounds are also involved in the appearance and texture of rots. In colour, apple rots may vary from the pale green of the typical *Penicillium* rot to the dark brown of *S. fructigena*. In texture the former is soft, whereas the latter is firm. Cole and Wood (1961) showed that in tissue rotted by *S. fructigena*, there is little (10–20%) breakdown of the water-insoluble pectin in cell walls and that extracts of rotted tissue contain very little polygalacturonase. Oxidized polyphenols are known to be able to inactivate this enzyme (Cole, 1956) and it was concluded that breakdown of cell walls was limited by this inhibitory action. In contrast, tissue rotted by *P. expansum* loses about 70% of the pectic substances and contains abundant polygalacturonase. Wilson *et al.* (1973) found that lesions on McIntosh apples inoculated with *P. expansum* after controlled atmosphere storage were darker than those inoculated after storage in air. Thus after controlled atmosphere storage there was little effect on the *o*-diphenol oxidase system by the soluble inhibitors associated with decay by *Penicillium*. Walker (1969) also found that there is a substantial reduction in *o*-diphenol oxidase activity in tissue rotted by *P. expansum*. Presumably, the iso-enzymes responsible for browning in McIntosh apples stored in a controlled atmosphere are less susceptible to inhibitors present in rotten tissue than those responsible for browning in fruit stored under refrigerated conditions in air.

F. Orchard Management

Studies of the relationship between cultural factors and storage behaviour have been in progress for many years. Many of these sought to relate manurial treatment to fruit composition and the incidence of storage disorders but were of little value to the pathologist because the organisms responsible for wastage were not identified. The direct effect of such treatment on the susceptibility of the trees themselves to infection by fungi capable of causing fruit rots after harvest was also neglected. Thus when the sources of inoculum are cankers and die-backs, the spore load dispersed on to the fruit during the growing season must be considerably influenced by tree vigour.

The earliest studies in England (Gregory and Horne, 1928) indicated that susceptibility to rotting was increased by high nitrogen content and decreased by high potash. Other workers sought to influence the composition of the fruit by experimenting with different combinations and levels of applied fertilizer and were able to demonstrate that added nitrogen increased the nitrogen content of the fruit and in so doing assisted the radial spread of rotting following inoculation with *Cytosporina ludibunda* (Muskett *et al.*, 1938).

A resurgence of interest in these investigations started in the 1950s when the use of refrigerated and controlled atmosphere storage became more widespread and growers were concerned that methods should be found to extend the length of storage life. These developments brought with them problems associated with long-term storage, e.g. the incidence of low temperature and senescent breakdown, and also the emergence of lenticel rotting by *Gloeosporium* spp. as a major disease of stored apples. Since the amount of rotting was closely associated with maturity and the length of storage, study of the diseases was related to cultural practices that affected susceptibility.

In a long-term study of the effects of nitrogen, phosphorus and potassium fertilizers between 1953 and 1958, Montgomery and Wilkinson (1962) observed consistent effects of applied nitrogen increasing wastage due to *G. album*.

Trees grown under sward were found to produce fruit that was relatively resistant to rotting. Hulme (1956) has shown that fruit grown under these conditions contains less nitrogen. Other effects detected in these comparisons include that of potash, which increased rotting in some years, but this effect decreased with time, and of phosphate, which in the absence of nitrogen applications resulted in fewer rots and in the presence of nitrogen was associated with increased rotting. Wilkinson (1957) found that the competitive effect of sward became less in the later years of his experiment and the level of nitrogen was sometimes higher than in fruit from corresponding plots under cultivation. In spite of this there was less rotting of fruit grown under sward. Sharples (1968b) has suggested that the influence of nitrogen may also be indirect in affecting the vigour of the tree and causing an imbalance in the K/Ca ratio in the fruit. Comparisons of the influence of orchard treatments were carried out in a number of countries at about this time and, in general, apart from minor differences reached similar conclusions (Martin, 1953; Ostrowski *et al.*, 1959; Tiller *et al.*, 1959; Nyhlen, 1960; Landfald, 1962).

However, in spite of these long-standing associations between mineral composition and rotting, no causal connection between the two has been established.

Since the completion of these comparisons, the development of successful fungicidal treatments for rotting has resulted in few data being collected of the effect of orchard treatments on rotting. This appears mainly to be because rotting was an unavoidable feature of storage experiments in the 1950s and early 60s and once it could be virtually eliminated no attempt was made to include it in routine assessments of the effect of mineral composition on storage behaviour. The current levels of susceptibility of commercially grown fruit to rotting are largely unknown as is the amount of inoculum currently being produced in orchards. Nevertheless, new cultural practices must inevitably alter fruit composition. The use of herbicides to kill part or all of the ground cover in orchards is a case in point. Apples from trees growing in orchards employing the "overall herbicide" technique (i.e. no ground cover) have been found to be larger and to have higher concentrations of potassium and lower concentrations of calcium. The content of nitrogen in the fruit has also been raised in some years by the "overall herbicide" treatment (Sharples *et al.*, 1975). While post-harvest fungicide treatments remain effective, orchard practices can and will be introduced without reference to their possible effect on the incidence of rotting. If such post-harvest use becomes less effective because of the emergence of resistant strains of the causal organisms or discontinued for other reasons, e.g. changing views on their potential as health hazards, then this situation must change.

G. Conditions of Storage

Fruit is stored to extend the period over which it may be marketed. It is necessary to determine the most suitable conditions of storage for each cultivar of apple and pear and this is done by varying the temperature and composition of the store atmosphere until the optimum combination is established. The later stages of storage life are accompanied by the development of physiological disorders and rotting, and accordingly the assessment of each combination of temperature and storage atmosphere must take into account the effect on the development of each type of wastage. The development of storage techniques in the last 20 or so years is a reflection not only of the continuing need to improve storage performance but also of the availability of more accurate instruments for the measurement of carbon dioxide and oxygen. As a result, apples which have been the main subject of experimentation may now be kept under near-anaerobic conditions.

1. *Temperature*

The minimum temperature for growth of most of the moulds that cause rotting in store is below 0°C, so their development cannot be prevented by the use of refrigeration. Low temperatures do, however, greatly reduce the rate of rotting by retarding the growth of the causal organisms and also the rate of ripening of the fruit. In tropical regions, fungi with higher minimum temperatures for growth are encountered and, in India, for example, infection of apples by *Gliocephelotrichum bulbitium* (Ellis and Heseltine) does not take place below 10°C and can therefore be controlled by storage below this temperature (Jamaluddin *et al.*, 1973).

2. *Carbon Dioxide*

The simplest form of controlled atmosphere storage is achieved by restricting the ventilation of the store and allowing the concentration of carbon dioxide to increase to a predetermined level. A number of English cultivars, e.g. Bramley's Seedling, may be stored in this way, but in many instances the method causes an increase in the incidence of physiological disorders (Tomkins, 1960).

Any increase in concentration above the optimum may lead to increased rotting and this effect may be greater as temperature is reduced (Tomkins, 1968).

3. *Oxygen*

The use of restricted ventilation will lead not only to an increase in the concentration of carbon dioxide present but to a reduction in the oxygen level as it is consumed by respiration. If the carbon dioxide produced is absorbed, the effect of sub-atmospheric concentrations of oxygen on rotting can reduce the incidence of rotting (Tomkins, 1966).

4. *The Combined Effect of Oxygen and Carbon Dioxide*

Many cultivars that derive no benefit from storage in CO_2 levels above about 5% or which are injured at higher concentrations can be stored in an atmosphere with a lower concentration of oxygen than that resulting from absorption by the fruit. Absorption of CO_2 by passing the gas mixture through a "scrubber" enables the oxygen level to be reduced without a corresponding rise in the CO_2 level within the store. The resultant reduction in store pressure is compensated for by admitting air. The most commonly used gas mixture obtained in this way is 5% CO_2 + 3% O_2 and gives good results with many cultivars grown all over the world. However, the presence of 5% CO_2, although beneficial in retarding softening, colour and rotting, has been found to be responsible in some instances for increasing susceptibility to physiological disorders (Fidler and North, 1963). Reducing the concentration of carbon dioxide has been shown to reduce the incidence of these disorders for cv. Cox's Orange Pippin and has not been accompanied by an increase in rotting (Edney, 1964a). Subsequent work by Fidler and North (1964) showed that in the presence of low concentrations of oxygen, carbon dioxide could have a damaging effect which leads to more rotting.

Investigations in many countries have led to the determination of the optimum conditions for storage of a large range of apples and pears (Fidler, 1973). Recommended conditions for the same cultivar vary from one country to another and reflect differences in growing conditions and consumer preference for the type of ripened fruit.

The relationship between the incidence of rotting and maturity is not entirely straightforward and it is not known which of the changes accompanying ripening have a direct effect on rotting. Ground colour of the skin and hardness of the fruit are commonly used as an indication of maturity. In a comparison of the incidence of rotting of Cox's Orange Pippin by *G. perennans*, Edney (1964a) found that although fruit stored in 0% CO_2 + 3% O_2 was yellower and softer than fruit stored in 5% CO_2 + 3% O_2, no effect on rotting was observed.

Although it is likely that the main effect of controlled atmosphere storage on rotting is due to the retarding effect on ripening, there is some evidence that the influence of storage conditions on the growth and physiology of the causal organism is also a contributory factor. Edney (1964a) has shown that the production of pectolytic enzymes by *G. perennans* and *G. album* was reduced by the gas mixtures used in the storage of Cox's Orange Pippin, the effect being greater for the latter species. This reduction may be a contributory factor to the finding reported by Montgomery (1958) and Olsson (1965b) that samples of Cox's Orange Pippin infected with *G. album* had less rotting in 5% CO_2 + 3% O_2 at 3.9°C after $4\frac{1}{2}$ months than those kept in air at 2.8°C for $3\frac{1}{2}$ months. Where *G. perennans* was the species concerned, there was more rotting after the same period and conditions of storage. However, Bompeix (1978b) stated that although controlled atmosphere storage reduces the development of lesions of *G. album*, it does not have much effect on mycelial growth and he concluded that the constraint on development comes entirely from the influence of storage conditions on the host tissue. Lockhart (1967) found that the growth of *G. album in vitro* was reduced by decreasing the concentration of oxygen but the addition of carbon dioxide to the low oxygen concentrations significantly stimulated growth. This author could not reconcile these *in vitro* data with the known reduction of rotting of apples in storage. These problems of relating the response of the fungus *in vitro* to the development of rotting in store are still in need of resolution.

V. Control Measures

The choice of treatment for post-harvest disease of top fruit is influenced by many considerations. The nature of each disease and the source(s) of infection are the main aspects to be considered, but the problem of residues and the regulations governing these vary considerably from one country to another and may be as important as any of the biological considerations. The amount of fungicide used for post-harvest purposes is relatively small compared with the quantities required for field application and as a result manufacturing companies do not produce chemicals specifically for post-harvest use. All those currently used have been developed for the control of major diseases of a number of crops and this is likely to continue in the future. Nevertheless, suitable materials are currently available for most of the serious diseases and, as a result, more fruit is treated after harvest nowadays than at any previous time.

So far as the use of liquid-phase fungicides is concerned, there are two main types of disease to be considered. The lenticel rots are caused very largely by pathogens present in the orchard as wood infections and this affords the opportunity to treat these infections directly in addition to protecting the fruit by applying orchard sprays. They may also be treated by post-harvest fungicidal dips or by a combination of the two. The "soft rots", e.g. *Penicillium* and *Botrytis*, have wind-borne spores with no localized sources of inoculum. They attack mainly through damage of various types, some of which is incurred during harvest operations. Thus although in some countries programmes of orchard sprays are used to contribute to their control, they are for the most part treated after harvest.

A. Orchard Sprays

1. *Protectant and Curative*

The first post-harvest pathogen for which programmes of orchard sprays were developed was *Gloeosporium*. Early work in the U.S.A. before the introduction of organic fungicides did not result in complete success, control not always being adequate and phytotoxicity also being a problem (Kienholz, 1956). This was followed by the use of captan and ziram which were an advance on anything previously used and protected fruit without causing any damage. The same worker had shown that *G. perennans* can infect fruit early in the growing season and as a result there was a need to spray from May to harvest to cover the total period when infection could take place. Later work with Cox's Orange Pippin and captan under English conditions (Moore and Edney, 1959) showed that the timing of individual sprays was related to the rainfall and that variation in the effect of a spray could be evaluated in terms of the amount of rain before and after application.

In 1955, applications in July and September were the most effective, each spray being preceded by a relatively dry period and followed by a wetter period. The next year the converse was true and the July spray had no effect, probably because rain before treatment had already induced a heavy dispersal of inoculum.

Although captan was a considerable advance on previously used materials, it possessed no eradicant ability and its successful use was dependent on an adequate cover being maintained on the fruit surface. It was replaced in the late 1960s by the benzimidazole and thiophanate methyl fungicides, which by virtue of their ability to control developing infections were also adapted for post-harvest use. A further attribute of these materials is that their ability to penetrate the tissues prolongs their effectiveness so that fewer applications are necessary. Burchill and Edney (1977) found that single sprays of benomyl (0.025%) applied in either June, July or August controlled rotting of apples (Cox's Orange Pippin) by *Gloeosporium* spp. This property is of special value on farms where spraying late in the season is difficult or impossible because branches laden with mature fruit are bent down and impede the passage of spray machinery. In some instances the use of benzimidazole materials in the orchard is discouraged because of the possibility that in so doing resistant strains might develop which would not be controlled by post-harvest treatment. There is no evidence of this happening with *Gloeosporium* spp. and where post-harvest use is forbidden, application before harvest has to be used. Recent work on these materials also indicates that orchard sprays can make a contribution to the control of pathogens with wind-dispersed spores such as *Penicillium*, *Botrytis* and *Alternaria* (Katschinski, 1974; Kristeva, 1974; Borecka, 1977; Novobranova and Gudkovskii, 1978).

2. Spore Suppressant

A property possessed by some chemicals is the ability to suppress the sporulation of fungi. Byrde *et al.* (1952) showed that spore production from cankers caused by *Nectria galligena* could be reduced by spraying during the winter using materials previously shown to have eradicant properties. Sharples and Somers (1959) subsequently showed that 0.03% phenyl mercuric chloride reduced the sporulation of wood infections of *G. perennans*. Attempts to employ this ability to reduce the fruit rot caused by *Gloeosporium* spp. were first made in England by Edney *et al.* (1961) using a proprietary phenyl mercury acetate compound applied during the dormant season since the chemical damages the tree if used in the growing season. Subsequent work by Burchill and Edney (1963) showed that 0.10% phenyl mercuric chloride reduced the sporulation of *G. album* and that in consequence there was less rotting of the fruit in store. Brook (1959), working in New Zealand, also used phenyl mercuric chloride and found that this enhanced the effect of protectant sprays of captan. As the effect of these dormant season sprays wears off, sporulation may be enhanced (Burchill and Edney, 1963) and this may mean that there are more spores available for re-infecting the wood when pruning operations are in progress in the autumn. Accordingly, efforts were made to find spore suppressant materials that could be applied during the growing season. Of many materials tried on a small scale, dichlorophen (5,5-dichloro-2,2-dihydroxy-diphenyl-methane) markedly reduced sporulation and this resulted in a significant reduction in rotting.

Corke (1967) found that assessment of the relative numbers of spores released after spore suppressant treatment was not sufficient in every case to evaluate the total effect of the chemical and observed that the viability of spores released must also be taken into account. He found that there were wide variations depending on the time of application and the type of treatment. Although spore suppressants have never been used on any scale, it is interesting to note that in later work, Corke and Sneh (1979) found that benomyl exhibited a considerable ability to suppress sporulation of *G. perennans*. It seems likely therefore that this property contributes to the total effect of orchard sprays of this material when applied for its eradicant effect.

B. Post-Harvest Treatment

1. *Dips*

Fungicidal treatment applied after harvest must be capable of controlling spores and latent infections on the skin and in the lenticels. The disparity in performance of pre- and post-harvest treatments was noted by Kienholz (1956) and many subsequent attempts have been made to find a fungicide that possessed the properties necessary for successful use after harvest. Some success was achieved in America with sodium hypochlorite (Baker and Heald, 1934) when used specifically for the control of *Penicillium* on apple, and later, aqueous salts of 2-amino-butane were used for the same problem on apples and pears (Eckert and Kolbezen, 1964; Maclean and Dewey, 1964). It was not, however, until the benzimidazoles became available that satisfactory treatment was available for most of the post-harvest disease of apple and pear.

Their success in treating *Penicillium* and *Botrytis* has been reported by a number of workers (Blanpied and Purnasiri, 1968; Daines and Snee, 1969; Spalding *et al.*, 1969; Cargo and Dewey, 1970). Recently, however, benzimidazole-resistant strains of *Penicillium* have been reported on pears in Australia (Wicks, 1977) and Oregon, U.S.A. (Bertrand and Saulie-Carter, 1978), on apples in New York (Rosenberger and Meyer, 1979) and on apples and pears in Australia (Koffman *et al.*, 1978). There are also indications of a similar situation developing with regard to *Botrytis* (Bertrand and Saulie-Carter, 1978). This, of course, necessitates further screening of fungicides that will control both resistant and sensitive strains (Rosenberger *et al.*, 1979; Burton and Dewey, 1981).

The control of lenticel rotting caused by *Gloeosporium* spp. is perhaps the greatest contribution made by the benzimidazoles. Early experimental results were variable (Bondoux, 1967b; Gorini *et al.*, 1969; Edney, 1970), but they have been widely adopted in countries that have permitted their use after harvest with considerable success. It is their ability to eradicate infections that have penetrated the flesh of the fruit (Edney, 1970) which sets them apart from unsuccessful materials which can provide control if used as protectant sprays. This characteristic is probably associated with the ability of the chemicals to penetrate sufficiently deeply into the host tissue. Estimates of how far this penetration extends in apple vary from 1 cm (Leroux *et al.*, 1975) to "mainly in the peel" (Goedicke and Riebel, 1978). However, since the depth of the lenticels is contained within the peel, this penetration within the peel tissue would be adequate to control rotting. On pear, Ben-Arie (1975) found that the concentration of benomyl and thiabendazole was highest in the peel with a declining gradient towards the core. Degradation of these residues proceeds throughout the storage period and was found by Kepczynska and Borecka (1979) to be slower in 5% CO_2 + 3% O_2 than in air.

The problem of infection by *Phytophthora* spp. places very high demands on any fungicide that might be used after harvest. The fungus is capable of rapid invasion of fruit before it is picked and treatment after harvest has to contend with an infection that has been developing rapidly for several days. The ability to spread by contact during the storage period is also a major source of wastage (Edney, 1978) and treatment must be able to prevent this happening. The acylalanine fungicide metalaxyl has shown considerable promise for this purpose (Edney and Chambers, 1981b), but its success on a commercial scale has not yet been established.

It is probably true to say that every new fungicide is screened for its possible use in a post-harvest role. This applies not only to the major diseases but also to those with a more localized importance as shown, for example, by the use of imazalil for *Alternaria* on apple and pear in Israel

(Prusky and Ben-Arie, 1981) and of antibiotics for *Clathridium* on apple in India (Thind, 1979). In commercial practice, post-harvest treatments also currently include the use of an anti-oxidant for scald and calcium salts for bitter pit. The latter will also decrease susceptibility to rotting (Gorini and Mori, 1976; Sharples and Johnson, 1976; Pratella *et al.*, 1979). The use of mixtures of two or more chemicals raises potential problems, both in terms of their effectiveness in controlling the post-harvest disorders for which they are used and of the possibility of producing a combination that will be toxic either to the fruit or the consumer. Little *et al.* (1980) have studied the use of various mixtures of fungicides and other chemicals which they term "multiformulation dips". This would not appear to be a very satisfactory trend although it may be necessary in the short term. It seems quite possible that so many permutations will be available that their assessment will become virtually impossible and in any event the effect on health of the consumer must be the paramount consideration. Some rationalization of the situation is long overdue.

2. Volatile Fungicides

In theory, it could be argued that the use of a volatile fungicide is the ideal way of controlling post-harvest rotting, in particular the lenticel rots which present difficulties of access for liquid-phase fungicides. The penetration of a fungicidal vapour into natural openings would overcome these problems. Although success was achieved in the control of *Penicillium digitatum* on oranges in the 1930s by the use of diphenyl incorporated into paper tissue wraps (Tomkins, 1936), attempts to find a similar type of treatment for pome fruits have so far been unsuccessful. The main problems encountered are those of phytotoxicity and taint. Spencer and Wilkinson (1960), for example, found that 2:4:5-(trichlorophenoxy-thio) trichloromethane would suppress the development of *G. perennans* on apple but could not overcome the problem of phytotoxicity. The method continues to attract further experimentation; compounds recently tried include carbon monoxide (Sommer *et al.*, 1981) and acetaldehyde (Stadelbacher and Prasad, 1974): none of these has yet been adopted for commercial use.

3. Warm Water Treatment

The use of warm water as a means of reducing post-harvest rotting of apple was first reported by Burchill (1964) following earlier work on other crops (Akamine and Arisumi, 1953; Smith, 1962). Immersion of Cox's Orange Pippin apples in water at between 40 and 45°C for periods of 6 min or more markedly reduced the incidence of rotting caused by *Gloeosporium* spp.

Steam/air mixtures were subsequently used to supply the heat and had the additional merit of facilitating the use of air at ambient temperature to cool the fruit (Edney and Burchill, 1967). However, it was also found that when treated in this way, a higher level of physiological disorders was induced (Edney *et al.*, 1966; Sharples, 1967) and as a result there was no overall benefit of the treatment. No such complication was encountered by Olsson (1965c) using cv. Ingrid Marie and it is possible that other cultivars could also be treated without any undesirable side-effects. Sharples and Johnson (1976) found that the incorporation of calcium nitrate in the water went some way to minimizing the risk of increased physiological breakdown. Although the method has not as yet been adopted for use on a large scale, it remains as a possible alternative to fungicidal treatment should one be wanted.

References

Agarwal, G. P. and Bisen, P. S. (1977). Post-infection changes in apple due to *Aspergillus niger* van Tiegh. III. Phenolics. *Phytopath. mediterr.* **16,** 137–139.

Akamine, E. K. and Arisumi, T. (1953). Control of post-harvest storage decay of fruits of papaya (*Carica papaya* L.) with special reference to the effect of hot water. *Proc. Am. Soc. hort. Sci.* **16,** 270–274.

Ames, A. (1915). The temperature relations of some fungi causing storage rots. *Phytopathology* **5,** 11–19.

Atkinson, D. (1974). Root growth and effects of competition in tree spacing experiments. *Rep. E. Malling Res. Stn 1973*, 70–71.

Ayub, A. N. B. H. and Swinburne, T. R. (1970). Observations on the infection and rotting of apples var. Bramley's Seedling by *Diaporthe perniciosa. Ann. appl. Biol.* **66,** 245–255.

Babovic, M., Perisic, M., Markovic, S., Stofanovic, S. and Pantellic, Z. (1979). Investigation of apple fruits in cold storage. *Zašt. Bilja* **30,** 83–87.

Baker, K. F. and Heald, F. D. (1934). An investigation of factors affecting the incidence of lenticel infection of apples by *Penicillium expansum. Bull. Wash. agric. Exp. Stn*, 298.

Bangerth, F. (1973). Investigations upon Ca related physiological disorders. *Phytopath. Z.* **77,** 20–37.

Ben-Arie, R. (1967). Control of storage rot of Grand Alexander apples in Israel. *1st Israel Congr. Pl. Path.*, 122.

Ben-Arie, R. (1975). Benzimidazole penetration, distribution and persistence in post-harvest treated pears. *Phytopathology* **65,** 1187–1189.

Bertrand, P. F. and Saulie-Carter, J. L. (1978). The occurrence of benomyl-tolerant strains of *Penicillium expansum* and *Botrytis cinerea* in the mid-Colombia region of Oregon and Washington. *Pl. Dis. Reptr* **62,** 302–305.

Blanpied, G. D. and Purnasiri, A. (1968). Thiabendazole control of *Penicillium* rot of McIntosh apples. *Pl. Dis. Reptr* **52,** 867–871.

Bolay, A. (1977). *Phytophthora syringae*, agent d'une grave pourriture des pommes en conservation. *Revue suisse Vitic. Hort. Arb.* **9,** 161–169.

Bompeix, G. (1966). Contribution à l'étude de la maladie des taches lenticellaires des pommes "Golden Delicious" en France. *Mem. Fac. Sci. Rennes Univ., Brest,* 84–88.

Bompeix, G. (1967). Maladie des taches lenticellaires des pommes "Golden Delicious" en France, 31–34. Thesis, University of Rennes.

Bompeix, G. (1978a). Some physiological aspects of host–parasite relationships during apple storage. *Fruits d'outre mer* **33**, 23–26.

Bompeix, G. (1978b). The comparative development of *Pezicula alba* and *P. malicorticis* on apples and in vitro (air and controlled atmosphere). *Phytopath. Z.* **91**, 97–109.

Bondoux, P. (1964). Les principales maladies cryptogamiques des pommes et des poires en conservation. *Arboric. fruit.* **11**, 17–25.

Bondoux, P. (1967a). Les maladies cryptogamiques des poires et des pommes au cours de l'entreposage. *Annls Épiphyt.* **18**, 509–550.

Bondoux, P. (1967b). L'eau chaude et le thiabendazole en traitement curatif contre les parasites lenticellaires des pommes. *C. r. hebd. Séanc. Acad. Agric. Fr.* **53**, 1314–1321.

Borecka, H. (1977). Control of apple rot caused by *Penicillium* spp. *Acta agrobot.* **30**, 229–238.

Borecki, Z. and Profic, H. (1962). The fungi *Trichothecium roseum* Link, *Pezicula malicorticis* (Cord.) Nannf. and *Penicillium expansum* (Link) Thom. as primary or secondary pathogens of apples. *Acta agrobot.* **12**, 72–94.

Brook, P. J. (1959). Experiments on the control of ripe spot of Sturmer apples. *N.Z. Jl. agric. Res.* **2**, 463–474.

Brooks, C. and Cooley, J. S. (1917). Temperature relations of apple-rot fungi. *J. agric. Res.* **8**, 139–164.

Burchill, R. T. (1964). Hot water as a possible post-harvest control of *Gloeosporium* rots of stored apples. *Pl. Path.* **13**, 106–107.

Burchill, R. T. and Edney, K. L. (1963). The control of *Gloeosporium album* rot of stored apples by orchard sprays which reduce sporulation of wood infections. *Ann. appl. Biol.* **15**, 379–387.

Burchill, R. T. and Edney, K. L. (1977). The control of *Gloeosporium* storage rot of apple by reduced spray programmes. *Ann. appl. Biol.* **87**, 51–56.

Burton, C. L. and Dewey, D. H. (1981). New fungicides to control benomyl-resistant *Penicillium expansum* in apples. *Pl. Dis.* **65**, 881–883.

Bykova, T. D., Davydova, M. A., Ozeretskova, O. L. and Moiseeva, N. A. (1977). Antifungal substances in apples. *Mikol. Fitopat.* **11**, 116–121.

Byrde, R. J. W., Crowdy, S. H. and Roach, F. A. (1952). Observations on apple canker. V. Eradicant spraying and canker control. *Ann. appl. Biol.* **39**, 581–587.

Cargo, C. A. and Dewey, D. H. (1970). Thiabendazole and benomyl for the control of post-harvest decay of apples. *Hortscience* **5**, 259–260.

Ceponis, M. J., Kaufman, J. and Butterfield, J. E. (1969). Moldy carpels in Delicious apples on the Greater New York market. *Pl. Dis. Reptr* **53**, 136–138.

Cole, J. S. (1956). The pathogenicity of *Botrytis cinerea, Sclerotinia fructigena* and *Sclerotinia laxa*, with special reference to the part played by pectolytic enzymes. *Ann. Bot., Lond.* **20**, 15–38.

Cole, M. and Wood, R. K. S. (1961). Types of rot, rate of rotting and analysis of pectic substances in apples rotted by fungi. *Ann. Bot., Lond.* **25**, 417–434.

Colhoun, J. (1938). Fungi causing rot of apple fruits in storage in Northern Ireland. *Ann. appl. Biol.* **15**, 88–99.

Cooper, J. W. and Padfield, C. S. (1965). Storage rots of apples. *Orchard. N.Z.* **38**, 135.

Corke, A. T. K. (1967). Screening trials of materials for suppressing spore production by *Gloeosporium perennans* on apple trees. *Ann. appl. Biol.* **60**, 241–249.

Corke, A. T. K. and Sneh, B. (1979). Antisporulant activity of chemicals towards fungi causing cankers on apple branches. *Ann. appl. Biol.* **91**, 325–330.

Daines, R. H. and Snee, R. D. (1969). Control of blue mold of apples in storage. *Phytopathology* **59**, 792–794.

Doesburg, J. J. (1957). Relation between the solubilization of pectin and the fate of organic acids during maturation of apples. *J. Sci. Fd Agric.* **8**, 206–216.

Eckert, J. W. and Kolbezen, M. J. (1964). 2-Amino-butane salts for control of post-harvest decay of citrus, apple, pear, peach and banana fruits. *Phytopathology* **59**, 978–986.

Edney, K. L. (1958). Observations on the infection of Cox's Orange Pippin apples by *Gloeosporium perennans*. *Ann. appl. Biol.* **46**, 622–629.

Edney, K. L. (1964a). The effect of the composition of the storage atmosphere on the development of rotting of Cox's Orange Pippin apples and the production of pectolytic enzymes by *Gloeosporium* spp. *Ann. appl. Biol.* **54**, 327–334.

Edney, K. L. (1964b). Some factors affecting the rotting of stored apples by *Gloeosporium* spp. *Ann. appl. Biol.* **53**, 119–127.

Edney, K. L. (1970). Some experiments with thiabendazole and benomyl post-harvest treatments for the control of storage rots of apples. *Pl. Path.* **19**, 189–193.

Edney, K. L. (1976). An investigation of persistent infection of stored apples by *Gloeosporium* spp. *Ann. appl. Biol.* **82**, 355–360.

Edney, K. L. (1978). The infection of apples by *Phytophthora syringae*. *Ann. appl. Biol.* **88**, 31–36.

Edney, K. L. and Burchill, R. T. (1967). The use of heat to control the rotting of Cox's Orange Pippin apples by *Gloeosporium* spp. *Ann. appl. Biol.* **59**, 389–400.

Edney, K. L. and Chambers, D. A. (1981a). Post-harvest treatments for the control of *Phytophthora syringae* storage rot of apples. *Ann. appl. Biol.* **97**, 237–241.

Edney, K. L. and Chambers, D. A. (1981b). The use of metalaxyl to control *Phytophthora syringae* rot of apple fruits. *Pl. Path.* **30**, 167–170.

Edney, K. L., Austin, W. G. L., Corke, A. T. K. and Hamer, P. S. (1961). Effect of winter spraying on rotting of stored apples by *Gloeosporium* spp. *Pl. Path.* **10**, 10–13.

Edney, K. L., Fidler, J. C., Mann, G. and North, C. J. (1966). Pre-storage heat treatment for lenticel rotting of apples. *Bull. Inst. int. Froid, Annexe 1966–1*, 243–249.

Edney, K. L., Tan, A. M. and Burchill, R. T. (1977). Susceptibility of apples to infection by *Gloeosporium album*. *Ann. appl. Biol.* **86**, 129–132.

Fawcett, C. H. and Spencer, D. M. (1967). Antifungal phenolic acids in apple fruits after infection with *Sclerotinia fructigena*. *Ann. appl. Biol.* **60**, 87–96.

Fawcett, C. H. and Spencer, D. M. (1968). *Sclerotinia fructigena* infection and chlorogenic acid content in relation to antifungal compounds in apple fruits. *Ann. appl. Biol.* **61**, 245–253.

Fidler, J. C. (1973). Appendix: storage conditions for apples and pears. The biology of apple and pear storage. *Res. Rev. Commonw. Burea Hort. Plantat. Crops*, No. 3, 48–51.

Fidler, J. C. and North, C. J. (1963). Core flush in apples. *Atti Congr. int. Conserv. Distrib. Prod. Ortofruttic., Bologna* **1**, 303–308.

Fidler, J. C. and North, C. J. (1964). Controlled atmosphere storage of apples. *A. R. Ditton Covent Garden Labs 1963–64*, 8–9.

Fuller, M. M. (1976). The ultrastructure of the outer tissues of cold stored apple fruits of high and low calcium content in relation to cell breakdown. *Ann. appl. Biol.* **83**, 299–304.

Goedicke, H.-J. and Riebel, A. (1978). Toxicological residue problems following the application of fungicides in apple production. *NachrBl. PflSchutz D.D.R.* **32**, 85–88.

Gorini, F. L. and Mori, P. (1976). Old and new fungicides for the prevention of pome fruit rots. *Inftore Agrario* **32**, 23849–23852.

Gorini, F. L., Mori, P. and Mariotti, A. (1969). Lotta contro i marciumi e la butteratura amara delle mele con trattamenti post-raccolta. *Inform. agrar. Verona* **25**, 1403–1404.

Gregory, F. G. and Horne, A. S. (1928). A quantitative study of the course of fungal invasion of the apple fruit and its bearing on the nature of disease resistance. Part I. A statistical method of studying fungal invasion. Part II. The application of the statistical method to certain specific problems. *Proc. R. Soc. Lond., Ser. B* **102**, 427–466.

Guthrie, E. J. (1959). The occurrence of *Pezicula alba* sp. nov. and *P. malicorticis*, the perfect stages of *Gloeosporium album* and *G. perennans* in England. *Trans. Br. mycol. Soc.* **42**, 502–506.

Harris, D. C. (1979). The occurrence of *Phytophthora syringae* in fallen leaves. *Ann. appl. Biol.* **91**, 309–312.

Horne, A. S. (1934). Infection in relation to disease in stored apples. *Rep. Fd Invest. Bd 1933*, 228–245.

Hulme, A. C. (1956). The nitrogen content of Cox's Orange Pippin apples in relation to manurial treatments. *J. hort. Sci.* **31**, 1–7.

Jackson, J. E., Palmer, J. W., Perring, M. A. and Sharples, R. O. (1977). Effects of fruit growth, chemical composition and quality at harvest and after storage. *J. hort. Sci.* **52**, 267–282.

Jamaluddin, Tandon, M. P. and Tandon, R. N. (1973). Studies on fruit rot of apple (*Malus sylvestris*) caused by *Gliocephalotrichum bulbilium* Ellis. *Proc. natn Acad. Sci. India, B* **43**, 154–158.

Katschinski, K. H. (1974). The occurrence and control of fungal storage diseases of apple. *NachrBl, PflSchutz D.D.R.* **28**, 226–228.

Kavanagh, J. A. and Glynn, A. M. (1966). Brown rot of apples caused by *Nectria galligena*. *Ir. J. agric. Res.* **5**, 143–14.

Kepczynska, E. K. and Borecka, H. (1979). Dynamics of disappearance of benzimidazole derivates in stored apples. *Fr. Sci. Rep. Skierniwice* **6**, 45–55.

Kidd, M. N. and Beaumont, B. A. (1924). Apple rot fungi in storage. *Trans. Br. mycol. Soc.* **10**, 98–118.

Kidd, M. N. and Beaumont, B. A. (1925). An experimental study of the fungal invasion of apples in storage with particular reference to invasion through the lenticels. *Ann. appl. Biol.* **12**, 14–33.

Kienholz, J. R. (1956). Control of bull's-eye rot on apple and pear fruits. *Pl. Dis. Reptr* **40**, 872–877.

Knee, M. (1973). Polysaccharides and glycoproteins of apple fruit cell walls. *Phytochemistry* **12**, 637–653.

Knee, M. (1977). Cell-wall degradation in senescent fruit tissue in relation to pathogen attack. *In* "Cell Wall Biochemistry" (B. Solheim and J. Raa, Eds), 259–262. University of Tromso, Norway.

Knee, M., Fielding, A. H., Archer, S. A. and Laborda, F. (1975). Enzymic analysis of cell wall structure in apple fruit cortical tissue. *Phytochemistry* **14**, 2213–2222.

Koffman, W., Penrose, L. J., Menzies, A. R., Davis, K. C. and Kaldor, J. (1978).

Control of benzimidazole-tolerant *Penicillium expansum* in pome fruit. *Sci. Hort.* **9**, 31–39.

Krapf, B. (1961). Entwicklung und Bau der Lentizellen des Apfels und ihre Bedeutung für die Lagerung. *Promotionsarb. Zurich*, No. 3059, 389–441.

Krasztev Kemenes, M. (1977). *Penicillium* rot on stored Golden Delicious apples. *In* "Ujabb Kutatási Eredmények a Gyümölcstermesztésben", 159–168. Gyümölcs-és Diszövénytermesztési Kutató Intezet, Budapest, Hungary.

Kristeva, M. (1974). Trials of fungicides against apple fruits in cold storage. *Gradinarska Lozarka Nauka* **11**, 49–54.

Lafferty, H. A. and Pethybridge, G. H. (1922). On a *Phytophthora* parasite on apples which has both amphigynous and paragynous antheridia; and on allied species which show the same phenomenon. *Scient. Proc. R. Dubl. Soc.* **17**, 29–43.

Landfald, R. (1962). Virninger av dyrikingsvilkar. Frokstørrelse og temperature ved lagring av James Grieve. *Frukt Baer* **15**, 57–61.

Leroux, P., Casanova, M. and Dachaud, R. (1975). Study on penetration and persistence of benomyl and carbendazim in apples treated after harvest. *Phytiat.-Phytopharm.* **24**, 49–56.

Little, C. R., Taylor, H. J. and Peggie, I. D. (1980). Multiformulation dips for controlling storage disorders of apples and pears. I. Assessing fungicides. *Sci. Hort.* **13**, 213–219.

Lockhart, C. L. (1967). Influence of controlled atmospheres on the growth of *Gloeosporium album in vitro. Can. J. Pl. Sci.* **47**, 649–651.

Maclean, D. C. and Dewey, D. H. (1964). Reduction of decay of pre-packaged apples with 2-aminobutane. *Q. Bull. Mich. St. Univ. agric. Exp. Stn* **47**, 225–230.

Marchal, El. and Marchal, Ea. (1921). Contribution à l'étude des champignons fructicole de Belgique. *Bull. Soc. r. Bot. Belg.* **4**, 109.

Martin, D. (1953). Variation between apple fruits and its relation to keeping quality. *Aust. J. agric. Res.* **4**, 235–248.

Martin, D. and Lewis, T. L. (1952). The physiology of growth in apple fruits. III. Cell characteristics and respiration activity of light and heavy crop fruits. *Aust. J. scient. Res., Ser. B* **5**, 315–317.

Mezzetti, A. (1958). Les principales pourritures des pommes et des poires pendant l'entreposage frigarifique. *Rapp. gén. Congr. pomol. Florence-Ferrare*, 54–65.

Montgomery, H. B. S. (1958). Effect of storage conditions on the incidence of *Gloeosporium* rots of apple fruits. *Nature* **182**, 737–738.

Montgomery, H. B. S. and Wilkinson, B. G. (1962). Storage experiments with Cox's Orange Pippin apples from a manurial trial. *J. hort. Sci.* **37**, 150–158.

Moore, M. H. and Edney, K. L. (1959). The timing of spray treatments for the control of storage rot of apple caused by *Gloeosporium* spp. *Rep. E. Malling Res. Stn 1958*, 106–109.

Moreau, C., Moreau, M. and Bompeix, G. (1965). Nouvelles techniques pour infecter les lenticelles des Pommes Golden par le *Pezicula alba* (Guthrie). *C.r. Acad. Sci., Paris* **261**, 521–523.

Mudretsova-Viss, K. A., Kolesnik, S. A. and Grinyuk, T. I. (1975), The resistance of Jonathan apples to fungal diseases in storage. *Mikol. Fitopat.* **9**, 414–417.

Muskett, A. E., Horne, A. S. and Colhoun, J. (1938). The effect of manuring upon apple fruits. *Ann. appl. Biol.* **25**, 50–67.

Ndubizu, T. O. C. (1976). Relation of phenolic inhibitors to resistance of immature apple fruits to rot. *J. hort. Sci.* **51**, 311–319.

Novobranova, T. I. and Gudkovskii, V. A. (1978). Effect of benomyl treatment on mycoflora and preservation of apples and pears. *Vest. sel.'khoz. Nauki, Kazakhstania* **21**, 35–39.

Nyhlen, Å. (1960). Kyllagring av olika appelsorter 1952–1956. *Sver. pomol. För. Årsskr. 1959* **60**, 91–110.

Ogilvie, L. (1931). A fruit rot of apples and pears due to a variety of *Phytophthora syringae*. *Rep. agric. hort. Res. Stn. Univ. Bristol 1930*, 81.

Ogilvie, L. (1935). The fungus flora of apple twigs and branches and its relation to apple fruit spots. *J. Pomol.* **13**, 140–148.

Olsson, K. (1965a). A study of the biology of *Gloeosporium album* and *G. perennans* on apples. *Medd. St. Vaxtskyddsanst. Stockholm* **13** (104), 189–259.

Olsson, K. (1965b). Om olika gloeosporiumarters upptradande pa appelen i olika typer av lager. *Vaxtskyddsnotiser* **29**, 59–61.

Olsson, K. (1965c). Varmvatten behandling mot *Gloeosporium* pa apple i olika typer av lager. *Vaxtskyddsnotiser* **29**, 59–61.

Olsson, K. (1967). On the occurrence of *Gloeosporium* on apple leaves. *Medd. St. Vaxtskyddsanst. Stockholm* **14** (116), 33–50.

Osterwalder, A. (1907). Zur *Gloeosporium* faule des Kernobstes. *Zbl. Bakt. Abt. 2* **18**, 825.

Ostrowski, W. (1971). Fruit rots and their control. *Mezimár Zeměd Čas* No. 4, 58–60.

Ostrowski, W., Guzewska, I. and Gdowski, J. (1959). Wplyw nawozenia mineralnego na wartosc przechowalnicza Iablek admiany Antonovka. *Pr. Inst. Sadow. Skierniew.* **4**, 365–375.

Perring, M. A. (1979). The effects of environment and cultural practices on calcium concentration in apple fruit. *Commun. Soil Sci. Pl. Anal.* **10**, 279–293.

Pratella, G. C. (1960). Avversita della pomacee conservate. *Infme. fitopatol.* **10**, 273–282.

Pratella, G. C., Bertolini, P. and Maccaferri, M. (1979). Calcium therapy of Abbe zitel pears in controlled atmosphere storage. *Frutticoltura* **41**, 41–43.

Preece, T. F. (1967). Losses of Cox's Orange Pippin apples during refrigerated storage in England, 1961–1965. *Pl. Path.* **16**, 176–180.

Prusky, D. and Ben-Arie, R. (1981). Control by imazalil of fruit storage rots caused by *Alternaria alternata*. *Ann. appl. Biol.* **98**, 87–92.

Rosenberger, D. A. and Meyer, F. W. (1979). Benomyl-tolerant *Penicillium expansum* in apple packing houses in eastern New York. *Pl. Dis. Reptr* **63**, 37–40.

Rosenberger, D. A., Meyer, F. W. and Cecelia, C. V. (1979). Fungicide strategies for control of benomyl-tolerant *Penicillium expansum* in apple storages. *Pl. Dis. Reptr* **63**, 1033–1037.

Schneider-Orelli, O. (1912). Versuche uber die Wachstumsbedingungen und Verbreitung der Faulnispilze des Lagerobstes. *Zbl. Bakt.* **32**, 161–169.

Sharples, R. O. (1964). The effects of fruit thinning on the development of Cox's Orange Pippin apples in relation to the development of storage disorders. *J. hort. Sci.* **39**, 224–235.

Sharples, R. O. (1967). The effect of post-harvest heat treatment on the storage behaviour of Cox's Orange Pippin apple fruits. *Ann. appl. Biol.* **59**, 401–406.

Sharples, R. O. (1968a). Fruit-thinning effects on the development and storage quality of Cox's Orange Pippin apple fruits. *J. hort. Sci.* **43**, 359–371.

Sharples, R. O. (1968b). The structure and composition of apples in relation to storage quality. *Rep. E. Malling Res. Stn 1967*, 185–189.

Sharples, R. O. and Johnson, D. S. (1976). Post-harvest chemical treatments for the control of storage disorders of apples. *Ann. appl. Biol.* **83**, 157–167.

Sharples, R. O. and Little, R. C. (1970). Experiments on the use of calcium sprays for bitter pit control in apple. *J. hort. Sci.* **45**, 49–56.

Sharples, R. O. and Somers, E. (1959). Control of *Gloeosporium perennans* with formulations of eradicant fungicides. *Pl. Path.* **8**, 8–12.

Sharples, R. O., Perring, M. A. and Johnson, D. S. (1975). Soil management/rate of nitrogen trial on Cox. *Rep. E. Malling Res. Stn 1974*, 74–75.

Smith, W. L., Jr (1962). Reduction of post-harvest brown rot and *Rhizopus* decay of eastern peaches with hot water. *Pl. Dis. Reptr* **46**, 861–865.

Sommer, N. F., Fortlage, R. J., Buchanan, J. R. and Kader, A. A. (1981). Effect of oxygen on carbon monoxide suppression of post-harvest pathogens of fruit. *Pl. Dis.* **65**, 347–349.

Spalding, D. N., Vaught, H. C., Day, R. H. and Brown, G. A. (1969). Control of blue mold development in apples treated with heated and unheated fungicides. *Pl. Dis. Reptr* **53**, 738–742.

Spencer, D. M. and Wilkinson, E. H. (1960). Further experimental work on the control of *Gloeosporium* in store. *Pl. Path.* **53**, 738–742.

Stadelbacher, G. J. and Prasad, K. (1974). Post-harvest decay control of apple by acetaldehyde vapour. *J. Am. Soc. hort. Sci.* **99**, 364–368.

Swinburne, T. R. (1970). Fungal rotting of apples. I. A survey of the extent and cause of current fruit losses in Northern Ireland. *Rec. agric. Res. North. Ire.* **18**, 15–19.

Tan, A. M. and Burchill, R. T. (1972). The infection and perennation of the bitter rot fungus, *Gloeosporium album*, on apple leaves. *Ann. appl. Biol.* **70**, 199–206.

Terui, M. (1959). Diseases of apples in storage. *Bull. Fac. Agric. Hirosaki Univ.* **5**, 58–62.

Tetley, U. (1930). A study of the anatomical development of the apple and some observations on the "pectic constituents of the cell walls". *J. Pomol.* **8**, 153–172.

Thind, T. S. (1979). Control of an apple fruit rot with antibiotics. *Hindustan Antibiot. Bull.* **22**, 14–17.

Tiller, L. W., Roberts, H. S. and Bollard, E. G. (1959). The Appleby experiments. A series of fertilizer and cold storage trials with apples in the Nelson district New Zealand. *Bull. N.Z. Dep. scient. ind. Res.* 129.

Tindale, G. B. (1966). Golden Delicious in cool storage. *Vict. Hort. Dig.* **10**, 23–25.

Tomkins, R. C. (1936). Wraps for the prevention of rotting of fruit. *Rep. Fd Invest. Bd 1935*, 129.

Tomkins, R. G. (1960). The conditions for the gas storage of certain fruits and vegetables obtained by the use of a simple small-scale method. *Proc. 10th int. Congr. Refrig., Copenhagen 1959* **3**, 189–192.

Tomkins, R. G. (1966). The effect of controlled atmosphere conditions on the extent of wastage caused by fungal rotting. *Bull. Inst. int. Froid, Annexe 1966–1*, 83–90.

Tomkins, R. G. (1968). Small scale storage trials. *A. R. Ditton Lab. 1967–68*, 12–18.

Tweedy, B. G. and Powell, D. (1963). The taxonomy of *Alternaria* and species of this genus reported on apples. *Bot. Rev.* **29**, 405–412.

Vyas, S. C., Singh, D. and Sharma, N. D. (1976). Some new fungi causing post-harvest diseases of apple (*Malus sylvestris*) *Pl. Dis. Reptr* **60**, 988–990.

Walker, J. R. L. (1969). Inhibition of the apple phenolase system through infection with *Pencillium expansum*. *Phytochemistry* **8**, 561–566.

Wallace, J., Kúc, J. and Draudt, H. N. (1962). Biochemical changes in the water insoluble material of maturing apple fruit and their possible relationship to disease resistance. *Phytopathology* **52**, 1023–1027.

Wicks, T. (1977). Tolerance to benzimidazole fungicides in blue mold (*Penicillium expansum*) on pears. *Pl. Dis. Reptr* **61**, 447–449.

Wilkinson, B. G. (1957). The effect of orchard factors on the chemical composition of apples. 1. Some effects of manurial treatments of grass. *J. hort. Sci.* **32**, 74–84.

Wilkinson, B. G. and Fidler, J. C. (1973). Physiological disorders. The biology of apple and pear storage. *Res. Rev. Commonw. Bureau Hort. Plantat. Crops*, No. 3, 63–131.

Wilkinson, B. G. and Sharples, R. O. (1967). The relation between time of picking and storage disorders in Cox's Orange Pippin apple fruits. *J. hort. Sci.,,* **42**, 67–82.

Wilkinson, E. H. (1942). Dry-eye rot of apples caused by *Botrytis cinerea* Pers. *J. Pomol.* **20**, 84–88.

Wilson, D. M., Nuovo, G. J. and Darby, W. B. (1973). Activity of o-phenol oxidase in postharvest apple decay in *Penicillium expansum* and *Physolospora obtusa*. *Phytopathology* **63**, 1115–1118.

Wright, T. R. and Smith, E. (1954). Relation of bruising and other factors to blue mold decay of Delicious apples. *Circ. U.S. Dep. Agric.* **935**, 15.

Zeller, S. M. and Childs, L. (1925). Perennial canker of apple trees (a preliminary report). *Bull. Ore. agric. Exp. Stn*, 217.

Zschokke, A. (1897). Uber der Bau der Haut und die Ursachen der verschiedener Haltbarkeit unserer Kernokstfuichten. *Landw. Jb. Schweiz.* **11**, 153–194.

4

Onions

R. B. MAUDE

I. Introduction

Man has used the onion (*Allium cepa* Linn.) as a food plant since early Egyptian times (3200–2780 B.C.) (Tackholm and Drar, 1954). In addition to culinary purposes, the Egyptians used onions in votive offerings, in embalming the dead, and may even have treated them as gods (Tackholm and Drar, 1954). The place of origin of *Allium cepa* is largely uncertain, but is thought to have been in the vicinity of West Pakistan and the mountainous region to its north (Jones and Mann, 1963). Onions are now grown in various cultivar forms world-wide. Jones and Mann (1963), in their review of onions, which included details of morphology, cultivation, breeding and disease problems, stated that world production of bulb onions was 6.5 million tonnes in 1960; this increased to an estimated 19.5 million tonnes in 1979 (Anon, 1980). British output increased over the same period (Fig. 4.1), largely as a result of advances in crop technology. Home production of bulbs, which had increased during the war years, slumped drastically during the 1950s due to high production costs because the crop was labour intensive and storage systems were poor. After harvest, bulb onions were often sold directly from the field or stored briefly before sale. This downward trend in cropping was reversed by the end of the 1950s when, with the introduction of a bed system of onion growing, effective pre-emergence selective herbicides and greater mechanization, the crop became less labour intensive. In addition, the introduction of new varieties based on the Dutch cultivar Rijnsburger with improved skin finish and better storage capabilities, the application of maleic hydrazide to increase the dormancy of stored bulbs (Tucker, 1971a; Whitwell *et al.*, 1973; Ward and Tucker, 1976) and the development of improved drying and storage facilities (Tucker, 1971b) combined to make it possible to store bulb onions for up to 6 months after harvest.

As production increased in Britain there was, however, an increase in field and storage pathogens. The incidence and effect of these pathogens on stored onions are reviewed here in relation to the world onion situation.

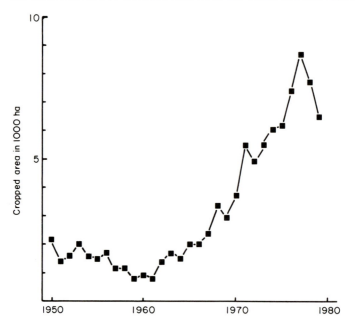

Fig. 4.1 Onion cropping 1950–1979. (From Ministry of Agriculture, Fisheries and Food agricultural statistics 1978–1979, 1968–1969 and 1957–1958.)

II. The Distribution and Identity of the Main Post-Harvest Diseases

Most post-harvest pathogens of onions occur widely throughout the world, but their severity is dictated by factors such as climate, crop rotation practices, drying and storage procedures and disease control measures. Although they appear mainly as storage problems, the majority of infections occur pre-harvest at some stage during the growth of the crop. They are caused mainly by fungi and bacteria; at present, virus problems are not common.

The main disease of stored bulb onions in temperate parts of the world is neck rot. It has caused considerable losses in most countries including America (Munn, 1917; Walker, 1925a, 1926; Hoyle, 1948; Vaughan et al., 1964; Kaufman and Lorbeer, 1967), Australia (Samuel, 1931; Hughes, 1970a), Britain (Massee, 1894; Croxall, 1945; Maude and Presly, 1977a, 1977b), Canada (McLarty, 1929; Jarvis, 1976), Czechoslovakia (Rod,

1978), Denmark (Hellmers, 1943), Germany (Sorauer, 1876; Brawtigam, 1977), Holland (van Doorn *et al.*, 1962), Israel (Netzer and Dishon, 1966), Japan (Hemmi and Niwa, 1937), Lithuania (Shidla, 1971), New Zealand (Brien, 1939; Harrow and Harris, 1969), Norway (Røed, 1951), Poland (Siemaszko, 1929) and Yugoslavia (Perisic *et al.*, 1977).

The main causal pathogen is *Botrytis allii* Munn, which is prevalent on all bulb onion types such as brown-, white- and silver-skinned. *Botrytis byssoidea* (Walker), mycelial neck rot, though much rarer also occurs on all bulb types. *Botrytis squamosa* (Walker), small sclerotial neck rot, does not normally attack brown-skinned bulb onions but it can affect white- and silver-skinned types. *Botrytis cinerea* (Fr.), grey mould, attacks the necks of silver-skinned onions only, but it causes a staining of the wrapper scales of a range of stored bulb onion cultivars (Clark and Lorbeer, 1973a, b).

Aspergillus and *Penicillium* spp. also cause post-harvest rotting of onion bulbs grown in hot climates, for example, in Egypt and Texas, U.S.A.; these fungi may also be a problem where high temperature drying systems are used. *Aspergillus niger* van Tiegh is a common post-harvest organism in these situations, but *B. allii* is less frequent.

Rotting caused by bacteria including *Pseudomonas*, *Erwinia* and occasionally *Lactobacillus* spp. occurs, but this is often masked by the primary disease.

Other fungal pathogens that can cause post-harvest diseases of onions in certain circumstances and situations include smudge (*Colletotrichum circinans*), pink root (*Pyrenochaeta terrestris*), bulb basal rot (*Fusarium oxysporum* f. sp. *cepae*) and white rot (*Sclerotium cepivorum*).

This·chapter reviews most of the post-harvest organisms mentioned but greatest emphasis is placed on bulb neck rot (*B. allii*), because of its over-riding importance in world onion production.

A. Onion Neck Rot

1. *Taxonomy of the Main Organisms*

In 1917, Munn investigated a neck rot disease in onions from Michigan and New York, U.S.A. He found a distinctive pathogen which he named *Botrytis allii* n. sp. Munn. Munn compared *B. allii* with species of *Sclerotium* (shape and size of sclerotia) and *Botrytis* (shape and size of conidia), but failed to find a positive similarity. On this basis he excluded *B. aclada* Fres., *B. cana* Kze et Schon, *B. cinerea* Pers., *B. fulva* Link, *B. parasitica* Cav. and *B. vulgaris* Fr. var. *interrupta* Fr. as being synonymous with *B. allii* n. sp. Walker (1925a, 1926) confirmed Munn's findings on the morphology and pathogenicity of *B. allii* but isolated another *Botrytis* which he named *B. byssoidea* n. sp. which caused a mycelial neck rot. This fungus was the main pathogen of onions in areas of Illinois and Wisconsin; unlike *B. allii* it spored poorly in culture and generally had larger conidia. He also named *B. squamosa* as a distinct and new species confined to the outer scales of white-skinned onions on which it produced a small sclerotial neck rot but caused little damage.

The two main neck rot pathogens *B. allii* and *B. byssoidea* were thought at one time to be conspecific, but Owen *et al.* (1950a) were unable to prove this relationship. Selected mycelial forms of *B. allii* quickly reverted to the normal, profusely sporing types typical of that fungus. Recently, Morgan (1971) upheld the distinction between *B. allii* and *B. byssoidea* when applying biometrics to the study of the taxonomy of *Botrytis* species.

In the majority of cases, *B. allii* appears to be the main cause of the disease. Van Poeteren (1939) recovered *B. allii* from 69% and *B. byssoidea* and *B. cinerea* each from 2% of onions affected with neck rot in Holland. Moore (1948) in Britain reported that in samples taken between 1943 and 1946 *B. allii* caused 72% of neck rot and *B. byssoidea* 26%. *Botrytis allii* was also the main pathogen of Norwegian onions (Røed, 1951).

Hennebert (1963), in reviewing the literature and herbarium specimens of *Botrytis* spp. occurring on onions, concluded that Fresenius' (1850) description of *B. aclada* was similar to Munn's for *B. allii*. Corrected spore measurements for *B. aclada* put it within the range of variation of conidia of *B. allii*. In a more recent publication (Hennebert, 1973), the two species are regarded as synonymous, with *B. aclada*, as the older description, being the preferred name. *Botrytis allii* Munn, however, is retained in this text because all fundamental and applied research refers to the pathogen by that name.

2. *The Disease and Causal Organisms*

Descriptions of neck rot disease of onions are given by Munn (1917) and Walker (1925a, 1926) and of the causal fungi by Munn (1917), Walker (1925a, 1926), Hickman and Ashworth (1943) and Ellis (1971).

Generally, when placed in store, bulbs infected with neck rot (*B. allii*) appear sound and healthy, but 8–10 weeks later symptoms appear. Necks of affected bulbs soften and removal of the brown wrapper scales reveals a mass of black sclerotia, 1–5 mm in diameter, encrusting the necks. Sometimes, a grey mould is seen below the sclerotia on the sides of the bulb. This comprises a dense mat of conidiophores usually less than 1 mm high, single stemmed and terminating in a head of conidia-bearing branches. Conidia are single celled, narrow and ellipsoidal.

If infected bulbs are sliced vertically, the affected parts are brown and discoloured in contrast to the white appearance of the unaffected storage tissue. This describes the disease in brown-skinned onions in Britain (Anon, 1981) and is comparable to Munn's (1917) description for American onions. Munn (1917) also described the occurrence of sclerotia on the sides and bases of bulbs. These symptoms have sometimes been seen on British onions by the author, especially where mechanical harvesting had damaged the wrapper scales of the bulbs. The root rot phase of the disease (Munn, 1917) has not been observed by the author in British crops. Mycelial neck rot (*B. byssoidea* Walker) differs in that affected bulbs carry more surface mycelium but sporulation is sparse (Walker, 1926).

3. *Disease Sources and Biology*

Neck rot is a post-harvest disease whose symptoms mainly become evident in stored bulbs. Because of this it proved extremely difficult to trace the primary source or sources of infection. Wood (1961), in reviewing the biology of *Botrytis* spp., noted the great discrepancy between neck rot, an important disease of bulb onions, and the almost total lack of symptom development in the field, thereby making it difficult to identify the source of infection. Viewed against this background the relatively recent information on seed-borne infection as a major source of the disease in certain countries is extremely relevant and is examined first in this account because it gives perspective to previous information on sources of this disease.

(a) *Disease transmission by infected seeds.* In the earlier literature, Munn (1917), Yarwood (1938) and Blodgett (1946) all reported that *B. allii* caused blasting of onion seed inflorescences in America resulting in loss of umbels and seed failure. More recently, Netzer and Dishon (1966) isolated *B. allii* from the umbels of onions grown for seed production in Israel. Although Munn (1917) washed *B. allii* spores from onion seeds, transmission of the disease by the contaminated seeds was not tested. The fungus was isolated from a single seed sample in Britain (Moore, 1948). More recently, Tichelaar (1970), in Holland, isolated *B. allii* from Dutch seed samples and showed that such seeds treated with benomyl produced onions with reduced levels of neck rot in store (Tichelaar, 1971). There was little evidence, however, to substantiate the relationship between onion seeds bearing the fungus and neck rot disease of the stored bulbs. This was supplied by Maude and Presly (1977a) who confirmed the occurrence of seed-borne infection in many commercial onion seed samples and who demonstrated seed transmission of the disease. Seedling infection was direct, with mycelium from the seed coat penetrating the tip of the attached cotyledon leaf (Maude and Presly, 1977a). Thereafter, the fungus progressed basipetally in the cotyledon and sometimes invaded the base of the first true leaf and its remaining tissues. In such cases the first leaf tissues were green and active and not senescing. Conidiophores were only produced on senescent and necrotic leaf tissues (Maude and Presly, 1977a). In plants grown from infected seeds in the field the fungus progressed by attacking the tips of the youngest leaves, and by growing downwards in their tissues as these became senescent. On the senescent leaf bases conidiophores bore conidia which when released attacked the tips of the next youngest leaves and the leaf invasion pattern was repeated. Ultimately, and late in the growing season, the leaves at the centre of the plant, contiguous with the bulb storage tissues, were invaded, causing neck rot (Maude and Presly, 1977a). Although these authors observed *B. allii* on the surface of the dry wrapper skins of developing bulbs the fungus did not penetrate the sides of the neck and bulb. This too was the experience of earlier workers (Munn, 1917; Walker, 1926). Although Maude and Presly (1977a) obtained some evidence for fungal growth within green leaves, Tichelaar (1967) considered that the fungus remained quiescent in the green leaf epidermis, only invading the mesophyll intracellularly when the leaves senesced. Latent infections in young leaves were also observed by Bochow and El Mosallamy (1979).

The disease was progressive in onion crops (Maude and Presly, 1977a), spreading rapidly in wet summers. It did not occur in all crops but was strictly source related, being essentially confined to those crops grown from infected seeds. The pathogen spread downwind from a diseased crop to infect healthy crops, but at a distance of 270 m from the source its effect was greatly diluted.

Because the disease was symptomless on onion leaves, healthy and infected crops were indistinguishable, but crops grown from seeds free from *B. allii* produced healthy bulbs, whereas those grown from infected seeds produced bulbs infected with neck rot. The importance of infected seeds as causal agents of neck rot in Britain was confirmed when direct relationships were established between the incidence of seed-borne infection and the percentage of neck rot infected stored bulbs (Fig. 4.2) and between the percentage of plants infected in the field and the disease in store (Maude and Presly, 1977a, b). Contrary to the views of earlier authors (Munn, 1917; Walker, 1926), *B. allii* was shown to be an active pathogen in onion crops (Maude and Presly, 1977a).

The importance of infected seeds as a source of neck rot is now recognized in European countries including Czechoslovakia (Janyska and Rod, 1979), East Germany (Bochow and Bottcher, 1978) and Yugoslavia (Perisic *et al.*, 1977).

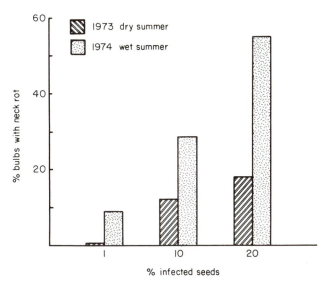

Fig. 4.2 Relationship between infected onion seeds and post-harvest neck rot.

In America, Ellerbrock and Lorbeer (1977a) isolated *B. allii*, *B. squamosa* and *B. cinerea* from blighted onion flowers, and agreed that *B. allii* was seed transmitted (Ellerbrock and Lorbeer, 1977b). The authors concluded, however, that infected seeds were not a major source of neck rot in New York State. The writer has tested many American seed samples but has isolated *B. allii* only occasionally from them.

(b) *Field infection and sources of inoculum.* Munn (1917) suggested that infection of onion plants occurred on dying leaves when the weather was wet during field drying before harvest. Walker (1952) reported that the growing plant was seldom affected but that the pathogen existed saprophytically on old dead leaves during the growing season, becoming parasitic and invading the necks of onions at or close to harvest. Various field sources of the pathogen have been suggested and are described here.

(i) *Soil-borne debris and sclerotia.* Soil-borne onion debris (Munn, 1917; Walker, 1926) and sclerotia (Walker, 1926, 1952) were suggested as potential overwintering sources of the neck rot pathogens (*B. allii* and *B. byssoidea*) from which crops might become affected if grown in contaminated soil during the following spring. *Botrytis allii* has been isolated from soil cropped with onions in America (Lorbeer and Tichelaar, 1970) and Israel (Kritzman and Netzer, 1978) by the use of selective media. Soil-borne sources may be important where onion crops are grown without or with minimal rotation as is reputed to happen in certain parts of America, such as on the organic soils of New York State.

They are, however, less likely to be of importance where extended rotation intervals severely reduce or eliminate soil inoculum by allowing decomposition of debris and sclerotia to occur. Thus the 3- to 4-year rotations practised in Britain (Anon, 1978) may have this effect, because *B. allii* survives on debris in sandy loam soil for two successive years only and sclerotia, though recovered from soil for up to 2 years, fail to germinate and produce conidiophores after 6 months (Maude *et al.*, 1982).

(ii) *Other sources.* Dumps such as cull piles of affected onions (Munn, 1917 and Walker, 1926 in America; van Doorn *et al.*, 1962 in Holland), overwintered seed production crops and overwintered bulb onion crops may act as sources of infection for spring-sown crops. Tichelaar (1967) observed *B. allii* sporing profusely on the upper layers of onion cull piles in June in Holland. The spring bulb crop is normally sown in March in the Netherlands. Ellerbrock and Lorbeer (1977b) found that cull piles and overwintering seed production fields in New York State were primary sources of the related pathogen *B. squamosa* for infection of onion fields in the spring. By implication, it is possible that *B. allii* might be disseminated from similar sources. Maude (1976) demonstrated experimentally that overwintered bulb onions could act as an infection bridge for *B. allii* between one spring-sown crop and the next.

In Britain, and to a certain extent in Europe, the primary source of post-harvest neck rot is infected seeds. Disinfection of the seeds and wide crop rotation virtually eliminates neck rot. Field sources of inoculum may be significant, however, where crop rotations are narrow or non-existent and where there is a substantial acreage of overwintered onions, such as bulb or seed production crops, causing a green bridge effect.

4. *The Disease in Stored Bulbs*

In the years prior to the use of seed treatment in Britain, losses in stored onion bulbs in bad seasons ranged from 15 to 50% (Derbyshire and Shipway, 1978).

Because neck rot increased with length of storage, growers sometimes believed that the disease spread by contact in store (Maude and Presly, 1977b). Munn (1917), however, found little evidence of this, and recent research (Maude and Presly, 1977b) confirmed this view (Fig. 4.3). The dry outer wrapper scales of bulb onions did not support the growth of *B. allii* (Munn, 1917; Walker, 1926) and were a barrier to contact spread of the disease. It has since been shown that increase of the disease with prolonged storage resulted not from contact spread of the disease in store but from the progressive rotting of bulbs which were already infected symptomlessly in their neck tissues when they were placed in store (Maude and Presly, 1977b).

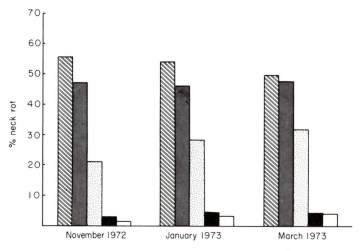

Fig. 4.3 Effect of length of storage on incidence of neck rot. (Histograms indicate neck rot levels in five different bulb onion crops.)

Bruising during lifting may increase the incidence of neck rot in store (Vaughan *et al.*, 1964), presumably by contamination of the damaged areas with the fungus during the lifting operation. Spread of the disease in store may also occur from a combination of mechanical damage to the bulbs during harvesting, resulting in breakage of the outer wrapper scales exposing the food tissues and where storage is under very humid conditions (Janyska and Rod, 1979).

5. *Disease Control*

In the past, neck rot was regarded as an intractable disease, difficult to control, because the lack of symptom expression in growing bulb crops meant that the source or sources and onset of the disease could not be recognized. Prophylactic sprays were not effective because their application was not timed correctly to achieve control. Strategies have been more successful, however, where they have been directed at an identifiable source of the disease, such as infected seeds, or where they have been more broadly based and designed to make the onion host less vulnerable to the pathogen such as the use of post-harvest drying systems.

(a) *Seed treatment control of neck rot.* Maude and Presly (1977a, b) demonstrated that infected onion seeds produced diseased but symptomless plants in the field which gave rise to bulbs with neck rot in store. This linked the seed source to post-harvest neck rot and produced a reasonable explanation for Tichelaar's (1971) success in reducing the disease in Dutch onion stores by seed treatment. The disease was virtually eliminated (less than 1% in dry and 3% in wet summers) in the U.K. by treating seeds for sowing with benomyl (2 g Benlate kg seed^{-1}) or benomyl plus thiram (3.3 g Benlate T kg seed^{-1}) (Maude and Presly, 1977b). Seed treatment with benomyl plus thiram was introduced commercially in Britain in 1974 and continues to give effective control of the disease (Maude and Presly, 1978, 1979; Maude et al., 1980). Recently, in Czechoslovakia (Janyska and Rod, 1980), seed treatment with benomyl plus thiram reduced post-harvest neck rot from 57.5 to 3.5% and increased yields by 114%. Bochow and Bottcher (1978) in East Germany also reported control of neck rot by seed treatment with benomyl and gained a further improvement in disease control by the addition of a single field spray of the fungicide just before the death of the crop foliage.

The choice of fungicide for seed treatment may vary from country to country but it is mainly based on the use of one of the benzimidazole-derived systemic fungicides singly or formulated with a broad-spectrum contact fungicide, such as thiram. Systemic fungicides that were ineffective as seed treatments (at similar application rates to benomyl) included thiabendazole, tridemorph and dodemorph (Maude et al., 1975). The non-systemic dicarboximide fungicide iprodione was effective when applied as a seed treatment (Maude and Presly, 1976, 1978). Later, Rod (1981) demonstrated that the dicarboximide fungicides iprodione, vinclozolin and procymidone were active in vitro against B. allii, but none were as toxic as the benomyl–thiram mixtures. Dips or surface applications of benomyl plus thiram (Benlate T) (Rod, 1980; Rod and Janyska, 1980) and dusting with iprodione (M. Hims, personal communication) has given effective control of the pathogen in onion sets.

Seed treatment alone has been used in Britain to control neck rot in onion crops grown to maturity and wind-rowed (field-dried) for 7–14 days. Such crops retain their foliage and are dried further with forced air at 15°C and stored at ambient temperatures (Tucker, 1971b). From 1975, however, quality requirements for onions stipulated by the European Economic Community (Anon, 1974) stimulated interest in the development of direct harvesting methods in Britain. The foliage (tops) of maturing onions were cut 5–8 cm above the bulbs in the field and the green bulbs were undercut, collected and taken directly into store. A rapid drying system was developed by Shipway (Anon, 1977) which removed excess moisture from the bulbs and sealed the necks by forcing air (425 m^3 air h^{-1} tonne onions^{-1}) through onion stacks at a temperature of 30°C for 3–5 days followed by a temperature reduction and storage at ambient temperatures.

Failure to achieve high airflows and temperatures in the post-harvest drying of direct harvested onion bulbs resulted in local outbreaks of neck rot in some stores in Britain in the late 1970s. The situation was redressed in subsequent years by strict adherence to the correct drying procedures (Anon, 1977).

At the present time, slightly more than 50% of bulb onions are direct harvested in Britain. Neck rot is controlled by seed treatment in all crops but higher airflow and drying temperatures are used for direct harvested onions.

(b) *Crop treatments.* Dithiocarbamate sprays or dusts were successful in controlling leaf rot (blast—*B. squamosa*) and increasing onion yields in America (Newhall and Rawlins, 1952). This was not the case, however, when fungicides, such as maneb, mancozeb and nabam, were tested as weekly sprays (June to August) to control neck rot in onions in New York State (Kaufman *et al.*, 1964). The fungicides had the opposite effect and increased the incidence of neck rot in the stored bulbs. Field sprays of benomyl did not control neck rot in stored bulbs in Britain (Maude and Presly, 1974). This failure may have been due to inaccurate timing of sprays in the absence of foliar symptoms.

Foliar sprays of desiccant chemicals (neo-decanoic acid mixtures) have been applied to accelerate the drying of leaves of late harvested crops and thereby possibly reduce the incidence of neck rot in the stored bulbs (Pendergrass *et al.*, 1969). Although the treatment desiccated the leaves, neck rot was increased in two of the three cultivars tested.

(c) *Post-harvest measures.* (i) *Fungicide treatments.* Post-harvest chemical dips and sprays applied to onion bulbs prior to storage proved ineffective in controlling neck rot in New York State (Lorbeer and Kawamoto, 1963). The application of fungicide dusts and desiccants to the topped necks of onions was more successful in experimental tests (Kaufman and Lorbeer, 1967). The fungicides dicloran and captan and the desiccant calcium chloride substantially reduced neck rot if applied to the cut necks of onions 1 h after inoculation with spores of *B. allii*. However, the use of chemicals on onion plants or harvested bulbs was superseded in New York State by the development of effective post-harvest drying systems.

(ii) *Drying treatments.* The concept of post-harvest drying of onions to reduce neck rot is not new. In 1925 Walker (1925b) demonstrated the importance of drying the necks of "topped" onions at 32–34°C to seal them against invasion by *B. allii*. One result of drying was that the onion necks no longer supported hyphal growth (Walker, 1925b) and furthermore there was a direct effect on the fungus in that growth of *B. allii in vitro* was prevented by exposure to temperatures of 35°C (Munn, 1917) and 38°C (van Doorn *et al.*, 1962). Harrow and Harris (1969) demonstrated that the fungus was killed by exposure to 36–38°C for 4 days. The rate of infection of onion leaf tissue inoculated with *B. allii* spores was severely reduced at 30°C (100% relative humidity), and, although 5% of leaves became infected, none produced sporing colonies (Presly, 1978). A reduction of 5°C restricted infection and asexual reproduction.

In a warm, dry climate the sun's radiant energy is sufficient to dry onions in the wind-row. In Oregon, U.S.A., supplementary artificial heat is probably necessary only 1 year in 5 (Vaughan *et al.*, 1964), whereas in temperate climates, for example, in Britain, post-harvest drying is necessary in most years.

In America, drying temperatures in excess of 43°C have been tested (Hoyle, 1948; Guzman and Hayslip, 1962; Vaughan *et al.*, 1964) and 40°C has been tried in the Netherlands (van Doorn, 1955; van Doorn *et al.*, 1962). In New Zealand, however, Harrow and Harris (1969) found that bulbs were severely damaged when dried at temperatures in excess of 38°C. There was no damage to bulbs and neck rot was considerably reduced in America by forcing air (224 m^3 air h^{-1} tonne onions^{-1}) at 34–35°C through bulk-stored onions over a 10- to 14-day curing period (Gunkel *et al.*, 1973). Similar effects were achieved in New York State when the method was adapted for use with onions stored in pallet boxes (Gunkel *et al.*, 1976): 46% of the crop in that state is stored in boxes. This post-harvest drying system is in commercial use in America. A similar system was developed independently in Britain (Anon, 1977), where air at 30°C is forced (425 m^3 air h^{-1} tonne onions^{-1}) through direct harvested, bulk-stored onion bulbs. Because the incidence of neck rot in most British stored bulbs was low as a result of seed treatment (Maude and Presly, 1977b), measuring the efficacy of this system in reducing the disease further was difficult (Shipway, 1980). In cooperative experiments with the Agricultural Development and Advisory Service, topped onions were inoculated in the field with spores of *B. allii* immediately before harvest. Post-harvest drying gave the greatest reduction of disease where inoculated bulbs were taken directly into store and dried at 30°C and where they were removed from the field no longer than 48 h after inoculation with the fungus (Maude *et al.*, 1980, 1981). In tests with infra red irradiation (Rosberg and Johnson, 1958), neck rot was reduced if bulbs were treated within 24 h of inoculation with spores of *B. allii*.

Forced air drying at temperatures of 30°C and above is the main basis for control of neck rot where infection sources include contaminated soil, onion dumps and overlapping seed production crops. Where infected seeds are important and seed treatment is employed, forced air drying provides a useful additional disease reduction facility for direct harvested onion bulbs.

(d) *Resistance to neck rot*. Resistance in coloured bulb onions to the neck rot fungi *B. allii* and *B. byssoidea* was shown to be related to the presence of phenolic compounds, protocatechuic acid (Angell *et al.*, 1930) and catechol (Link and Walker, 1933), in the dry wrapper scales. Under severe infection pressure, however, Hatfield *et al.* (1948) failed to establish a correlation between skin colour (red, yellow and white cultivars) and resistance to *B. allii*. They found that *B. allii* was least affected (by comparison with *Colletotrichum circinans* (smudge) and *Aspergillus niger* (black mould)) by volatile and non-volatile substances in onion storage tissues. Later, Owen *et al.* (1950b) reported that neck rot was less on pungent than mild cultivars and that coloured, pungent cultivars were less diseased than white cultivars. These discoveries appear not to have been incorporated into breeding programmes in America. In Europe where pungent or strong bulb onions command a large market, van der Meer *et al.* (1970) produced resistance testing techniques for assaying the reaction of seedlings, leaves and bulbs to *B. allii*. They reported that most seedlings and bulbs of *Allium cepa* were susceptible, but that occasional bulbs of the variety Rijnsburger were resistant. Seedlings and bulbs of *Allium* species such as *A. porrum* (leek) and *A. schroenoprasum* (chives), bearing a pseudostem instead of a bulb, were moderately resistant. The susceptible reaction of *A. cepa* to *B. allii* also applied to the related pathogen *B. squamosa* (leaf rot or blast). Bergquist and Lorbeer (1971) found no resistance to *B. squamosa* in 67 cultivars of *A. cepa*. Resistance was demonstrated, however, in *A. fistulosum* (Japanese bunching onion) and in interspecific hybrids (*A. cepa* × *A. fistulosum*). Resistance to the neck rot fungus, *B. allii*, was also obtained in *A. fistulosum* and in interspecific crosses (*A. cepa* × *A. fistulosum*) (Maude *et al.*, 1980).)

Non-destructive infection tests are being devised (Currah *et al.*, 1982) as an aid to this breeding programme. In common with other workers, no resistance to *B. allii* was found in the bulbs of common European onion cultivars (Maude *et al.*, 1980) nor in seedlings of cultivars from the Middle East, India and Russia (Currah *et al.*, 1982).

B. *Aspergillus* and *Penicillium*

The storage mould fungi *Aspergillus niger*, *A. alliaceus* and *Penicillium* spp. affect stored onion bulbs in hot climates, for example, in Texas, U.S.A. (Miller and Dillon, 1979a), Egypt (Hussein *et al.*, 1977) and India

(Venkatarayan and Delvi, 1951) or where onions are dried at high temperatures and then stored under cool conditions, e.g. Britain (Stow, 1975).

1. *Black* (Aspergillus niger) *and Yellow* (A. alliaceus) *Moulds*

(a) *The diseases and causal organisms.* Black mould (*A. niger*) affects the necks of onions and occasionally the base or sides of bulbs where the wrapper scales have been damaged, thereby exposing the underlying storage tissue (Chupp and Sherf, 1960). The fungus invades downwards in the storage scales, causing water soaking. White mycelium develops on the affected scales and through sporulation produces masses of black conidia: hence the name sooty or black mould. Machacek (1929) considered that *A. niger* was a primary pathogen causing a minor rot of imported Spanish onions. It became more extensive if secondary bacteria were involved.

Yellow mould (*A. alliaceus*) was isolated by Walker and Murphy (1934) in the U.S.A. from garlic imported from Italy. Infection of the storage tissue progressed in the manner described for *A. niger* with the difference that *A. alliaceus* produced black sclerotia and yellow conidia. Both fungi caused bulbs to shrivel. *Aspergillus alliaceus* has been reported on bulbs imported into Sweden (Ekstrand, 1948) and Thailand (Anon, 1961). *Aspergillus niger* and closely related species, however, more commonly affect stored bulbs.

(b) *Disease sources and biology.* Wayne (1917) found *A. niger* on onion seeds and sets and considered them to be potential sources of the disease in store. *Aspergillus niger* occurred on onions in field crops in New South Wales (Anon, 1939), but caused storage rotting only if bulbs were stored for long periods. It was, however, a major cause of rot of stored yellow onions in Texas between 1974 and 1976 (Miller and Dillon, 1979a), causing losses of 22–76% (1974, 1975) in onions stored for 3 months at 26.7°C and 70% relative humidity. Losses of 33% (1979) and 61% (1980) in onion bulbs due to *A. niger* were recorded during summer storage in Japan (Tanaka and Nonaka, 1981). Miller and Dillon (1979a) considered that the main source of the pathogen was the soil and that the outer scales of onion bulbs became contaminated by conidia of *Aspergillus* and thereafter progress of infection depended on the amount of rain before harvest and high relative humidity post-harvest in store. Hatfield *et al.* (1948) found *A. niger* to be a weak pathogen of onion bulb scales, but phenolic substances present in brown-skinned bulbs increased, rather than decreased, their susceptibility to the fungus. Walker and Murphy (1934) failed to infect leaves of growing onions with *A. alliaceus* and concluded that it was mainly a bulb pathogen. Bulbs inoculated with that fungus rotted completely if stored at 36°C for 21 days.

Infection progressed almost as rapidly at 28 and 32°C, but lower (20°C) and higher (40°C) temperatures caused only slight decay. At 16°C no rotting had occurred after 3 months storage. Thus disease development and expression was related to high temperatures but not to temperatures such as 16°C, which are more likely to be used in Britain for post-harvest drying of wind-rowed, non-topped onions.

Venkatarayan and Delvi (1951) considered that spread of *A. niger* in Indian stores might be due to the movement of mites which had fed on conidia of the fungus and that mite-infested stores might act as sources of infection for healthy bulbs.

(c) *Disease control.* Control of these storage moulds of onions in Britain may become more difficult with the increased use of high temperature drying (Stow, 1975). To some extent the promotory effects of high temperature on these diseases is limited because temperatures are reduced to ambient levels (on average 7°C over winter in Britain) in a relatively short space of time. This is not normally possible in hot climates where ambient temperatures are high during crop production and subsequently throughout storage. Miller and Dillon (1979b) reported that similar numbers of dried (at 37.8°C) and undried onions were infected with *A. niger* in Texas. Black mould was reduced only if air at 36% relative humidity was circulated during high temperature, post-harvest drying. Aycock and Jenkins (1960) reported significant reductions in *Aspergillus* mould of shallots if bulbs were field-dried for 6 days, then dipped in fungicide (2, 4, 5-trichlorophenoxide) for 15 min before or after being dried at 37.8°C for 6 days. Dusting topped onions with calcium carbonate reduced post-harvest bulb rot by 16–17% (Tanaka and Nonaka, 1981). Pryor (1950) reported that black mould in transported onions could be significantly reduced by container treatments with the gas nitrogen trichloride at 430 mg m^{-3}.

In contrast to bulb treatments, Wayne (1919) suggested that black mould in stored onions might be controlled by in-furrow seed treatment with formaldehyde.

2. *Blue Moulds* (Penicillium *spp.*)

(a) *The disease and causal organisms.* Blue moulds caused by various *Penicillium* spp. occur on fruits, bulbs, roots and seeds of a wide range of plants throughout the world (Chupp and Sherf, 1960). Cut surfaces of onion bulbs dipped in copper sulphate were rapidly colonized by *Penicillium* spp. (Dillon-Weston and Taylor, 1943); these fungi also rot many types of stored flower bulbs (Moore, 1949) in Britain. *Penicillium* moulds generally cause a soft watery rot on the surface of which are borne broom-like conidiophores

covered by masses of blue-green conidia. These symptoms were caused by
P. cyclopium Westling in stored English, Spanish and Dutch onions in the
U.K. (Wijeratnam and Lowings, 1978). The neck of onions and damaged
bulb tissue may be attacked. Blue mould (*P. corymbiferum* Westling) is a
serious field and post-harvest disease of garlic in California, U.S.A. (Smalley and Hansen, 1962).

(b) *Disease sources and biology.* In garlic, primary inoculum is probably
provided by infected cloves in the planting material. *Penicillium
corymbiferum* survives poorly in soil. Contamination of the garlic bulbs
during maturation and field-drying (wind-rowing) provided inoculum for
post-harvest infection of the cloves many of which were diseased when the
bulbs were opened after 3 months storage (Smalley and Hansen, 1962).
Wounding predisposed cloves to infection and certainly the practice of
breaking open the bulbs to release the cloves prior to planting provided
damaged areas that were suceptible to infection by *Penicillium* spp. Garlic
plants grown from cloves stored at 25°C were more severely diseased than
those grown from cloves stored at 5°C.
    It seems probable that the damage incurred in releasing the garlic cloves
for planting ensures the continuity of the disease from season to season.
This is unlikely to be the case with bulb onions which are grown from seeds.
In the author's experience and from the research of others (Kudrina, 1967),
however, *Penicillium* and *Aspergillus* spp. are often found contaminating
onion seed samples but the relevance of this observation to the occurrence
of moulds in stored onion bulbs is not known.

(c) *Disease control.* Smalley and Hansen (1962) suggested a combination
of post-harvest dipping of garlic bulbs in a mercurial fungicide followed by
in-store drying as a starting point for the development of practical
control measures.

C. Bacteria

A number of bacterial species are associated with post-harvest decay of
onion bulbs, as primary pathogens and as secondary invaders following
fungal, nematode or insect attack. Where fungal and bacterial pathogens
occur together in bulbs the bacteria are usually masked. Thus, when onion
neck rot (*B. allii*) was controlled in Britain, the full extent of bacterial
rotting became apparent (Taylor *et al.*, 1980). In circumstances where
bacteria are the primary disease agents they may cause considerable
damage (Cother *et al.*, 1976).
    In general, bacterial decay of bulbs is favoured by high temperatures,

and, because of this, problems may be more severe in warmer climates. The types of organism involved may differ in different climatic situations although many bacteria are widely distributed.

In the temperate climate of Britain, yellow-colonied *Erwinia herbicola*-like bacteria and a *Lactobacillus* sp. were the predominant organisms from decaying, commercially stored onions (Taylor, 1975). These bacteria were pathogenic when re-inoculated into healthy onion bulbs. *Pseudomonas alliicola*, the cause of slippery skin disease, has been isolated but is not common in British onions (Taylor, 1975); it has also been detected in bulbs imported from Spain (Roberts, 1973). It was the first bacterial pathogen of onions to be recognized and was described by Stewart (1899) in America. Classic research on this organism was done by Burkholder (1942) and Starr and Burkholder (1942). *Pseudomonas cepacia* Burk., the cause of sour skin disease, is also responsible for problems of stored onions in America (Burkholder, 1949), Canada (Creelman, 1967) and in Italy (Bazzi, 1979), but it has not been reported in British bulbs. In Australia, Cother *et al.* (1976) identified *Pseudomonas aeruginosa* to be the cause of an internal brown rot similar to slippery skin of bulb onions.

In warmer climates, bacterial soft rots (*Erwinia* and *Pseudomonas* spp.) caused up to 10.7% bulb losses after 3 months storage of Texan yellow onions where *A. niger* was the main pathogen (Miller and Dillon, 1979a). *Pseudomonas alliicola*, *P. cepacia* and *Erwinia aroidea* occurred on average in 9% of isolations from Egyptian onions (Hussein *et al.*, 1977).

1. *The Diseases and Causal Organisms*

Onions affected with slippery skin disease appear sound but when cut open some of the inner scales are brown and water-soaked and have a cooked appearance (Burkholder, 1942). Pressure applied at the base of the bulb may cause the centre core to slip out (Chupp and Sherf, 1960): hence the name "slippery skin". Infection progresses downwards in the storage tissue; eventually bulbs become soft and rotten. Mature bulbs are very susceptible and may rot completely within 10 days at room temperature (Burkholder, 1942). *Pseudomonas alliicola* is a Gram-negative motile bacterium with one to several polar flagella. Information on cultural and biochemical identification tests is given by Burkholder (1942) and Starr and Burkholder (1942). Unlike *P. alliicola*, *P. cepacia* (sour skin) attacks only the outer storage scales (Burkholder, 1949). Infected scales appear yellow and slimy, and bulbs may have a sour, vinegary odour: hence the name "sour skin". The bacterium is a Gram-negative, non-spore-forming rod with rounded ends. It is motile with 1–3 polar flagella and is on average 1.9

× 0.8 μm. Information on cultural and biochemical identification tests is given by Burkholder (1949).

The *Lactobacillus* sp. isolated by Taylor (1975) caused severe water-soaking of the internal tissues of onion bulbs, but only at high temperatures (35°C). Serological and biochemical tests (Taylor and Holden, 1977) showed that the organism was similar to *L. plantarum*, a homofermentative bacillus on herbage silage (Keddie, 1959). An *Erwinia*-like organism was the commonest of the pathogens isolated from British onions (Taylor, 1975) and caused severe water-soaking and discolouration of onion bulbs. Bio-chemically, it was similar to *E. herbicola*, but strong serological rela-tionships were not secured (Taylor and Holden, 1977).

2. *Disease Sources and Biology*

The field sources of bacteria causing post-harvest rotting are not well defined. *Pseudomonas alliicola* and *P. cepacia* have been isolated from soil in which onions had grown (Kawamoto and Lorbeer, 1967) and *P. cepacia* has been obtained from organic muck soil (Ballard *et al.*, 1970) and from irrigation water (Irwin and Vaughan, 1974). In the crop situation, Hevesi and Viranyi (1975) found necrotic lesions caused by *P. alliicola* on the seed stalks and leaves of onion plants in Hungary. When re-inoculated the bacterium produced elongated, sunken white, tongue-shaped lesions simi-lar to those which occurred naturally. Abundant rainfall in June and July increased the disease in the field and bulb crop (Vitanov, 1976). In India, disease outbreaks in seed production crops caused a severe stalk rot, inflorescences rotted and failed to set seed (Swarup *et al.*, 1974). Vitanov (1974), however, failed to demonstrate seed and inflorescence infection in inoculation tests with *P. alliicola*. *Pseudomonas alliicola* clearly affects the growing crop which increases in susceptibility with age. Burkholder (1942) considered that infection of the bulb probably occurred close to harvest time.

Pseudomonas cepacia is not naturally aggressive in the field crop situation. Kawamoto and Lorbeer (1974) failed to establish infection by inoculation of unwounded plants, but the tissue at the junction of the leaf blade and the sheath was particularly susceptible when stab-inoculated and kept moist (Kawamoto and Lorbeer, 1972a, 1974). This circumstantially supports Burkholder's (1949) suggestion that the bacterium may gain entry during field topping of onions. Populations of *P. cepacia* declined rapidly in stab-inoculated leaves, suggesting that air-dried, diseased leaves were not sour-ces of inoculum in the field (Kawamoto and Lorbeer, 1972b). *Pseudomonas aeruginosa* (Cother *et al.*, 1976), like *P. cepacia*, did not infect unwounded plants and it was concluded that special conditions such as high rainfall, low temperature and leaf abrasion were necessary for wound invasion to occur. *Lactobacillus* and the *Erwinia herbicola*-like organisms generally occur as leaf surface saprophytes in field crops and it is presumed that they act as secondary pathogens only when climatic factors and physiological conditions of the host tissues are favourable.

3. *Disease Control*

Bacteria causing post-harvest diseases present a range of organisms, some specific to onions (*P. alliicola*, *P. cepacia*), whereas others (*Erwinia*, *Lactobacillus* spp.) have a wide host range.

Post-harvest drying treatments appear to give little control of these organisms with some such as *Lactobacillus* spp. being increased by high temperature drying regimes. It would be perhaps more effective to prevent maturing onions from becoming infected by these bacteria, but this requires a greater understanding of their life-cycles. For example, in the case of *P. phaseolicola* (halo-blight of bean), a detailed investigation of seed transmission (Taylor *et al.*, 1979a) and field spread of the pathogen (Taylor, 1972) has resulted in the development of effective strategies for its control (Taylor *et al.*, 1979b). With the possible exception of *P. alliicola*, however, this approach may not be feasible for the storage bacteria, which are not predictable disease-causing organisms.

Chemicals including the antibiotic streptomycin and the organophosphorus insecticide chlorothion have been shown to be inhibitory to *P. alliicola in vitro* (El-Helaly *et al.*, 1962).

III. Other Diseases of Stored Bulbs

There are soil-borne fungi that infect the roots and cause plant losses in growing crops of bulb onions. By invading the roots and base plates of maturing bulbs, they may also present post-harvest problems when affected crops are harvested and stored. Such diseases include white rot (*Sclerotium cepivorum* Berk.), *Fusarium* basal rot (*F. oxysporum* f. sp. *cepae* (Hanz.) Snyder and Hansen) and pink root (*Pyrenochaeta terrestris* (Hansen) Gorenz). Smudge (*Colletotrichum circinans* (Berk.) Vogl.), however, only affects the necks and outer bulb scales of onions. Other sources of these fungi include contaminated sets and possibly infected seeds (*C. dematium* f. *circinans*, Behr 1963; *F. oxysporum* f. sp. *cepae*, Szatala, 1964). The general biology of these pathogens is described by Walker (1952) and Chupp and Sherf (1960). They are reviewed briefly here.

1. *White rot* (Sclerotium cepivorum)

The disease is widespread. It occurs mainly in salad (green) onions in Britain but in recent years has appeared more frequently in bulb onion fields. It has not caused post-harvest problems in the U.K., but these are reported from Australia where the fungus has caused breakdown and rotting of stored bulbs (Hughes, 1970b).

Typical white rot is distinguishable from neck rot (*B. allii*) in that it affects the base of onion bulbs, where it usually produces a white mycelium bearing a mass of small black sclerotia (0.3–0.5 mm in diameter) (Anon, 1981). Field treatments by the application of dicarboximide fungicides to infected organic soils (Canada: Utkhede and Rahe, 1979) and sandy loam soils (Britain: Entwistle *et al.*, 1980) offer a rational approach to pre- and post-harvest control of the disease. Post-harvest treatments have been tried and Hughes (1970b) reported the efficacy of the fungicide dicloran applied as a dust in the control of white rot in stored bulbs.

2. Fusarium *basal rot* (Fusarium oxysporum *f. sp.* cepae)

The causal pathogen is *Fusarium oxysporum* f. sp. *cepae* (Hanz.) Snyder and Hansen, but the disease was previously ascribed to various *Fusarium* species. Basal rot occurs in many countries, particularly in soils where onions have been grown for many years with little crop rotation.

The fungus infects roots, the base plate, and the storage tissues of the lower halves of onion bulbs, producing a whitish mould. Infection moves upwards in the bulb scales. No sclerotia are produced on the mould, thereby distinguishing basal rot from white rot (*S. cepivorum*). Lorbeer and Stone (1965) reviewed the mechanism of field infection in the light of previous research and identified two onion cultivars with some resistance. They found no evidence to support Walker and Larson's (1959) report of bulb to bulb spread of *Fusarium* in store and concluded that post-harvest infection of bulbs was of minor importance. It was evident, however, that bulbs which became infected in the field continued to rot in store. In this respect the post-harvest drying systems advocated for the treatment of neck rot (*B. allii*) should have a similar effect on basal rot and may be of value in places such as Georgia, U.S.S.R where *Fusarium* and *Botrytis* post-harvest rots of onion are equally important (Sardzhveladze *et al.*, 1976). Some reduction of *Fusarium* storage rot by fumigation has been reported in India (Dang and Thakur, 1973).

3. *Pink root* (Pyrenochaeta terrestris)

The pathogen (*Pyrenochaeta terrestris* (Hansen) Gorenz) is widely distributed (Walker, 1952; Chupp and Sherf, 1960) and affects a number of monocotyledonous and dicotyledonous plants. It is a wound invader, causing roots to turn pink, shrivel and die with the result that bulb yields may be reduced. It is not prominent in stored onions, but it is not uncommon to find it in association with *Fusarium* in affected onion base plates (Davis and Henderson, 1937; Hess, 1962). In this context, it may exacerbate post-harvest spoilage of onions. Resistance to *P. terrestris* (Walker, 1952; Marlatt and McKittrick, 1958; Chupp and Sherf, 1960) and additionally to *F. oxysporum* f. sp. *cepae* has been reported (Mehr *et al.*, 1962; Wooliams, 1966).

4. *Smudge* (Colletotrichum circinans)

The disease is caused by *Colletotrichum circinans* (Berk.) Vogl. It was discovered in Britain by Berkeley (1851) and now occurs widely in Europe and North America. The fungus infects through the cuticle of the neck and outer bulb scales of maturing bulbs, forming minute black stromata which produce acervuli which may be concentrically arranged. The blemish or "smudge" so produced on the outer scales of fresh marketed or stored onions devalues the produce. White-skinned cultivars are particularly susceptible (Walker, 1952; Chupp and Sherf, 1960). The disease is rarely seen on the brown-skinned Rijnsburger cultivars, which constitute the bulk of stored onions in Britain. The post-harvest drying systems in current use in Britain and America should also reduce smudge incidence.

IV. Conclusions

Post-harvest diseases may place a major restriction on the long-term storage of onions. Improvements in the control of these problems, however, have resulted in a commensurate increase in bulb health and storage life. Such improvements have been made possible by the recognition that post-harvest diseases have their origins in the field and not in the store. In the case of neck rot caused by *B. allii* (the main storage disease of bulb onions), it has been possible to break the life-cycle in some situations by a simple seed treatment and in others to render the host tissues less susceptible to infection by the adoption of post-harvest drying systems. Such treatments have wider effects in that they may also be used to control other soil-borne field diseases of onion such as pink root, white rot, basal rot and smudge, some of which can persist into the storage phase. Generally, however, field remedies are necessary for these diseases and can usually be provided (Utkhede and Rahe, 1979; Entwistle *et al.*, 1980; Katan, 1980; Katan *et al.*, 1980). In many situations, hygienic field practices including extended rotations, the burial or covering of onion debris with soil, and avoidance of overlapping crops are beneficial in the control of post-harvest diseases of onions.

Our ability to control moulds (*Aspergillus* and *Penicillium* spp.) that develop at high temperatures and bacteria (*Pseudomonas*, *Lactobacillus* and *Erwinia* spp.), however, is more limited. Some methods of control have been suggested, but essential supportive information on the biology of these pathogens is fragmentary and a greater research input in this direction is required if practical remedies are to be developed.

Acknowledgements

The author is grateful to Dr R. T. Burchill for his helpful comments on the chapter and to Dr J. D. Taylor for his criticism of the section on bacterial pathogens. I wish to express my thanks to Mrs N. C. Wood, who typed the script; to Mrs A. Spencer, who prepared the references; and to Mrs J. M. Bambridge, who produced the diagrams.

References

Angell, H. R., Walker, J. C. and Link, K. P. (1930). The relation of protocatechuic acid to disease resistance in the onion. *Phytopathology*, **20**, 431–438.

Anon (1939). Plant diseases. Notes contributed by the Biological Branch. *Agric. Gaz. N.S.W.* **1**, 199–203.

Anon (1961). *Aspergillus alliaceus. Q. Rep. Pl. Prot. Comm. S.E. Asia Pacific Reg.*, Jan–Mar & April–June.

Anon (1974). "E.E.C. Standards for Onions." Ministry of Agriculture, Fisheries and Food, London.

Anon (1977). Dry bulb onions. Effects of various handling and drying practices on storage v7/40. *Rep. Kirton Exp. hort. Stn* **14**, 53–57.

Anon (1978). "Dry Bulb Onions", Hort. Enterpr. Booklet I. Ministry of Agriculture, Fisheries and Food, London.

Anon (1980). Onions. *F.A.O. Prod. Ybk 1979* **33**, 155–156.

Anon (1981). "Onion Neck Rot", Leaflet 779. Ministry of Agriculture, Fisheries and Food, London.

Aycock, R. and Jenkins, J. M., Jr (1960). Methods of controlling certain diseases of shallots. *Pl. Dis. Reptr* **44**, 934–939.

Ballard, R. W., Palleroni, N. J., Doudoroff, M., Stanier, R. Y. and Mandel, M. (1970). Taxonomy of aerobic Pseudomonads: *Pseudomonas cepacia, P. marginata, P. alliicola* and *P. caryophylli. J. gen. Microbiol.* **60**, 199–214.

Bazzi, C. (1979). Identification of *Pseudomonas cepacia* on onion bulbs in Italy. *Phytopath. Z.* **95**, 254–258.

Behr, L. (1963). On a total loss of germinating onions caused by *C. dematium* f. *circinans. Zbl. Bakt.* **116**, 552–561.

Bergquist, R. R. and Lorbeer, J. W. (1971). Reaction of *Allium* spp. and *Allium cepa* to *Botryotinia* (*Botrytis*) *squamosa*. *Pl. Dis. Reptr* **55**, 394–398.

Berkeley, M. J. (1851). *Gdnrs' Chron.* **11**, 595.

Blodgett, E. C. (1946). Observations on blasting of onion seed heads in Idaho. *Pl. Dis. Reptr* **30**, 77–81.

Bochow, H. and Bottcher, H. (1978). Zur Bekampfung von *Botrytis allii* Munn durch Einsitz von Fungizicken. *NachrBl. dt. PflSchutzdienst, Berl.* **32**, 135–137.

Bochow, H. and El Mosallamy, H. M. (1979). Investigations on the infection behaviour of *Botrytis allii* Munn in *Allium cepa* L. *Archiv. Phytopath. PflSch.* **15**, 103–112.

Brawtigam, S. (1977). *Botrytis allii* Munn an Zwiebelsaatgiet. *NachrBl. dt. Pfl-Schutzdienst, Berl.* **31**, 195.

Brien, R. M. (1939). List of plant diseases recorded in New Zealand. *Bull. N.Z. Dep. scient. ind. Res.*, No. 67, 39 pp.

Burkholder, W. H. (1942). Three bacterial plant pathogens: *Phytomonas carophylli* sp. n., *Phytomonas alliicola* sp. n., and *Phytomonas manihotus* (Arthand—Berthet et Bondar) Vilgas. *Phytopathology* **32**, 141–149.

Burkholder, W. H. (1949). Sour skin, a bacterial rot of onion bulbs. *Phytopathology* **40**, 115–117.

Chupp, C. and Sherf, A. F. (1960). "Vegetable Diseases and their Control." Ronald Press, New York.

Clark, C. A. and Lorbeer, J. W. (1973a). Reaction of onion cultivars to *Botrytis* brown stain. *Pl. Dis. Reptr* **57**, 210–214.

Clark, C. A. and Lorbeer, J. W. (1973b). Symptomatology, etiology and histopathology of *Botrytis* brown stain of onion. *Phytopathology* **63**, 1231–1235.

Cother, E. R., Darbyshire, B. and Brewer, J. (1976). *Pseudomonas aeruginosa*: cause of internal brown rot of onion. *Phytopathology* **66**, 828–834.

Creelman, D. W. (1967). Diseases of vegetable crops. *Can. Pl. Dis. Surv.* **47**, 44–53.

Croxall, H. E. (1945). Some factors influencing loss of onion bulbs during storage. *Rep. Long Ashton Res. Stn 1945*, 143–147.

Currah, L., Maude, R. B., Presly, A. H., Ockendon, D. J. and Bolland, C. (1982). *Botrytis* diseases of onions. *Rep. natn. Veg. Res. Stn 1981*, 60–61.

Dang, J. K. and Thakur, D. P. (1973). Controlling storage rot on onion bulbs caused by *Fusarium solani* by fumigation. *Indian Phytopath.* **25**, 359–361.

Davis, G. N. and Henderson, W. J. (1937). The interrelationship of the pathogenicity of a *Phoma* and a *Fusarium* on onions. *Phytopathology* **27**, 763–772.

Derbyshire, D. M. and Shipway, M. R. (1978). Control of post-harvest deterioration in vegetables in the UK. *Outl. Agric.* **9**, 246–252.

Dillon-Weston, W. A. R. and Taylor, R. E. (1943). Development of *Pencillium* on the cut surfaces of certain vegetables. *Nature, Lond.* **151**, 54–55.

Ekstrand, H. (1948). A fungal disease new to Sweden on imported eating onions. *Vaxtskyddsnotiser* **4**, 63–64.

El-Helaly, A. F., Abo-El-Dahab, M. K. and Zeitoun, F. M. (1962). Effect of organic phosphorus pesticides on certain phytopathogenic bacteria. *Phytopathology* **53**, 762–764.

Ellerbrock, L. A. and Lorbeer, J. W. (1977a). Etiology and control of onion flower blight. *Phytopathology* **67**, 155–159.

Ellerbrock, L. A. and Lorbeer, J. W. (1977b). Sources of primary inoculum of *Botrytis squamosa*. *Phytopathology* **67**, 363–372.

Ellis, M. B. (1971). "Dematiaceous Hyphomycetes." Commonwealth Mycological Institute, Kew.

Entwistle, A. R., Munasinghe, H. L., Jonas, L. B. and Haynes, A. J. (1980). White rot disease of onions—autumn sown bulb onions. *Rep. natn Veg. Stn 1979*, 67.

Fresenius, G. (1850). "*Botrytis aclada* Fresen", Beitrage zur Mykologie, 16–18. Heinrich Ludwig Bronner, Frankfurt.

Gunkel, W. W., Lorbeer, J. W., Kawamoto, S. O., Kaufman, J. and Smith, H. A., Jr (1973). Recent development on artificial heating: a method of control of *Botrytis* neck rot in bulk stored onions. *Prog. Rep. 30th N.Y. Farm Electrif. Coun.*, 57–61.

Gunkel, W. W., Lorbeer, J. W., Bensin, R. F. and Smith, H. A., Jr (1976). Application of the artificial heating method to control *Botrytis* neck rot in pallet box stored onions. *Prog. Rep. 33rd N.Y. Farm Electrif. Coun.*, 88–93.

Guzman, V. L. and Hayslip, N. C. (1962). Effect of time of seeding and varieties on onion production and quality when grown in two soil types. *Proc. Fla. St. hort. Soc.* **75**, 156–162.

Harrow, K. M. and Harris, S. (1969). Artificial curing of onions for control of neck rot (*Botrytis allii* Munn). *N.Z. Jl agric. Res.* **12**, 592–604.

Hatfield, W. C., Walker, J. C. and Owen, J. H. (1948). Antibiotic substances in onion in relation to disease resistance. *J. agric. Res.* **77**, 115–135.

Hellmers, E. (1943). *Botrytis* on *Allium* species in Denmark. *Botrytis allii* Munn and *B. globosa* Raabe. *Medd. Vet. Højsk. plantepat. Afd., Kbh.* **25**, 51 pp.

Hemmi, T. and Niwa, S. (1937). On gray mould neck rot of stored onions. *Forsch. Plfkr., Kyoto* **3**, 234–249.

Hennebert, G. L. (1963). Les *Botrytis* des *Allium*. *15th Int. Symp. Phytopharm. Phyt.*, 851–876.

Hennebert, G. L. (1973). *Botrytis* and *Botrytis* like genera. *Persoonia* **7**, 183–204.

Hess, W. M. (1962). Pink root of onion caused by *Pyrenochaeta terrestris*. *Diss. Abstr.* **23**, 1478–1479.

Hevesi, M. and Viranyi, F. (1975). An unknown symptom on onion plants caused by *Pseudomonas alliicola* (Burkholder) Starr et Burkholder. *Acta Phytopath. Acad. Sci. Hung.* **10**, 281–286.

Hickman, C. J. and Ashworth, D. (1943). The occurrence of *Botrytis* spp. on onion leaves with special reference to *B. squamosa*. *Trans. Br. mycol. Soc.* **26**, 153–157.

Hoyle, B. J. (1948). Onion curing—a comparison of storage losses from artificial, field, and non-cured onions. *Proc. Am. Soc. hort. Sci.* **52**, 407–414.

Hughes, I. K. (1970a). Onion diseases in Queensland. *Qd agric. J.* **96**, 607–612.

Hughes, I. K. (1970b). Onion storage rot control. *Qd J. Agric. Animal Sci.* **27**, 391–392.

Hussein, F. N., Abd-Elrazik, A., Darweish, F. A. and Rushdi, M. H. (1977). Survey of storage diseases of onions and their incitants in upper Egypt. *Egypt J. Phytopath.* **9**, 15–21.

Irwin, R. D. and Vaughan, E. R. (1974). Bacterial soft rot of onions—role of irrigation water and identification of the causal organisms. *Proc. Am. Phytopath. Soc.* **1**, 47 (Abstract).

Janyska, A. and Rod, J. (1979). The causes of spreading of *Botrytis allii* Munn in onions under storage conditions. *Bull. Vyzk. slecht. Ust. zelin, Olomouc* **21 /22**, 70–83.

Janyska, A. and Rod, J. (1980). The effect of the treatment of onion (*Allium cepa* L.) seed on the yields and on the occurrence of neck rot. *Zahradnictvi* **7**, 197–208.

Jarvis, W. R. (1976). "Onion Neck Rot", Fact Sheet 019. Ministry of Agriculture and Food, Ontario.

Jones, H. A. and Mann, L. K. (1963). "Onions and their Allies." Interscience Publishers, New York.

Katan, J. (1980). Solar pasteuration of soils for disease control status and prospects. *Pl. Dis.* **64**, 450–454.

Katan, J., Rotem, I., Daniel, J. and Finkel, Y. (1980). Solar heating of the soil for the control of pink root and other soil-borne diseases in onions. *Phytoparasitica* **8**, 39–50.

Kaufman, J. and Lorbeer, J. W. (1967). Control of *Botrytis* neck rot of onions by fungicidal dusts and desiccant chemicals. *Pl. Dis. Reptr* **51**, 696–699.

Kaufman, J., Lorbeer, J. W. and Friedman, B. A. (1964). "Relationship of Fungicides and Field Spacings to *Botrytis* Neck Rot of Onions Grown in New York", Publ. ARS 51–1. United States Department of Agriculture, Agricultural Research Service, Washington, D. C.

Kawamoto, S. O. and Lorbeer, J. W. (1967). Soft rot bacteria associated with onion decay. *Phytopathology* **57**, 341.

Kawamoto, S. O. and Lorbeer, J. W. (1972a). Histology of onion leaves infected with *Pseudomonas cepacia*. *Phytopathology* **62**, 1266–1271.

Kawamoto, S. O. and Lorbeer, J. W. (1972b). Multiplication of *Pseudomonas cepacia* in onion leaves. *Phytopathology* **62**, 1263–1265.

Kawamoto, S. O. and Lorbeer, J. W. (1974). infection of onion leaves by *Pseudomonas cepacia*. *Phytopathology* **64**, 1440–1445.

Keddie, R. M. (1959). The properties and classification of Lactobacilli isolated from grass and silage. *J. appl. Bact.* **22**, 403–416.

Kritzman, G. and Netzer, D. (1978). A selective medium for isolation and identification of *Botrytis* spp. from soil and onion seed. *Phytoparasitica* **6**, 3–7.

Kudrina, V. N. (1967). Changes in the microflora and germination of vegetable seeds during storage. *Izv. timiryazev sel'.-khoz. Akad.* **2**, 175–183.

Link, K. P. and Walker, J. C. (1933). The isolation of catechol from pigmented onion scales and its significance in relation to disease resistance in onions. *J. biol. Chem.* **100**, 379–383.

Lorbeer, J. W. and Kawamoto, S. O. (1963). "Annual Progress Report on Research Supported by the Orange County Vegetable Improvement Co-operative Association Incorporated." Cornell University, Ithaca, New York.

Lorbeer, J. W. and Stone, K. W. (1965). Reaction of onion to *Fusarium* basal rot. *Pl. Dis. Reptr* **49**, 522–526.

Lorbeer, J. W. and Tichelaar, G. M. (1970). A selective medium for the assay of *Botrytis allii* in organic and mineral soils. *Phytopathology* **60**, 1301 (Abstract).

Machacek, J. E. (1929). The black mold of onion caused by *Aspergillus niger*. *Phytopathology* **19**, 733–739.

Marlatt, R. B. and McKittrick, R. T. (1958). Pink root resistant onions in Arizona. *Pl. Dis. Reptr* **42**, 1310–1311.

Massee, G. (1894). An onion disease. *Gdnrs' Chron.* **16**, 120.

Maude, R. B. (1976). Neck rot (*Botrytis allii*) in the autumn sown bulb onion crop. *Rep. natn. Veg. Res. Stn 1976*, 94–95.

Maude, R. B. and Presly, A. H. (1974). Neck rot of onions. *Rep. natn. Veg. Res. Stn 1973*, 95.

Maude, R. B. and Presly, A. H. (1976). Neck rot of onions. *Rep. natn. Veg. Res. Stn 1975*, 97.

Maude, R. B. and Presly, A. H. (1977a). Neck rot (*Botrytis allii*) of bulb onions. I. Seed-borne infection and its relationship to the disease in the onion crop. *Ann. appl. Biol.* **86**, 163–180.

Maude, R. B. and Presly, A. H. (1977b). Neck rot (*Botrytis allii*) of bulb onions.

II. Seed-borne infection in relationship to the disease in store and the effect of seed treatment. *Ann. appl. Biol.* **86**, 181–188.

Maude, R. B. and Presly, A. H. (1978). Neck rot (*B. allii*) in the spring-sown bulb onion crop. *Rep. natn. Veg. Res. Stn 1977*, 93–94.

Maude, R. B. and Presly, A. H. (1979). Neck rot of bulb onions. *Rep. natn. Veg. Res. Stn 1978*, 69.

Maude, R. B., Presly, A. H. and Walker, J. A. (1975). Neck rot of onions. *Rep. natn. Veg. Res. Stn 1974*, 109.

Maude, R. B., Presly, A. H., Miller, J. M. and Large, A. (1980). Neck rot of bulb onions. *Rep. natn. Veg. Res. Stn 1979*, 60–63.

Maude, R. B., Presly, A. H., Bambridge, J. M. and Spencer, A. (1981). Neck rot of bulb onions. *Rep. natn. Veg. Res. Stn 1980*, 64.

Maude, R. B., Bambridge, J. M. and Presly, A. H. (1982). The persistence of *Botrytis allii* in field soil. *Pl. Path.* **31**, 247–252.

McLarty, H. R. (1929). Report of the Dominion Field Laboratory of Plant Pathology, Summerland, B. C. *In* "Report of the Dominion Botanist for the Year 1928", 142–162. Division of Botany, Canadian Department of Agriculture.

Mehr, A. E., O'Brien, M. J. and Davis, E. W. (1962). Pathogenicity of *Fusarium oxysporum* f. sp. *cepae* and its interaction with *Pyrenochaeta terrestris* on onion. *Euphytica* **11**, 197–208.

Miller, M. E. and Dillon, R. C., Jr (1979a). Survey of bulb diseases on onions in South Texas. *J. Rio Grande Vall. hort. Soc.* **33**, 25–28.

Miller, M. E. and Dillon, R. C., Jr (1979b). Effects of artificial drying and controlled atmosphere storage on control of black mold (*Aspergillus niger*) of onions. *Phytopathology* **69**, 530.

Moore, W. C. (1948). Diseases of crop plants (1943–1946). *Bull. Minist. Agric. Fish. Fd, Lond.* **139**, 90 pp.

Moore, W. C. (1949). Diseases of bulbs. *Bull. Minist. Agric. Fish. Fd, Lond.* **117**, 176 pp.

Morgan, D. J. (1971). Numerical taxonomic studies of the genus *Botrytis* II. Other *Botrytis* taxa. *Trans. Br. mycol. Soc.* **56**, 327–335.

Munn, M. T. (1917). Neck-rot disease of onion. *Bull. N.Y. St. agric. Exp. Stn* **437**, 455 pp.

Netzer, D. and Dishon, I. (1966). Occurrence of *Botrytis allii* in onions for seed production in Israel. *Pl. Dis. Reptr* **50**, 21.

Newhall, A. G. and Rawlins, W. A. (1952). Control of onion blast and mildew with carbamates. *Phytopathology* **42**, 212–214.

Owen, J. H., Walker, J. C. and Stahmann, M. A. (1950a). Variability in onion neck rot fungi. *Phytopathology* **40**, 749–768.

Owen, J. H., Walker, J. C. and Stahmann, M. A. (1950b). Pungency, color and moisture supply in relation to disease resistance in the onion. *Phytopathology* **40**, 292–297.

Pendergrass, A., Isenberg, F. M., St John, L. E., Jr and Lisk, D. J. (1969). A foliage drying harvest aid for onions. *Hort. Sci.* **4**, 294–297.

Perisic, M., Babovic, M. and Markovic, S. (1977). Contribution to the study of *Botrytis allii*, the parasite of onion. *Arh. poljopr. Naute Teh.* **30**, 143–149.

Presly, A. H. (1978). *Botrytis* species on overwintered salad onions. Ph.D. Thesis, University of Birmingham.

Pryor, D. H. (1950). Reduction of post-harvest spoilage in fresh fruits and vegetables destined for long distance shipment. *Fd Technol.* **4**, 57–62.

Roberts, P. (1973). A soft rot of imported onions caused by *Pseudomonas alliicola* (Burkh.) Starr and Burkh. *Pl. Path.* **22**, 98.

Rod, J. (1978). Evaluation of the storability of Czechoslovak cultivars of onion (*Allium cepa* L.). *Sbornik UVTIZ, zhradnictvi, 5–6, 1978–1979,* 99–103.

Rod, J. (1980). Onion (*Allium cepa*) seed treatment. *Ochr. Rost.* **16**, 111–119.

Rod, J. (1981). The efficacy of some fungicides on fungi *Botrytis allii* and *Botrytis cinerea* under *in vitro* conditions. *Ochr. Rost.* **17**, 113–117.

Rod, J. and Janyska, A. (1980). The control of *Botrytis allii* in onion (*Allium cepa* L.) sets. *Zahradnictvi* **7**, 279–288.

Røed, H. (1951). Botrytis (gray mold) on *Allium cepa* and *Allium ascalonicum* in Norway. *Acta Agric. scand.* **1**, 20–39.

Rosberg, D. W. and Johnson, H. B. (1958). Post-harvest infra red irradiation for the control of gray mold neck rot disease of onions. *Phytopathology* **48**, 345.

Samuel, G. (1931). Summary of plant disease records in South Australia for the two years ending June 30th, 1930. *J. Dep. Agric. S. Aust.* **34**, 746.

Sardzhveladze, S., Dolidze, M. and Kirimelashvili, N. (1976). The development of diseases of onion during storage under different conditions. *Jr. Nil Zemledeliya Gruz. S.S.R.* **23**, 141–145.

Shidla, L. A. (1971). Fungi of the genus *Botrytis* in Lithuania, Part 1. Species and their distribution on agricultural and medicinal plants. *Tr. Akad. Nank. Lit. S.S.R. Ser. B* **1**, 23–30.

Shipway, M. R. (1980). Developments in onion drying and storage. *A. Rev. Kirton Exp. hort. Stn 1979,* 33–39.

Siemaszko, W. (1929). Grey mould of onion—*Botrytis allii* Munn. *Roczn. Nauk roln.* **21**, 12 pp.

Smalley, E. B. and Hansen, H. N. (1962). Penicillium decay of garlic. *Phytopathology* **52**, 666–678.

Sorauer, P. (1876). Das Verschimmeln der Speiszwiebeln. *Osterr. Landw. Wehnbl.* 147.

Starr, M. P. and Burkholder, W. H. (1942). Lipolytic activity of phytopathogenic bacteria determined by means of spirit blue agar and its taxonomic significance. *Phytopathology* **32**, 598–604.

Stewart, F. C. (1899). A bacterial rot of onion. *Bull. N.Y. St. agric. Exp. Stn* **164**, 209–212.

Stow, J. R. (1975). Effects of humidity on losses of bulb onions (*Allium cepa*) stored at high temperature. *Expl Agric.* **11**, 81–87.

Swarup, J., Magarkoti, M. S. and Saksena, H. K. (1974). A new bacterial rot of onion caused by *Pseudomonas alliicola* in India. *Indian J. Mycol. Pl. Path.* **3**, 187–189.

Szatala, O. (1964). *Fusarium* rot of onions in Hungary. *Annls Inst. Prot. Plant. Hung.* **9**, 301–311.

Tackholm, V. and Drar, M. (1954). Flora of Egypt. *Bull. Fac. Sci. Egypt Univ.* **3**, xii + 644 pp.

Tanaka, K. and Nonaka, F. (1981). Studies on the rot of onion bulbs caused by *Aspergillus niger* and its control by lime applications. *Bull. Fac. Agric. Saga Univ.* **51**, 47–51.

Taylor, J. D. (1972). Field studies on halo-blight of beans (*Pseudomonas phaseolicola*) and its control by foliar sprays. *Ann. appl. Biol.* **70**, 191–197.

Taylor, J. D. (1975). Bacterial rots of stored onion. *Rep. natn. Veg. Res. Stn 1974,* 116–117.

Taylor, J. D. and Holden, C. M. (1977). Bacterial rots of stored onions. *Rep. natn. Veg. Res. Stn 1976,* 104.

Taylor, J. D., Dudley, C. L. and Presly (née Gray), L. (1979a). Studies of halo-blight seed infection and disease transmission in dwarf beans. *Ann. appl. Biol.* **93**, 267–277.

Taylor, J. D., Phelps, K. and Dudley, C. L. (1979b). Epidemiology and strategy for control of halo-blight of beans. *Ann. appl. Biol.* **93**, 167–172.

Taylor, J. D., Dudley, C. L. and Littlejohn, I. H. (1980). Bacterial rots of stored onions. *Rep. natn. Veg. Res. Stn 1979*, 76–77.

Tichelaar, G. M. (1967). Studies on the biology of *Botrytis allii* on *Allium cepa*. *Neth. J. Pl. Path.* **73**, 157–160.

Tichelaar, G. M. (1970). *Botrytis allii. Jversl. Int. plziektenk. Onderz. 1969*, 30.

Tichelaar, G. M. (1971). *Botrytis allii. Jversl. Inst. plziektenk. Onderz. 1970*, 32.

Tucker, W. G. (1971a). Understanding onions. 1. Physiology of bulb storage. *Comml Grow.*, No. 3919, 273 and 276.

Tucker, W. G. (1971b). Understanding onions. 2. Practicalities of storing the crop. *Comml Grow.*, No. 3920, 307 and 315.

Utkhede, R. S. and Rahe, J. E. (1979). Evaluation of chemical fungicides for control of onion white rot. *Pestic. Sci.* **10**, 414–418.

van Doorn, A. M. (1955). Control of onion neck rot by artificial curing. *Meded. Dir. Tuinb.* **18**, 250–258.

van Doorn, A. M., Koert, J. L. and Kreyger, J. (1962). Investigations on the occurrence and control of neck rot (*Botrytis allii* Munn) in onions. *Versl. landbouwk. Onderz. RijkslandbProefstn* **68.7**, 83 pp.

van der Meer, Q. P., Bennekom, J. L. van and van der Giessen, A. C. (1970). Testing onions (*Allium cepa* L.) and other *Allium* species for resistance to *Botrytis allii* Munn. *Euphytica* **19**, 152–162.

van Poeteren, N. (1939). Onderzoek over het koprot in de uien van de oogst 1938. *Versh. Meded. plziektenk. Dienst Wageningen*, 90.

Vaughan, E. K., Cropsey, M. G. and Hoffman, E. N. (1964). Effects of field-curing practices, artificial drying and other factors in the control of neck rot in stored onion. *Stn tech. Bull. Ore. agric. Exp. Stn* **77**, 21 pp.

Venkatarayan, S. V. and Delvi, M. H. (1951). Black mould of onions in storage caused by *Aspergillus niger*. *Curr. Sci.* **20**, 243–244.

Vitanov, M. (1974). Investigations on the pathogenicity of *Pseudomonas alliicola* (Burkh.) Starr and Burkholder. *Gradinar Lozar. Nauka.* **11**, 48–56.

Vitanov, M. (1976). Effect of harvest dates and storage on onion slippery skin infection (*Pseudomonas alliicola* Burkh.) on onion bulbs. *Gradinar Lozar. Nauka.* **13**, 63–71.

Walker, J. C. (1925a). Two undescribed species of *Botrytis* associated with the neck rot diseases of bulb onions. *Phytopathology* **15**, 708–713.

Walker, J. C. (1925b). Control of mycelial neck rot of onion by artificial curing. *J. agric. Res.* **30**, 365–373.

Walker, J. C. (1926). *Botrytis* neck rots of onions. *J. agric. Res.* **33**, 893–928.

Walker, J. C. (1952). "Diseases of Vegetable Crops." McGraw-Hill, New York.

Walker, J. C. and Larson, R. H. (1959). "Onion Diseases and their Control", Agriculture Handbook No. 208. United States Department of Agriculture, Washington, D.C.

Walker, J. C. and Murphy, A. (1934). Onion-bulb decay caused by *Aspergillus alliaceus*. *Phytopathology* **24**, 289–291.

Ward, C. M. and Tucker, W. G. (1976). Respiration of maleic hydrazide treated and untreated onion bulbs during storage. *Ann. appl. Biol.* **82**, 135–141.

Wayne, van Pelt (1917). A new fungus disease causing serious damage in storage houses. *Mon. Bull. Ohio agric. Exp. Stn* **1**, 152–156.

Wayne, van Pelt (1919). Onion diseases as found in Ohio. *Mon. Bull. Ohio agric. Exp. Stn* **1**, 70–76.

Whitwell, J. D., Frith, L. and Williams, J. H. (1973). Experiments on the use of maleic hydrazide as a sprout suppressant on spring sown bulb onions. *Expl Hort.* **25**, 87–96.

Wijeratnam, R. S. W. and Lowings, P. H. (1978). Watery rot of stored onions caused by *Penicillium cyclopium* Westling. *Pl. Path.* **27**, 100.

Wood, R. K. S. (1961). The biology and control of diseases caused by *Botrytis* spp. *Proc. 1st Br. Insectic. Fungic. Conf.*, 309–314.

Wooliams, G. E. (1966). Resistance of onion varieties to *Fusarium* basal rot and to pink root. *Can. Pl. Dis. Surv.* **46**, 101–103.

Yarwood, C. E. (1938). *Botrytis* infection of onion leaves and seed stalks. *Pl. Dis. Reptr* **22**, 428–429.

5

Carrots

B. G. LEWIS AND B. GARROD

I. Introduction

The cultivated carrot, *Daucus carota* L., is considered to have been selected from the anthocyanin form of *Daucus carota* ssp. *carota*, which probably has a centre of diversity in Afghanistan (MacKevic, 1929). Previous selection has concentrated on the phenotypic components of yield, root shape, retardation of inflorescence development, and quality characteristics such as colour, but has not included resistance to disease. Amongst the present range of commercial stocks, there do not appear to be obvious differences in storage potential between cultivars, though no detailed comparisons have been made. However, there is a good deal of variation within cultivars, and this needs to be taken into account when results of experiments are assessed.

The wild form of *Daucus carota* ssp. *carota* is distributed mainly in cool temperate regions where the biennial habit, involving a quiescent winter-survival period, has resulted in the development of the root and hypocotyl (Esau, 1940) as a storage organ. Although the cultivated carrot is grown in a very wide range of climates, including sub-tropical conditions (Banga, 1976), it is mainly in cool temperate regions that the crop is stored for long periods in the winter following summer and autumn production.

The length of the period between harvesting and marketing is very variable. A very large proportion of the crop is marketed or used for canning immediately and, in these situations, post-harvest infection is minimal. Disease problems during short-term storage have been confined to carrots pre-packed in polyethylene bags where high humidities and temperatures have allowed very rapid infection by pathogens such as *Erwinia carotovora* (Jones) Bergey *et al.* var. *carotovora* Dye (see Ch. 9) and *Chalaropsis thielavioides* Peyronel (Derbyshire and Shipway, 1978).

Long-term storage is achieved in many countries by storing in clamps. Usually, this means simply earthing-over heaps of carrots, but more elaborate systems of crating and ventilation for the initial cooling of the roots have been described by Djacenko (1971). However, when ambient temperatures rise in spring, sprouting and rapid infection of the stored roots often occur.

In most cool temperate countries where carrots are produced, controlled temperature storage is used, although as yet only a small proportion of the crops are stored in this way. These stores are maintained at temperatures just above freezing point and a humidity as near saturation as possible, conditions that are generally considered to give maximum storage potential (Lockhart and Delbridge, 1972; van den Berg and Lentz, 1973; Apeland and Hoftun, 1974; Krahn, 1974). Under these conditions, carrots can often be stored up to 8 months (Krahn, 1974). A recent development is the system of ice-bank storage coupled with positive ventilation, in which the humidity approaches saturation and the temperature can be very precisely controlled at $0.6 \pm 0.1°C$, some 2°C lower than other typical commercial cold stores (Hawkins *et al.*, 1978).

In cool temperate regions, the development of the carrot plant slows down and growth virtually ceases as the ambient temperature falls during autumn. Although metabolism of the root slows down, the tissues remain capable of such functions as wound repair and active defence for a period of several months under conditions that simulate a cool temperate soil environment in winter, viz. a high humidity and a temperature just above freezing point of the tissues. With the onset of higher temperatures under natural conditions, the plant would eventually become re-activated, producing new foliage and inflorescence. This essential function of the root as a survival and storage organ in the biennial cycle of the carrot plant helps in understanding its longevity and disease resistance in contrast with such organs as leaves, petioles or fruits, which have a shorter natural life span and a much more limited storage potential. By harvest, the older foliage of carrot plants has begun to senesce and abscission layers have begun to form in the petiole bases, but are incomplete (Davies *et al.*, 1981), and the root has developed all the secondary tissues shown in Fig. 5.1. The outer part of the root consists of pericyclic parenchyma covered with periderm, the outermost cells of which are dead (Fig. 5.2). If the roots remain in soil to be stored by "strawing over", these outer tissues presumably remain intact and afford substantial protection since there are no reports of serious disease problems except violet root rot caused by *Helicobasidium purpureum* Pat., which is also capable of infection during the growing season (Anon, 1980). Apart from seedling infection, attack by other species appears to be mainly associated with senescent foliage and only occasionally results in crown infection by this stage (Goodliffe and Heale, 1975; Geary, 1978; Wall and Lewis, 1980).

When roots are harvested, however carefully, the lateral roots are severed and the tap root is broken. With machine harvesting and bulk hand-

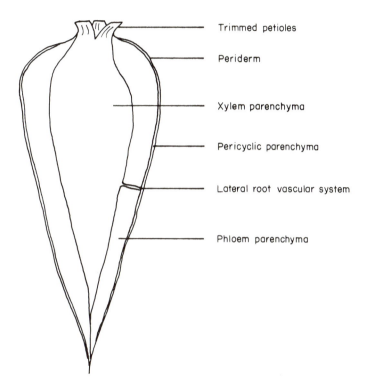

Fig. 5.1 Internal morphology of the carrot root (longitudinal section).

ling, damage is usually substantial. Tucker (1974) reported 20–30% visible damage of machine-harvested roots, and as much as 60% sometimes occurs (Derbyshire and Crisp, 1971; Anon, 1975). During harvest, machinery commonly causes deep flesh wounds as well as surface abrasion, and during subsequent bulk handling, more surface abrasion, as well as bruising and cracking of roots, usually occurs (Anon, 1975). Currently, these roots are usually transferred directly to clamps, barns or refrigerated stores without treatment to heal wounds or minimize infection in other ways.

II. Incidence of Various Carrot Pathogens in Storage

At present, information about the incidence and relative importance of storage disease of carrots is incomplete. There have been very comprehensive surveys (e.g. Rader, 1952; Mukula, 1957; Årsvoll, 1969) of storage diseases from only a small proportion of the main carrot-growing countries

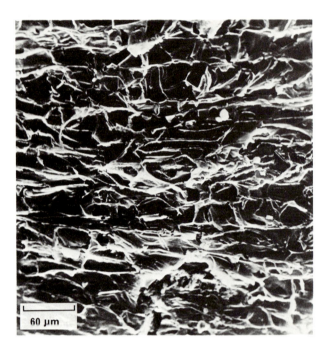

Fig. 5.2 Scanning electron micrograph of carrot root periderm tissue. (Courtesy of W. P. Davies, Harper Adams Agricultural College, Newport, Shropshire, U.K.)

of the world, and information about certain storage methods is much more comprehensive than others; in general, much more is known about refrigerated storage than clamp storage or field storage. Nevertheless, there is fairly general agreement about pathogens of major importance and the storage conditions where they are prevalent (Table 5.1). *Botrytis cinerea* Pers. ex Pers., *Sclerotinia sclerotiorum* (Lib.) De Bary and *Mycocentrospora acerina* (Hartig) Deighton appear to be most generally important. *Stemphyllium radicinum* (Meier *et al*.) Neergaard may be equally prevalent, but its incidence is difficult to assess because secondary pathogens, such as *B. cinerea* and *S. sclerotiorum* often invade the lesions and mask the primary infection (Rader, 1952); perhaps for this reason its importance has been underestimated. *Rhizoctonia carotae* Rader is potentially important and has sometimes caused major losses where it has remained endemic in stores (Jensen, 1971), but there is very little information about this pathogen. Under conditions where *Erwinia carotovora* becomes prevalent, it causes a very rapid, damaging epidemic. However, it only becomes important at high temperatures and it therefore only occurs occasionally in clamp storage when

Table 5.1 *Occurrence of the major pathogens in various storage systems*

Pathogen	Storage Method Where Disease is Prevalent			References
	F/S	Cl/S	C/S	
B. cinerea	+	+	+	Lauritzen (1932), Rader (1952), Årsvoll (1969), Derbyshire and Crisp (1971)
S. sclerotiorum	+	+	+	Rader (1952), Mukula (1957), Årsvoll (1969), Derbyshire and Crisp (1971)
M. acerina	+	+	+	Årsvoll (1969), Derbyshire and Crisp (1971)
R. carotae		Rarely	+	Rader (1952), Mukula (1957), Jensen (1971), Derbyshire and Shipway (1978)
S. radicinum		+	+	Lauritzen (1926), Neergaard (1945), Mukula (1957)
E. carotovora		+		Rader (1952)

F/S = field storage; Cl/S = clamp storage; C/S = cold storage.

temperatures rise and in unrefrigerated transit, particularly with pre-packed carrots (see Ch. 9).

Another group of pathogens of secondary importance includes *Rhizopus* spp., *Fusarium* spp., *Phytophthora cactorum* (L. & C.) Schroet, *Phytophthora megasperma* Dr. and *Pythium* spp. Although less prevalent than the main pathogens, they occasionally cause devastating diseases under rather unusual conditions (Rader, 1952). Rots of minor importance are caused by *Penicillium* spp. and *Cylindrocarpon* spp. (Rader, 1952; Mukula, 1957). Since there is relatively little information about the behaviour and incidence of these last two groups of pathogens, most of our ensuing comments are based on the pathogens of major importance.

III. Symptoms Caused by Major Fungal Pathogens

A comprehensive survey of symptoms is given by Rader (1952), Mukula (1957) and Årsvoll (1969). The following descriptions are based on those of the above authors, together with our own observations.

A. *Botrytis cinerea*

All parts of the root are susceptible to this fungus, although lesions are most frequent on either the crown or the tip. Infected tissues are light brown and water-soaked initially, appearing spongy or leathery in older lesions with indiscrete margins to healthy tissues. The surface of lesions usually becomes characteristically greyish-brown, covered with masses of conidiophores and conidia and, commonly, black sclerotia are formed in close proximity to the decayed tissues. At temperatures approaching 0°C the fungus occasionally develops a white aerial mycelium lacking conidia, and thus resembles *S. sclerotiorum*. Later, under humid conditions, a leathery layer of sclerotia may form.

B. *Sclerotinia sclerotiorum*

Lesions occur anywhere on stored roots and are typically characterized by the presence of rapidly spreading, white, cottony mycelium. Infected tissue is typically soft and watery and only slightly discoloured, and can be distinguished from bacterial soft rot by the absence of sliminess, provided the lesion has not been invaded by secondary organisms. Sclerotia form on the surface of lesions, the rinds of which are initially white, but later turn black.

C. *Mycocentrospora acerina*

Blackish, rapidly spreading lesions may occur on any part of the root where the tissues have been damaged but are most common in the crown and root-tip regions. This fungus may penetrate deeply into the root tissues, causing a soft and watery, dark brown, later black, decay. Macroscopically, it is often difficult to distinguish this disease from that caused by *Stemphylium radicinum*, but they are easily separated with the aid of a microscope since *M. acerina* produces characteristic chains of thick-walled, dark-coloured chlamydospores in the black lesion, whereas *S. radicinum* does not.

D. *Rhizoctonia carotae*

Lesions may occur anywhere on the root. In the first instance small, whitish hyphal knots are apparent on the root surface. Beneath these hyphal knots, small pits form which slowly enlarge into sunken craters lined with white flocculent mycelium. The superficial mycelium is usually more sparse than that of *S. sclerotiorum* but can be abundant under high humidity conditions where, macroscopically, the organisms can easily be confused. At microscopic level, they can readily be distinguished by the presence of abundant clamp connections on *R. carotae*.

E. *Stemphylium radicinum*

Black, often deep, lesions may occur on any part of the root and are characterized by the discrete black margins that separate the diseased from healthy tissues. Typically, lesion tissue is fairly firm and mealy. However, at high humidities and temperatures, the decayed tissue may be more soft and watery with a brownish advancing margin. Macroscopically, *S. radicinum* is very difficult to distinguish from *M. acerina* (see above).

IV. Origin and Form of Inoculum

Post-harvest infection often originates from inoculum that has built up on the growing crop (Rader, 1952; Goodliffe and Heale, 1975; Geary, 1978). Most post-harvest pathogens of carrot rarely infect the roots in the field, but are usually confined to the foliage. At harvest, some, such as *M. acerina* (Davies *et al.*, 1981), *S. sclerotiorum* (Geary, 1978) and *B. cinerea* (Goodliffe and Heale, 1975), have progressed from the lamina into the base of the older petioles and are, therefore, not removed when the rest of the foliage is trimmed. *Mycocentrospora acerina* is also abundant as resting mycelium and chlamydospores in the rhizosphere of growing roots; these are the main forms of inoculum of this fungus entering storage on the root surfaces and on soil particles amongst the petioles in the crown region (Wall and Lewis, 1980; Davies *et al.*, 1981). *Botrytis cinerea* and *S. sclerotiorum* are also present amongst foliage debris adhering to the roots during harvesting operations and this is thought to be an important source of primary infection (van den Berg and Lentz, 1966; Goodliffe and Heale, 1975; Geary, 1978). These forms of inoculum are usually inconspicuous on harvested roots and a superficial inspection would not be useful in indicating storage potential. The possibility of predicting storage potential by inspection of crops before harvest or by incubation tests on samples of roots after harvest has not been explored.

After entering the store in these various ways, inoculum of certain pathogens can remain endemic, surviving on crates and pallets, unless these are effectively sterilized. This form of survival is particularly important with *R. carotae* (Rader, 1952; Jensen, 1971), but a wide range of other species has been isolated from storage containers (Rader, 1952; Mukula, 1957).

V. Spread from Root to Root

The humidity within stores is usually sufficiently high to favour surface spread of fungi; *S. sclerotiorum* (Rader, 1952; Mukula, 1957; Derbyshire and Crisp, 1978; Geary, 1978) and *R. carotae* (Rader, 1952) usually spread in this way, forming characteristic foci of infection. *Botrytis cinerea* can also spread by mycelial growth from an infected crown to neighbouring roots (Goodliffe and Heale, 1975), but high levels of conidia have been commonly found in circulating air in stores (Rader, 1952; Mukula, 1957) and probably provide an alternate means of spread (Mukula, 1957). Subsequent levels of these diseases may reflect the effects of storage conditions on spread of the pathogen, rather than initial levels of inoculum. On the other hand, storage potential reflects the numbers of contaminated roots brought into storage with pathogens such as *M. acerina*, which does not spread significantly from root to root (Davies *et al.*, 1981); treatments that reduce inoculum are more likely to be effective with this organism than with those which spread from root to root.

VI. Infection and Resistance of Roots

Fungi that commonly infect foliage in the field often enter the root via the petioles. At harvest, the incomplete abscission layers, which are present even in the oldest remaining petiole bases, do not constitute an obvious structural impedance to infection (Davies *et al.*, 1981). From the petioles, the main storage pathogens, *B. cinerea*, *S. sclerotiorum* and *M. acerina*, can invade the crown, which consists of stem and hypocotyl (Esau, 1940), and thence the root region (Goodliffe and Heale, 1975; Geary, 1978; Davies *et al.*, 1981). Senescent petioles appear to be particularly susceptible (Davies *et al.*, 1981), but younger petioles and the remainder of the crown regions are often sufficiently resistant in early storage to localize infection by *M. acerina* and *B. cinerea* (Goodliffe and Heale, 1975; Davies *et al.*, 1981).

Infection rarely occurs through undamaged parts of the root despite the often patchy development of periderm on some roots at harvest (Davies, 1977a). Resistance of the root at this stage is very high (Day, 1972; Davies, 1977a; Goodliffe and Heale, 1978), the periderm and pericycle being particularly impenetrable (Davies *et al.*, 1981). Some infection occasionally occurs through parts of the root without obvious macroscopic damage; pathogens that are able to infect in this way include *S. sclerotiorum* (Geary, 1978) and occasionally *B. cinerea* (Goodliffe and Heale, 1977; Sharman and Heale, 1979) when a supply of exogenous nutrients is present. Infection by all carrot storage pathogens is considerably facilitated by wounds (Mukula, 1957; Derbyshire and Crisp, 1971), and some, such as *M. acerina*, are virtually restricted to wound infection (Davies *et al.*, 1981).

Few detailed quantitative studies of wound infection have been done, except with *M. acerina*. Here, the susceptibility of wounds varies considerably with depth (Davies *et al.*, 1981), indicating that there is a gradient of resistance, varying from a high level in the periderm and pericycle to a low level in the xylem parenchyma (Fig. 5.3). In early storage, a large proportion of lesions caused by *M. acerina* and *B. cinerea* remain localized (Davies, 1977a; Goodliffe and Heale, 1978; Davies *et al.*, 1981). This localized phase is particularly noticeable with *M. acerina*, occurring on all types of infection site on the root; a characteristic reaction of the neighbouring tissue is the accumulation of lignin and suberin, together with a fluorescence of the cells in ultraviolet light (Davies *et al.*, 1981). In the localized phase, the fungus is unable to cross this reaction zone.

The time when lesions become progressive varies considerably among different pathogens. It can apparently occur early in storage with *S. sclerotiorum* when inoculum and food-base levels are high (Geary, 1978), but is particularly durable, lasting up to 4–6 months, with *M. acerina* at a storage temperature of about 2°C. This phase of lesion restriction might be described as latent infection (Anon, 1973), but whether the fungus is truly quiescent (Gaeumann, 1950; Verhoeff, 1974) or whether it is progressing slowly is not known. Latent infections on a wide range of fruits appear to be localized until the fruits ripen or senesce (Simmonds, 1941, 1963; Edney, 1956, 1958; Meredith, 1964; Ayob and Swinburne, 1970; Swinburne, 1971, 1973; see Ch.

1). In the carrot root, also, physiological changes, such as declining ability to accumulate anti-fungal substances (Day, 1972; Goodliffe and Heale, 1978), may be regarded as an aspect of senescence, though it appears to be much less sudden than that described for many fruits that have a climacteric phase. Very little information is available about latent infections with other carrot pathogens, though the change from localized to progressive infection with *B. cinerea* has been attributed to a decline in ability of the root to accumulate 8-hydroxy-6-methoxy-3-methyl-3,4-dihy-droisocoumarin (6-methoxymellein) (Goodliffe and Heale, 1978). The period of restricted lesions with these two pathogens appears to be extended by storing plants at low temperature. Thus, the diseases have tended to become prevalent, primarily as a result of attempts to lengthen the storage period and not because low temperatures are more favourable for infection.

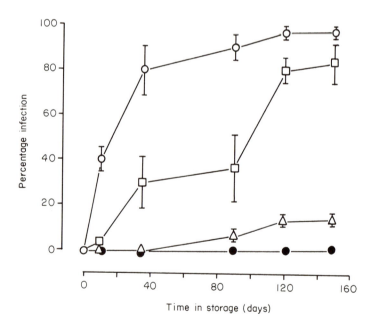

Fig. 5.3 Proportion of inoculated sites showing localized or progressive lesions on xylem parenchyma ○; phloem parenchyma □; pericyclic parenchyma △; periderm ●; with time in storage. Vertical bars represent the standard errors of the means. (Courtesy of *Transactions of the British Mycological Society* **77**, 139–151, 1981.)

Progressive lesions caused by all these pathogens are most common in the crown region, but can begin on any part of the root where a primary infection has already occurred. This phase of the diseases is very damaging, since *B. cinerea, S. sclerotiorum* and *M. acerina* cause unrestricted lesions with extensive maceration of the tissues, involving breakdown of cell walls by pectolytic and cellulolytic enzymes (Brown, 1915; Echandi and Walker, 1957; B. G. Lewis, unpublished). Development of these rots, which often destroy the whole root in the case of *S. sclerotiorum* and *B. cinerea*, is often very rapid and the pathogens can spread into adjacent roots; *S. sclerotiorum* usually forms characteristic foci of infection, involving large numbers of roots at this stage (Derbyshire and Shipway, 1978). With some pathogens the onset of progressive infection probably occurs as tissue resistance declines (Davies, 1977a; Goodliffe and Heale, 1978); this probably coincides with the phase in the biennial cycle of the plant when growth of new shoots develops and might be considered to be an indication of root senescence.

The decline in resistance is associated with a fall in levels of anti-fungal compounds. Carrot root tissues have been shown to contain a number of anti-fungal compounds. Some of them have been detected in extracts of uninoculated tissues (De Lacy and Virtanen, 1957; Chakravarty and Srivastava, 1967; Garrod *et al.*, 1978; Garrod and Lewis, 1979) and others were induced to accumulate in response to injury or the challenging of tissues with various micro-organisms (Condon and Kuć, 1960; Condon *et al.*, 1963; Menke *et al.*, 1964; Herndon *et al.*, 1966; Chalutz *et al.*, 1969; Coxon *et al.*, 1973; Jaworski *et al.*, 1973; Goodliffe and Heale, 1978; Harding and Heale, 1980, 1981). Unfortunately, most authors do not give details about the pre-experimental history of the roots and many of the experiments have been performed under environmental conditions, which are very dissimilar to commercial storage conditions. A further complication is that many experiments have been performed on transverse slices of root, which consist of a mixture of tissues, and which may vary in ability to accumulate certain anti-fungal compounds, such as *cis*-heptadeca-1,9-diene-4, 6-diyne-3,8-diol (falcarindiol). This polyacetylenic compound is strongly anti-fungal, inhibiting both germination and mycelial growth of *M. acerina* (Garrod *et al.*, 1978; Garrod and Lewis, 1982), as well as *B. cinerea* and a wide range of other fungi (Kemp, 1978). There is a close correlation between levels of this compound in the various healthy, uninoculated carrot root tissues and the ability of *M. acerina* to invade them. The levels of falcarindiol in the periderm and pericyclic parenchyma are well above the ED_{50}* value for inhibition of germination and mycelial growth of *M. acerina* (Garrod *et al.*, 1978; Garrod and Lewis, 1982); they are sufficiently high to account for the

* Estimated dose to give 50% inhibition.

very high level of resistance of this tissue early in storage and the requirement of a breach in the periderm and pericyclic parenchyma for infection (Davies *et al.*, 1981).

Of the anti-fungal compounds that are enhanced by inoculation or infection, 6-methoxymellein has received most attention. Sufficient quantities diffuse out of carrot roots to account for at least part of the inhibition of pre-penetration phases of *M. acerina* and localization of lesions in xylem parenchyma (Davies and Lewis, 1981b). The potential of roots to accumulate 6-methoxymellein falls with time in storage and may be involved with the general decline of resistance to *B. cinerea* (Goodliffe and Heale, 1978). Only in roots resistant to *B. cinerea* did the levels of 6-methoxymellein in the lesion exceed the ED_{50} for hyphal growth of this fungus (Goodliffe and Heale, 1977); these authors also claimed that the well-known increase in susceptibility to *B. cinerea* when roots lose turgor (Årsvoll, 1969; Djacenko, 1971; Heale *et al.*, 1977) is also correlated with a decrease in the potential for accumulating 6-methoxymellein. Levels of 6-methoxymellein found by various workers are strikingly variable; part of this might be explained by differences in tissue sampling procedure, but part might be explained by differences in the storage environment, since accumulation is known to be induced by low temperature treatment (Sondheimer, 1957), ethylene and various other chemical treatments (Chalutz *et al.*, 1969; Coxon *et al.*, 1973; Phan and Sarkar, 1975; Sarkar and Phan, 1975). It might be possible to increase the resistance of roots and improve storage potential by manipulating the storage environment in order to maximize 6-methoxymellein accumulation. Treatment with low levels of ethylene (Sarkar and Phan, 1975) merits further consideration in this respect, provided that there are no deleterious side-effects of this treatment.

The polyacetylenic compound *cis*-heptadeca-1,9-diene-4,6-diyne-3-ol (falcarinol) has also been found in carrot root tissue and levels were enhanced by inoculation with conidia of *B. cinerea* (Harding and Heale, 1980) to concentrations that these authors considered strongly inhibitory. However, there is considerable discrepancy between their estimations of sensitivity of *B. cinerea* to falcarinol and those of Kemp (1978). Nevertheless, the compound merits further attention.

Although *p*-hydroxybenzoic acid has been found in inoculated tissue (Davies, 1977b; Harding and Heale, 1980, 1981; Davies and Lewis, 1981a), the levels were too low to explain inhibition of either *M. acerina* or *B. cinerea*. In potato tissue, *p*-hydroxybenzoic acid is thought to be involved in lignin biosynthesis (Robertson *et al.*, 1968) and, if it is similarly involved in carrot, the compound may affect resistance in ways other than by direct toxicity.

VII. The Influence of Wound Healing on Storage Potential

Since wounds are an important avenue for invasion of carrot roots by storage pathogens, an understanding of wound infection and wound resistance is likely to yield useful information about storage potential. The treatment of other root crops, involving exposure to high temperature and humidity for a short period after harvest in order to "harden" the periderm and heal wounds, was originally developed for sweet potato (Weimer and Harter, 1921; Artschwager and Starrett, 1931) and is now used for potato (Burton, 1966), yams (Gonzalez and Collazo de Rivera, 1972; Passam et al., 1976) and cassava (Booth and Coursey, 1974). Such treatment also promotes healing of wounds, involving lignification, suberization and sometimes callus development in carrot roots (Sirtautaite and Sokolova, 1965; Davies, 1977a; Garrod et al., 1982). It is probable that under natural conditions, where the ambient temperature is reasonably high at harvest, some wound healing may occur before roots are cold-stored. For example, during studies of transit conditions Vakis et al. (1975) reported that a period of about 4 days was required for carrots to cool from an ambient temperature of 20–25°C to the temperature of the store at 5°C. However, it is currently recommended that roots intended for refrigerated storage are transferred to the store and the temperature is reduced as soon as possible to 0–1°C (Apeland and Hoftun, 1974) without a healing period, and a rapid cooling rate has also been recommended for ice-bank storage (Hawkins et al., 1978).

Although infection was not studied in detail in their experiments, Saburov and Sirtautaite (1966) reported a general decrease in rotting after wound repair. Treatment at high temperature and a humidity approaching saturation dramatically reduced infection of wounds by M. acerina (Lewis et al., 1981). A period as short as 12 h reduced infection of the normally susceptible xylem parenchyma to negligible levels (Fig. 5.4) and the beneficial effects were maintained for several months of storage. The overall result was to reduce the probability of infection of deep wounds to levels normally expected in pericyclic parenchyma.

The resistance of repaired phloem parenchyma tissue partly involves killing of inoculum by falcarindiol, which accumulates on the surface of the wound when the oil ducts in this tissue are severed during wounding, whereas on areas of wound surfaces where no oils are present, other compounds, probably including 6-methoxymellein, have a fungistatic effect (Lewis et al., 1981). Accumulation of these anti-fungal substances appears to be the main mechanism that reduces infection of wounds by M. acerina, since lignification and· suberization account for comparatively little impedance to this fungus (Garrod et al., 1982). Since falcarindiol is

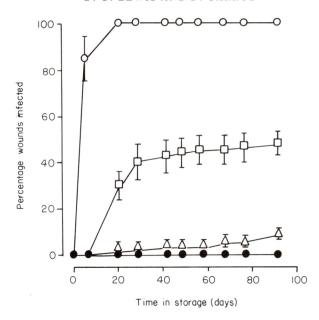

Fig. 5.4 The relationship between the healing period of xylem parenchyma tissue and subsequent development of infection (localized and progressive). ○, Unhealed tissue; □, tissue healed for 6 h; △, tissue healed for 12 h; ●, tissue healed for 2 days. Vertical bars represent the standard errors of the means. (Courtesy of *Annals of Applied Biology* **99**, 35–42, 1981.)

considered a wide spectrum (Kemp, 1978), pre-formed inhibitor (Garrod and Lewis, 1980), it would be particularly effective in protecting the carrot root, because its accumulation after wounding would not rely on metabolic activity and, therefore, would not be temperature dependent. Such a mechanism would clearly be of value in protecting this biennial storage organ during low temperature conditions which usually prevail in the winter between vegetative and reproductive phases of growth, and is, therefore, likely to be of significance in field storage as well as refrigerated storage.

Much more work on wound repair and the conditions that promote it needs to be done before it could be recommended for commercial practice. However, these preliminary results appear promising and the subject merits further attention. If these healing conditions are deemed worthwhile, there is an urgent need to check whether other diseases become prevalent under these short-term conditions of high temperatures and humidities and whether the beneficial effects of healing would be offset by an increase in other diseases, such as *Erwinia carotovora* and *Chalaropsis thielavioides* (Derbyshire and Shipway, 1978).

VIII. Interrelationships Between Age of Roots at Harvest and Storage Potential

When attempting to optimize storage potential, the timing of harvest is an important consideration with most crops, the aim being to effect a compromise between "ripeness" or "maturity" of the crop and its potential for resisting infection. In commercial practice, the harvest "maturity" of the carrot crop varies, depending on the outlet of the crop. It is usually determined mainly on the basis of root size, which can be varied by altering the length of the growing period (Kurdina, 1957; Fritz and Habben, 1975) or population density (Bleasdale, 1963). There are various reports that carrot crops of different ages also differ in susceptibility to storage diseases (Kurdina, 1957; Mukula, 1957; Sirtuataite and Sokolova, 1965; Saburov and Sirtautaite, 1966) and attempts to correlate other variables, such as carbohydrate, carotene and phenol composition of the roots with storage potential, have been made (Plantenius, 1934; Smith, 1948; Kurdina, 1957; Phan and Hsu, 1973; Weichmann and Ammerseder, 1974; Fritz and Habben, 1975). Disease incidence has also been correlated with concentrations of glucose and sucrose at harvest (Mukula, 1957; Sirtautaite, 1965). However, no causal relationship between these variables and resistance has been found in carrot roots and some of the correlations are poor (Weichmann and Kappel, 1977).

A particularly important factor in determining storage potential is the resistance of wounds, which appears to vary with root age (Davies and Lewis, 1980). When plants were sown at approximately monthly intervals, harvested simultaneously after growing periods varying between 4 and 7 months, then wounded and inoculated with *M. acerina*, numbers of lesions increased more slowly on the youngest roots; these differences were maintained for at least 5 months of storage. Increased susceptibility of the older roots was largely attributable to a decline in the healing potential of wounds (Lewis *et al.*, 1981). Such changes with age may explain some of the year to year variation in storage potential of crops (Davies and Lewis, 1980) and probably represent a decline in the rate of development of defence barriers. These studies need to be extended to include comparisons with the other main storage pathogens.

In view of the heterogeneity of roots and the seasonal variation in growing conditions, the absolute age would only give an approximate guide to healing potential. There is a need to identify the characteristics that have a predominant influence on wound healing potential, to find out how these change during development of the plant, and to devise practical ways of deciding when they have reached an optimum level.

IX. Pre-Storage Treatments

A number of fungicidal dips applied immediately after harvest have been found to be effective in reducing losses in storage. Of several fungicides tested against *M. acerina* by Derbyshire and Crisp (1971), methyl-l-(butylcarbamoyl)-2-benzimidazolecarbamate (benomyl) was the only effective one, reducing losses over a 6 month storage period from around 35% to around 10%. Benomyl also reduced infection by other pathogens in the United Kingdom; in one season, the overall rotting was reduced from around 40% to around 5% (Derbyshire and Crisp, 1978). At the time of these experiments (1971–73), there were also major reductions in rots caused by *B. cinerea*. In Canada, the structurally related fungicides of the thiabendazole group were equally effective in controlling storage rots, of which *B. cinerea* was the most prevalent (Lockhart and Delbridge, 1974). Although benomyl has not been tested on carrots more recently, it is probably less effective now that organisms such as *B. cinerea* have become insensitive to it (Dekker, 1976). In the United States, 2,6-dichloro-4-nitroaniline and sodium *o*-phenylphenate were as effective as benomyl in controlling post-harvest rots (Wells and Merwarth, 1973), but the effectiveness of sodium *o*-phenylphenate may depend on the spectrum of pathogens present, since it was not active against *M. acerina* (Derbyshire and Crisp, 1971).

Some fungicide studies have shown incidentally that, where carrot roots dipped in water as well as undipped roots were used as controls, a small proportion of the benefit is due to immersion in water (Derbyshire and Crisp, 1971, 1978). More striking reductions in infection have been achieved by washing roots prior to storage (Lockhart and Delbridge, 1972). Although these effects have been attributed to removal of inoculum (Lockhart and Delbridge, 1972), there may be other beneficial effects since Garrod *et al.* (1982) reported that more rapid and effective lignification of wounds occurs when roots are washed. Short periods (10–20 s) of dipping in water at 52°C were also beneficial (Wells and Merwarth, 1973). This effect is mainly a high temperature effect, but dipping in water at 24°C occasionally reduced rotting significantly.

X. Conclusions and Future Developments

One of the main interests in the post-harvest pathology of the carrot root arises from a need to improve the quality and extend the storage life of the crop. Traditional methods of storage—in field, clamp or barn—have limits in storage life which are largely imposed by variables other than infection;

these systems endeavour to store the roots for time spans that correspond approximately with the natural survival period of roots between foliage senescence in the first year and foliage regeneration in the second year of the plants' biennial cycle and coincide with a period when the root is highly resistant to infection.

Problems of infection appear to have arisen because technological developments in environmental control have removed most of the environmental constraints on storage potential and have extended the time span to a phase where natural resistance has become sufficiently low to allow disease levels to become the predominant limiting variable. Improvement in disease management is, therefore, one of the main developments that would extend the range of storage potential and allow the full potential of environmental technology to be exploited.

Some progress has already been made in diagnosing the main causal agents; elucidating the conditions favouring the establishment and spread of diseases in the store; improving the technology of handling crops to reduce damage and subsequent wound infection; and treating with fungicidal dips to retard disease development. By taking advantage of these findings, it is possible, on an experimental scale, with roots lifted with minimum damage and then treated with a fungicidal dip of benomyl and stored at 0–1°C, to store roots for up to 8 months with negligible losses.

At present, fungicidal dips are a necessary component of this success. However, it would be prudent to view this as a temporary expedient because of the possible hazard of human toxicity; realistically, also, the useful life of most fungicides is limited by the development of microbial resistance so that the range of useful fungicides needs to be extended.

Coupled with this on-going programme of fungicidal measures, other promising improvements, such as wound-healing treatments, and hitherto unexploited measures, such as breeding for disease resistance, might be explored. With the exploitation of male sterile lines in *Daucus carota* to develop F_1 hybrids for commercial use (Banga, 1976), there is a possibility of reducing heterogeneity in yield components and also, inadvertently, reducing heterogeneity in disease resistance. Thus, it would be desirable to include improvement of resistance as one of the aims in breeding programmes. Unfortunately, the genetic base of modern carrots is narrow, being founded upon a few eighteenth century Dutch varieties, but could be considerably extended since cultivated forms of *D. carota* cross freely with *D. carota* ssp. *carota* and probably most other wild forms.

References

Anon (1973). "A Guide to the Use of Terms in Plant Pathology", Phytopathological Papers No. 17. Commonwealth Mycological Institute, Kew.

Anon (1975). "Crop Damage in Mechanised Carrot Harvesting, 1972–74", Farm Mechanisation Studies No. 27. Ministry of Agriculture, Fisheries and Food, London.

Anon (1980). "Violet Root Rot of Carrots and Other Root Crops", Leaflet 346. Ministry of Agriculture, Fisheries and Food, London.

Apeland, J. and Hoftun, H. (1974). Effects of temperature-regimes on carrots during storage. *Acta Hort.* **38**, 291–308.

Årsvoll, K. (1969). Pathogens on carrots in Norway. *Scient. Reps agric. Coll. Norway* **48**, No. 2.

Artschwager, E. and Starrett, R. C. (1931). Suberization and wound-periderm formation in sweet potato and gladiolus by temperature and relative humidity. *J. agric. Res.* **43**, 353–364.

Ayob, N. and Swinburne, T. R. (1970). Observations on the infection and rotting of apples var. Bramley's seedling by *Diaporthe perniciosa. Ann. appl. Biol.* **66**, 245–255.

Banga, O. (1976). Carrot. *In* "Evolution of Crop Plants" (N. W. Simmonds, Ed.), 291–293. Longman, London and New York.

Bleasdale, J. A. K. (1963). "The Bed System of Carrot Growing", Short Term Leaflet No. 27. Ministry of Agriculture, Fisheries and Food, London.

Booth, R. H. and Coursey, D. G. (1974). Storage of cassava roots and related post-harvest problems. *In* "Cassava Processing and Storage", Proceedings of an Interdisciplinary Workshop, Pattaya, Thailand, April, 1974. Research Centre, IDRC-O13e.

Brown, W. (1915). Studies on the physiology of parasitism. 1. The action of *Botrytis cinerea. Ann. Bot.* **29**, 313–348.

Burton, W. G. (1966). "The Potato: A Survey of its History and Factors Influencing its Yield, Nutritive Value, Quality and Storage." European Association for Potato Research, Wageningen.

Chakravarty, D. K. and Srivastava, D. N. (1967). Studies on the resistance of carrot roots to *Pythium aphanidermatum. Ann. Bot.* **31**, 739–747.

Chalutz, E., DeVay, J. E. and Maxie, E. C. (1969). Ethylene-induced isocoumarin formation in carrot root tissue. *Pl. Physiol., Wash.* **44**, 235–241.

Condon, P. and Kuć, J. (1960). Isolation of a fungitoxic compound from carrot root tissue inoculated with *Ceratocystis fimbriata. Phytopathology* **50**, 267–270.

Condon, P., Kuć, J. and Draudt, H. N. (1963). Production of 3-methyl-6-8-hydroxy-3,4-dihydroisocoumarin by carrot root tissue. *Phytopathology* **53**, 1244–1250.

Coxon, D. T., Curtis, R. F., Price, K. R. and Levett, G. (1973). Abnormal metabolites produced by *Daucus carota* roots stored under conditions of stress. *Phytochemistry* **12**, 1881–1885.

Davies, W. P. (1977a). Infection of carrot roots by *Mycocentrospora acerina* (Hartig) Deighton. Ph.D. Thesis, University of East Anglia.

Davies, W. P. (1977b). Infection of carrots in cool storage by *Centrospora acerina. Ann. appl. Biol.* **85**, 163–164.

Davies, W. P. and Lewis, B. G. (1980). The inter-relationship between the age of carrot roots at harvest and infection by *Mycocentrospora acerina* in storage. *Ann. appl. Biol.* **95**, 11–17.

Davies, W. P. and Lewis, B. G. (1981a). Antifungal activity in carrot roots in relation to storage infection by *Mycocentrospora acerina* (Hartig) Deighton. *New Phytol.* **88**, 109–119.

Davies, W. P. and Lewis, B. G. (1981b). Behaviour of *Mycocentrospora acerina* on periderm and wounded tissues of carrot roots. *Trans. Br. mycol. Soc.* **77**, 369–374.

Davies, W. P., Lewis, B. G. and Day, J. R. (1981). Observations on infection of stored carrot roots by *Mycocentrospora acerina*. *Trans. Br. mycol. Soc.* **77**, 139–151.

Day, J. R. (1972). Host–parasite interactions between *Centrospora acerina* (Hartig) Newhall and *Daucus carota* L. Ph.D. Thesis, University of East Anglia.

Dekker, J. (1976). Acquired resistance to fungicides. *A. Rev. Phytopath.* **14**, 405–428.

De Lacy, P. and Virtanen, A. I. (1957). On antifungal factors in carrots. *Suom. kemistilehti* **30**, 218.

Derbyshire, D. M. and Crisp, A. F. (1971). Vegetable storage diseases in East Anglia. *Proc. 6th Br. Insectic. Fungic. Conf.*, 167–172.

Derbyshire, D. M. and Crisp, A. F. (1978). Studies on treatments to prolong the storage life of carrots. *Expl Hort.* **30**, 23–28.

Derbyshire, D. M. and Shipway, M. R. (1978). Control of post-harvest deterioration in vegetables in the U.K. *Outl. Agric.* **9**, 246–252.

Djacenko, V. S. (1971). Some biological peculiarities and storage methods of carrots. *Acta Hort.* **20**, 80–91.

Echandi, E. and Walker, J. C. (1957). Pectolytic enzymes produced by *Sclerotinia sclerotiorum*. *Phytopathology* **47**, 303–306.

Edney, K. L. (1956). The rotting of apples by *Gloeosporium perennans* Zeller and Childs. *Ann. appl. Biol.* **44**, 113–128.

Edney, K. L. (1958). Observations on the infection of Cox's Orange Pippin apples by *Gloeosporium perennans* Zeller and Childs. *Ann. appl. Biol.* **46**, 622–629.

Esau, K. (1940). Developmental anatomy of the fleshy storage organ of *Daucus carota*. *Hilgardia* **13**, 175–209.

Fritz, D. and Habben, J. (1975). Determination of ripeness of carrots (*Daucus carota* L.). *Acta Hort.* **52**, 231–238.

Gaeumann, E. (1950). "Principles of Plant Infection." Crosley, Lockwood, London.

Garrod, B. and Lewis, B. G. (1979). Location of the antifungal compound falcarindiol in carrot root tissue. *Trans. Br. mycol. Soc.* **72**, 515–517.

Garrod, B. and Lewis, B. G. (1980). Probable role of oil ducts in carrot root tissue. *Trans. Br. mycol. Soc.* **75**, 166–169.

Garrod, B. and Lewis, B. G. (1982). The effect of falcarindiol on hyphal growth of *Mycocentrospora acerina*. *Trans. Br. mycol. Soc.* **78**, 533–536.

Garrod, B., Lewis, B. G. and Coxon, D. T. (1978). *Cis*-heptadeca-1,9-diene-4, 6-diyne-3,8-diol, an antifungal polyacetylene from carrot root tissue. *Physiol. Pl. Path.* **13**, 241–246.

Garrod, B., Lewis, B. G., Brittain, M. J. and Davies, W. P. (1982). Studies on the contribution of lignin and suberin to the impedance of wounded carrot root tissue to fungal invasion. *New Phytol.* **90**, 99–108.

Geary, J. R. (1978). Host–parasite interactions between the cultivated carrot (*Daucus carota* L.) and *Sclerotinia sclerotiorum* (Lib.) De Bary. Ph.D. Thesis, University of East Anglia.

Gonzalez, N. A. and Collazo de Rivera, A. (1972). Storage of fresh yam (*Dioscorea alata* L.) under controlled conditions. *J. Agric. Univ. P. Rico.* **56**, 46–56.

Goodliffe, J. P. and Heale, J. B. (1975). Incipient infections caused by *Botrytis cinerea* in carrots entering storage. *Ann. appl. Biol.* **80**, 243–246.

Goodliffe, J. P. and Heale, J. B. (1977). Factors affecting the resistance of stored carrot roots to *Botrytis cinerea*. *Ann. appl. Biol.* **85**, 163 (Abstract).

Goodliffe, J. P. and Heale, J. B. (1978). The role of 6-methoxy mellein in the resistance and susceptibility of carrot root tissue to the cold-storage pathogen *Botrytis cinerea*. *Physiol. Pl. Path.* **12**, 27–43.

Harding, V. K. and Heale, J. B. (1980). Isolation and identification of the antifungal compounds accumulating in the induced resistance response of carrot root slices to *Botrytis cinerea*. *Physiol. Pl. Path.* **17**, 277–289.

Harding, V. K. and Heale, J. B. (1981). The accumulation of inhibitory compounds in the induced resistance response of carrot root slices to *Botrytis cinerea*. *Physiol. Pl. Path.* **18**, 7–15.

Hawkins, J. C., Messer, H. J. M. and Lindsay, R. T. (1978). Cooling of vegetables with positive ventilation and an ice bank cooling. *Agric. Res. Coun. Res. Rev.* **4**, 34–37.

Heale, J. B., Harding, V., Dodd, K. and Gahan, P. B. (1977). *Botrytis* infection of carrot in relation to the length of the cold storage period. *Ann. appl. Biol.* **85**, 453–457.

Herndon, B. A., Kuć, J. and Williams, E. B. (1966). The role of 3-methyl-6-methoxy-8-hydroxy-3,4-dihydroisocoumarin in the resistance of carrot root to *Cerotocystis fimbriata* and *Thielaviopsis basicola*. *Phytopathology* **56**, 187–191.

Jaworski, J. G., Kuć, J. and Williams, E. B. (1973). Effect of Ethrel and *Cerotocystis fimbriata* on the accumulation of chlorogenic acid and 6-methyoxy mellein in carrot root. *Phytopathology* **63**, 408–413.

Jensen, A. (1971). Storage diseases of carrots, especially Rhizoctonia Crater Rot. *Acta Hort.* **20**, 125–129.

Kemp, M. S. (1978). Falcarindiol, an antifungal polyacetylene from *Aegopodium podograria*. *Phytochemistry* **17**, 1002.

Krahn, R. (1974). Long term storage of perishable vegetables. *Acta Hort.* **38**, 443–450.

Kurdina, V. N. (1957). Effect of sowing dates on harvest, chemical composition and storage quality of carrots. *Artoreferat Kand. M.T.S.K.H.A.*, No. 28.

Lauritzen, J. I. (1926). The relation of black rot to the storage of vegetables. *J. agric. Res.* **33**, 1025–1041.

Lauritzen, J. I. (1932). Development of certain storage and transit diseases of carrots. *J. agric. Res.* **44**, 861–912.

Lewis, B. G., Davies, W. P. and Garrod, B. (1981). Wound healing in carrot roots in relation to infection by *Mycocentrospora acerina*. *Ann. appl. Biol.* **99**, 35–42.

Lockhart, C. L. and Delbridge, R. W. (1972). Control of storage diseases of carrots by washing, grading and postharvest fungicide treatments. *Can. Pl. Dis. Surv.* **52**, 140–142.

Lockhart, C. L. and Delbridge, R. W. (1974). Control of storage diseases of carrots with postharvest fungicide treatments. *Can. Pl. Dis. Surv.* **54**, 52–54.

MacKevic, V. I. (1929). The carrot of Afghanistan. *Bull. appl. Bot., Genet. Pl. Breed.* **20**, 517–557.

Menke, G. H., Patel, P. N. and Walker, J. C. (1964). Physiology of *Rhizopus stolonifer* infection on carrots. *Z. PflKrankh. PflPath. PflSchutz* **71**, 128.

Meredith, D. S. (1964). Appressoria of *Gloeosporium musarum* Cke and Massee on banana fruits. *Nature, Lond.* **201**, 214–215.

Mukula, J. (1957). On the decay of stored carrots in Finland. *Acta Agric. scand.*, Suppl. 2, 1–132.

Neergaard, P. (1945). "Danish Species of *Alternaria* and *Stemphylium:* Taxonomy, Parasitism, Economic Significance." E. Munksgaard, Copenhagen.

Passam, H. C., Read, S. J. and Rickard, J. E. (1976). Wound repair in yam tubers: physiological processes during repair. *New Phytol.* **77**, 325–331.

Phan, C. T. and Hsu, H. (1973). Physical and chemical changes occurring in the carrot root during growth. *Can. J. Pl. Sci.* **53**, 629–634.

Phan, C. T. and Sarkar, S. K. (1975). Studies on the biosynthesis of isocoumarins in carrot root tissues: induction by substances other than ethylene and effects of metabolic inhibitors. *Revue can. Biol.* **34**, 23–32.

Plantenius, H. (1934). Physiological and chemical changes in carrots during growth and storage. *Bull. Cornell Univ. agric. Exp. Stn* **161**, 1–18.

Rader, W. E. (1952). Diseases of stored carrots in New York State. *Bull. Cornell Univ. agric. Exp. Stn.* **889**, 64 pp.

Robertson, N. F., Friend, J., Aveyard, M., Brown, J., Huffee, M. and Homans, A. (1968). The accumulation of phenolic acids in tissue culture–pathogen combinations of *Solanum tuberosum* and *Phytophthora infestans*. *J. gen. Microbiol.* **54**, 261–268.

Saburov, N. V. and Sirtautaite, S. S. (1966). The storage life of carrots in relation to maturity. *Proc. Timiryazev agric. Acad.* **5**, 121–134.

Sarkar, S. K. and Phan, C. T. (1975). The biosynthesis of 8-hydroxy-6-methoxy-3-methyl-3,4-dihydroisocoumarin and 5-hydroxy-7-methoxy-2-methylchromone in carrot root tissues treated with ethylene. *Physiologia Pl.* **33**, 108–112.

Sharman, S. and Heale, J. B. (1979). Germination studies on *Botrytis cinerea* infecting intact carrot (*Daucus carota*) roots. *Trans. Br. mycol. Soc.* **73**, 147–154.

Simmonds, J. H. (1941). Latent infection in tropical fruits discussed in relation to the part played by species of *Gloeosporium* and *Colletotrichum*. *Proc. R. Soc. Qd* **52**, 92–100.

Simmonds, J. H. (1963). Studies on the latent phase of *Colletotrichum* species causing ripe rots of tropical fruits. *Qd J. agric. Sci.* **20**, 373–425.

Sirtautaite, S. S. (1965). Certain factors which indicate the quality of carrots during storage. *Dokl. T.S.K.H.A.* **114**, 139–147.

Sirtautaite, S. S. and Sokolova, N. P. (1965). A study of factors which determine the quality of carrots in storage. *Dokl. T.S.K.H.A.* **102**, 433.

Smith, W. H. (1948). Storage of carrots. *Agriculture, Lond.* **55**, 119–124.

Sondheimer, E. (1957). The isolation and identification of 3-methyl-6-methoxy-8-hydroxy-3,4-dihydroisocoumarin from carrots. *J. Am. chem. Soc.* **79**, 5036–5039.

Swinburne, T. R. (1971). The infection of apples cv. Bramley's Seedling, by *Nectria galligena* Bres. *Ann. appl. Biol.* **68**, 253–262.

Swinburne, T. R. (1973). The resistance of immature Bramley's Seedling apples to rotting by *Nectria galligena* Bres. *In* "Fungal Pathogenicity and the Plant's Response" R. J. W. Byrde and C. B. Cutting, Eds), 365–382. Academic Press, London.

Tucker, W. G. (1974). The effect of mechanical harvesting on carrot quality and storage performance. Symposium on Vegetable Storage, Weihenstephan, September 1973. *Acta Hort.* **38**, 359–367.

Vakis, N., Marriott, J. and Proctor, F. J. (1975). A study of transit conditions and packaging of carrots exported by sea from Cyprus. *J. Fd Sci. Agric.* **26,** 609–615.

van den Berg, L. and Lentz, C. P. (1966). Effect of temperature, relative humidity, and atmospheric composition on changes in quality of carrots during storage. *Fd Technol., Champaign* **20,** 104–107.

van den Berg, L. and Lentz, C. P. (1973). High humidity storage of carrots, parsnips, rutabagas and cabbage. *J. Ass. Soc. hort. Sci.* **98,** 129–132.

Verhoeff, K. (1974). Latent infection by fungi. *Rev. Phytopath.* **12,** 99–110.

Wall, C. J. and Lewis, B. G. (1980). Infection of carrot plants by *Mycocentrospora acerina. Trans. Br. mycol. Soc.* **74,** 587–593.

Weichmann, J. and Ammerseder, E. (1974). Influence of C.A. storage conditions on carbohydrate changes in carrots. *Acta Hort.* **38,** 339–344.

Weichmann, J. and Kappel, R. (1977). Harvesting dates and storage-ability of carrots. *Acta Hort.* **62,** 191–194.

Weimer, J. L. and Harter, L. L. (1921). Wound cork formation in sweet potato. *J. agric. Res.* **21,** 637–647.

Wells, J. M. and Merwarth, F. L. (1973). Fungicide dips for controlling decay of carrots in storage for processing. *Pl. Dis. Reptr* **57,** 697–700.

6

Brassicas

J. D. GEESON

I. Introduction

The term "brassicas" collectively refers to the various cultivated forms, which are usually described as botanical varieties, of *Brassica oleracea* (Nieuwhof, 1969; see Table 6.1), of which the stems, leaves, buds or inflorescences are eaten, normally in cooked form, as vegetables (Harrison *et al.*, 1969; Brouk, 1975). Also included in this chapter is Chinese cabbage or pe-tsai (*B. pekingensis*), one of a group of closely related *Brassica* species and varieties of oriental origin, which is increasingly grown in the U.K., Europe and North America for use as a cooked or salad vegetable (Harrison *et al.*, 1969).

Table 6.1 *The main cultivated forms of* Brassica oleracea L.

Latin Name	Common Name	Main Edible Parts
B. o. var. *gongylodes* L.	kohlrabi	stem
B. o. var. *capitata* L. f. *alba* DC.	white cabbage	leaves
f. *rubra* (L.) Thell.	red cabbage	
B. o. var. *sabauda* L.	savoy	
B. o. var. *acephala* DC.	kale	
subvar. *lacinata* L.	curly kale	
B. o. var. *gemmifera* DC.	Brussels sprouts	axillary buds
B. o. var. *botrytis* L.	cauliflower	inflorescences
B. o. var. *italica*	sprouting broccoli	
	calabrese	

After Nieuwhof (1969).

The total area of brassicas grown for human consumption in the U.K. is over 50 000 ha, which produces crops valued in excess of £100 million per annum. This is comprised mainly of three major crops: cabbages, cauliflowers and Brussels sprouts, which occupy *c.* 25 000, 15 000 and

13 000 ha, yielding *c.* 550 000, 300 000 and 200 000 tonnes, respectively (Anon, 1981). The utilization and therefore the practical requirement for storage of each of these vegetables varies, and the relevance of storage diseases to commercial storage will be discussed.

Many pathogens are common to several or all brassica crops. However, the storage potential of brassicas varies widely: at one extreme the group includes cabbage cultivars that have been specially developed for long-term storage over many months, and at the other extreme very ephemeral crops such as sprouting broccoli which has a post-harvest life of only a few days. Considering these differences in the time scale of infection and disease development, and also in the plant organs affected and symptoms produced, the diseases of each vegetable will be dealt with separately. This chapter is essentially concerned with post-harvest fungal diseases, and only brief mention will be made of bacterial diseases, as these are discussed in greater detail elsewhere in this volume (see Ch. 9). Some post-harvest fungal diseases of brassicas result from pre-harvest infections; field factors affecting these diseases and any relevant aspects of cultivation and harvesting that may predispose the vegetables to post-harvest infection are also discussed. Separate sections are devoted to the use of fungicides to control diseases of storage cabbage, and to the effects of controlled or modified storage atmospheres on the development of fungal spoilage, particularly of white cabbage.

II. Cabbage

Many different types of green or white cabbage are grown for harvesting almost throughout the year. Non-headed cultivars harvested in the spring and headed cultivars which mature in summer and early autumn, together with the winter-hardy "January King" types, are normally harvested as required for marketing or processing, although some buffer stocks may be stored to provide supplies when field or weather conditions preclude harvesting. However, some dense-headed types of white cabbage, which mature in late autumn, have been bred specifically for storage in order to extend the period of availability. In the U.K., *c.* 3000 ha, yielding *c.* 130 000 tonnes of winter white cabbage (sometimes referred to as Dutch white cabbage), mostly open-pollinated selections of Langedijk 4 or related F_1 hybrids (e.g. Hidena, Bartolo), are grown. Much of this crop is stored for up to 10 months to provide continuity of supply for the fresh market or for processing into coleslaw and other prepared salads. These or similar late-maturing storage cultivars are also grown and stored in Holland, West Germany and North America, either for the fresh market or for processing into coleslaw or sauerkraut (Isenberg, 1968; Nieuwhof, 1969). Late-maturing, dense-headed cultivars of red cabbage (*B. o. capitata* f. *rubra*) are also stored commercially, particularly in Holland, for subsequent marketing or processing (Nieuwhof, 1969; Roosenboom, 1978), although very little red cabbage is stored in the U.K.

Techniques for long-term storage of cabbage have developed in three stages. Underground clamp storage as used for root vegetables has been employed in the U.S.S.R. (Sofronov, 1959; Rodin, 1963), Germany (Vogel and Neubert, 1964) and Holland (Nieuwhof, 1969), and has been tried, though with limited success, in the U.K. (North and Gray, 1961; Whitwell, 1970; Kear and Symons, 1973). Storage of cabbage in barns or other suitable farm buildings, frequently with the facility of forced ventilation of ambient night air for cooling, has been practised in the U.S.A. (Isenberg, 1968), Holland (Nieuwhof, 1969), Germany (Harnack *et al.*, 1973) and Scandinavia (Suhonen, 1969). Barn storage is still widely used in the U.K. for storage of winter white cabbage over 3–4 months until late February or early March (Derbyshire and Shipway, 1978; Geeson and Kear, 1978). However, in most areas where storage cabbage is grown, purpose-built refrigerated stores in which the cabbages are normally stacked in palletized containers are now considered necessary for long-term storage of 6–10 months duration (Isenberg, 1968; van den Berg and Lentz, 1973; Geeson and Browne, 1978; Roosenboom, 1978). Most of the documented literature on post-harvest diseases of cabbage relates to storage cultivars maintained under refrigerated storage. Where other types of green or white cabbage are stored, this is typically for much shorter periods, although disease problems are similar and may be one of the factors limiting storage (Lutz and Hardenburg, 1968; Robinson *et al.*, 1975).

A. *Botrytis cinerea* (Grey Mould Disease)

Grey mould disease, caused by *Botrytis cinerea*, is the major cause of wastage and the factor that usually limits the duration of successful long-term storage of winter white cabbage crops in the U.K. (Derbyshire, 1973; Brown *et al.*, 1975; Geeson and Browne, 1978). It also occurs on stored red cabbage (J. D. Geeson and K. M. Browne, unpublished observation). The fungus is the most common pathogen of stored cabbage in Germany (Seidel and Baresel, 1978) and in the United States (Sherf, 1972; Yoder and Whalen, 1975a) and has also been recorded on stored cabbage in Canada (Lockhart, 1976), Holland (Nieuwhof, 1969) and Finland (Tahvonen, 1981).

Botrytis cinerea causes a rapidly spreading, brown soft rot of cabbage tissue, and the lesions soon develop a characteristic thick, felt-like surface growth of mycelium and spores, but there are no reports of the formation of

sclerotia on infected cabbage tissue. Although rots that are restricted to the
outer layers of leaves can be removed by trimming after removal from store,
infections of the cut stem and petiole bases frequently penetrate deep into
the "head" and result in complete loss of the cabbage (Geeson and Robinson,
1975).

The behaviour of *B. cinerea* as a wound pathogen and its ability to infect
damaged or moribund tissue is well known and has been reviewed by Jarvis
(1977). Initial infection of stored cabbage frequently takes place via damaged
tissue or on the margins of senescing outer wrapper leaves (Brown *et al.*, 1975;
Geeson and Robinson, 1975; Yoder and Whalen, 1975a; Wale and Epton,
1981). Yoder and Whalen (1975a) demonstrated that the ability of either
conidia or mycelium of *B. cinerea* to infect stored cabbage was dependent on a
supply of exogenous nutrients, and Yoder (1977) subsequently suggested
that this could be provided by debris or by wounds in the cabbage itself.

The importance of mechanical injury, such as bruises or splits, in providing
infection sites, particularly for *B. cinerea*, and the need for careful harvesting
and handling of storage cabbage to minimize this damage has been recognized by many authors (Sherf, 1972; Geeson and Robinson 1975; Lockhart,
1976; Derbyshire and Shipway, 1978). Shipway (1977a) also found that the
severe damage that occurred during machine-harvesting of winter white
cabbage resulted in a great increase in losses due to fungal decay. The
accelerated spread of *Botrytis* infection following mid-term trimming of
stored winter white cabbage (Anon, 1979; Shipway, 1981; J. D. Geeson and
K. M. Browne, unpublished observation) may be a further demonstration of
the ability of the fungus to invade wounds, although the exposure by peeling
of inner leaves, which are inherently more susceptible than outer wrapper
leaves to infection by *B. cinerea* (Yoder and Whalen, 1975b), may also play a
significant part.

Cabbage leaf tissue damaged by freezing when in the field especially in
crops harvested and handled while the outer wrapper leaves are still in a
frozen condition, or because of freezing due to poor control of store temperature is also very susceptible to infection by *B. cinerea*. Indeed, the fungus
can infect and cause complete decay of cabbages standing in the field during
winter months. Furry *et al.* (1977) have discussed in considerable detail the
temperatures and times of exposure necessary to cause freezing injury in the
headed cabbage cultivars growing in New York State, and both Derbyshire
and Shipway (1978) and Geeson and Browne (1978) have stressed the importance of early harvesting of winter white cabbage crops destined for storage
before serious damage has been caused by night frosts.

Another frequent site for infection by *Botrytis* on stored cabbage is on
tissue already damaged by other pathogens. Geeson and Browne (1978)
observed *B. cinerea* as a secondary organism infecting the lesions caused by

Alternaria spp. or by *Mycosphaerella brassicicola*, and Walkey and Webb (1978) reported that *B. cinerea* caused severe secondary infections of the large necrotic leaf lesions caused by turnip mosaic virus.

Once established on damaged or senescent tissue, *B. cinerea* spreads rapidly to colonize apparently healthy tissue (Wale and Epton, 1981), and much of this spread is due to infection of neighbouring heads and the development of "nests" of decay in the storage bins (Geeson and Browne, 1978). Poapst *et al.* (1979) have suggested that this organism, in common with some other storage pathogens, may actually produce an environment promoting senescence in, and parasitism on, the host tissue by its own formation of the plant senescence hormone, ethylene.

In a study of the susceptibility of stored cabbage tissues to infection by *B. cinerea*, Yoder and Whalen (1975b) demonstrated that for a given cultivar of cabbage, outer leaves were less susceptible than inner leaves, although susceptibility was not related to the location of the tissues within individual leaves. However, they found that the major factor in the susceptibility to infection by *Botrytis* was the cultivar of cabbage. In support of this conclusion, Geeson (1980) found that the susceptibility to *B. cinerea* and to other storage pathogens of ten cultivars of winter white cabbage varied considerably between cultivars and between growing sites. Rubin and his co-workers in the U.S.S.R. have made an extensive study of the complex role of phenolic compounds in the mechanism of resistance to *B. cinerea* in cabbage tissues, and this work has been reviewed by Rubin and Ivanova (1960) and more recently by Jarvis (1977). They concluded that the high peroxidase activity in resistant cultivars increased considerably when the tissue was infected by *B. cinerea* and that this was a major factor contributing to resistance. These findings were exploited by Brumshtejn and Metlickij (1963), who selected seedlings showing high peroxidase activity in their attempts to breed cabbage cultivars with resistance to *B. cinerea*. In a later study of resistance and susceptibility to *Botrytis* in cultivars of storage cabbage grown in Russia, Paliliov and Frolov (1974) found that cultivars with higher pigment content had better resistance and therefore greater storage potential. In addition to these differences in host susceptibility, single spore isolates of *B. cinerea* also differ considerably in their virulence towards stored cabbage tissue (Yoder, 1977).

Several studies have been made of the effects of inorganic fertilizer applications on the susceptibility of stored cabbages to *B. cinerea* and other spoilage fungi. Polegaev *et al.* (1979) reported that the incidence of *B. cinerea* during storage declined markedly with phosphorus/potassium and nitrogen/phosphorus/potassium fertilization. In contrast, Wale and Epton (1981) were unable to establish any consistent relationship between levels of nitrogen fertilizer applied and the development of *B. cinerea* and

Alternaria brassicicola. Shipway (1981) found that nitrogen levels up to 300 kg ha^{-1} had no effect on the extent of fungal spoilage in store, but that the highest level (350 kg ha^{-1}) tested resulted in a greater proportion of completely rotted heads.

Two of the most important post-harvest factors affecting infection and spoilage of cabbage by *B. cinerea* are temperature and relative humidity. Although most strains of the fungus have optimum temperatures between 20 and 25°C (Jarvis, 1977; Dennis and Geeson, 1980), the minimum temperatures for growth of some strains are below the 0–2°C generally used for long-term storage of cabbage. Dennis and Geeson (1980) found that strains of *B. cinerea* isolated from stored vegetables or from infected field and glasshouse crops were all able to grow on nutrient media at −2°C; minimum growth temperatures of −2 and −0.8°C have also been reported by van den Berg and Lentz (1968) and by Crüger (1978), respectively. Although serious spoilage of cabbage by *Botrytis* frequently occurs at commercial storage temperatures, grey mould cannot be completely prevented in practice by lowering these temperatures because of the danger of freezing injury to the cabbages themselves. However, the severity of the disease can be reduced by use of the lowest possible storage temperatures that do not damage the crop. Polegaev and Pastukhov (1976) found that the incidence of grey mould disease was reduced by decreases in storage temperatures to between 3 and −1.5°C, and Crüger (1978) reported a significant drop in the virulence of *B. cinerea* on stored cabbage below 4°C.

The influence of relative humidity (r.h.) within the store and the extent of desiccation of the outer wrapper leaves on spoilage of stored cabbage by *B. cinerea* and other pathogens is the subject of conflicting reports. Pendergrass and Isenberg (1974) observed that the turgid wrapper leaves of headed cabbage stored at 1°C and 100% r.h. developed less *Botrytis* than the more flaccid wrapper leaves on cabbage stored at this temperature and 85 or 75% r.h., and similarly van den Berg and Lentz (1973) found less fungal decay on cabbage stored at 98–100% than at 90–95% r.h. In contrast to these observations, Yoder and Whalen (1975a) stated that the optimum r.h. for the decay of healthy cabbage tissue was above 97% and that no decay occurred below 93%, and Polegaev and Pastukhov (1976) reported that the incidence of grey mould disease was reduced by lower humidities within the range of 99–80% r.h. Similarly, J. D. Geeson and K. M. Browne (unpublished) found that fungal spoilage, particularly by *B. cinerea*, of three cultivars of winter white cabbage stored at 2°C and 100, 95 or 85% r.h. was most severe at 100% and almost eliminated at 85% r.h., although this treatment resulted in unacceptably high weight loss and softening. In accordance with these results, storage of winter white cabbage in ice-bank cooled stores at 0.5°C and 98% r.h., although reducing evaporative loss, was found to increase fungal spoilage as compared with conventional brine-cooled or direct expansion cooled stores run at similar temperatures and 90–95% r.h. (Shipway, 1981; C. Dennis and K. M. Browne, unpublished). Shipway (1976) also made similar observations on the practice of polythene-lining of cabbage storage bins. Bunnemann and Hansen (1973), Geeson and Browne (1979a) and Shipway (1981) have all concluded that some dehydration of the outer wrapper leaves does reduce fungal attack and improves the storage life of white cabbage.

In addition to their ability to infect and decay cabbages at the low temperatures and high humidities (0–2°C, *c*. 95% r.h.) used for commercial storage, mycelium or ungerminated conidia of *B. cinerea* can survive without nutrients under these conditions for up to 12 months (van den Berg and Lentz, 1968; Geeson and Wallwork, 1981). Inoculum of the fungus surviving on walls, storage containers and other surfaces within the store could therefore play an important role in the infection of subsequent cabbage crops brought into the store.

B. *Alternaria* spp.

Two *Alternaria* spp., *A. brassicae* Berk. (Sacc.) and *A. brassicicola* (Schw.) Wiltshire, which are specific to hosts in the family Cruciferae (Ellis, 1971), are widespread pathogens of cabbage and other leafy brassicas. Both fungi cause seedling diseases and leaf spots in the field, and also post-harvest diseases variously described as "Alternaria spot", "black leaf spot" or "dark leaf spot" (Rangel, 1945; Ramsey and Smith, 1961; Changsri and Weber, 1963; Nieuwhof, 1969; Channon and Maude, 1971; Ryall and Lipton, 1972). The nomenclature of the genus *Alternaria* and of these two species in particular has been somewhat confused in the past. *Alternaria brassicae* was formerly described, especially by American authors as *A. herculea* (Ell. and Mart.) Elliot, and *A. brassicicola* was previously named as *A. oleracea* Milbraith and in some cases erroneously as *A. brassicae* Berk. (Sacc.) (see Wiltshire, 1947). In order to avoid further confusion, these two fungi are referred to as *A. brassicae* and *A. brassicicola*, respectively, throughout this chapter.

Both fungi are responsible for serious wastage of cabbage during storage and marketing, although their relative importance varies between different cabbage-growing areas. Spoilage by both species has been recorded in Holland (Nieuwhof, 1969) and the United States (Ramsey and Smith, 1961; Ryall and Lipton, 1972), although Ramsey and Smith (1961) stated that *A. brassicae* is the cause of most serious losses, and Sherf (1972) mentioned only *A. brassicae* as a commonly encountered storage disease of cabbages in New York State. In winter cabbage crops stored in the U.K., *A.*

brassicicola was considered to be an important spoilage organism by Brown *et al.* (1975) and by Kear *et al.* (1977), and has also been recorded by Shipway (1981) and by Wale and Epton (1981). In contrast, *A. brassicae* has only been reported once as an infrequent spoilage organism (Geeson and Robinson, 1975).

A third species, *A. alternata* (Fr.) Keissler, which has a much less restricted host range, is also a frequent spoilage organism on cabbage in cool storage in the U.K. (Geeson and Browne, 1978), and *A. tenuis* (= *A. alternata*, see Simmons, 1967) has been recorded on stored cabbage in the U.S.A. (Adair, 1971; Pendergrass and Isenberg, 1974).

The diseases caused by each of these species are very similar in appearance, particularly in the early stages, and the causal organism is best identified by isolation (Geeson and Browne, 1979a), and by microscopic examination of the spores (see Ellis, 1971). Infected tissue, which is frequently surrounded by a chlorotic margin (Nieuwhof, 1969), is discoloured dark brown or black with a dry or leathery texture, and later develops a sparse or sometimes fairly dense superficial growth of dark mycelium and conidia (Geeson and Robinson, 1975). Larger lesions caused by *A. brassicicola* are usually identifiable by uniform, dark olive-black surface sporulation, giving a sooty appearance (see Fig. 6.1), whereas the surface growth of *A. alternata* is typically dark grey to greyish-black, and that of *A. brassicae* brown or dark brown with characteristic concentric zonation (Ramsey and Smith, 1961; Changsri and Weber, 1963; Ryall and Lipton, 1972; Sherf, 1972).

Leaf spots caused by *A. brassicae* and *A. brassicicola* present on the outer wrapper leaves of the cabbage at harvest may continue to spread during storage (Ramsey and Smith, 1961; Ryall and Lipton, 1972), and both Ramsey and Smith (1961) and Sherf (1972) have stressed the importance of trimming off all infected leaves to prevent further spread during transit and storage. Although Ramsey and Smith (1961) reported that wounds are not necessary for post-harvest infection by *A. brassicae* or *A. brassicicola*, Geeson and Robinson (1975) found that storage rots in winter white cabbage caused by *Alternaria* spp. usually began by infection at damaged areas or on the margins of senescent outer leaves. Lesions on winter white cabbage are typically restricted in area (<5 cm diameter), but frequently penetrate through several layers of leaves and may require extensive trimming (Brown *et al.*, 1975; Geeson and Robinson, 1975; Geeson and Browne, 1978). Rots caused by these fungi rarely spread by contact to adjacent heads in store, but they are frequently followed by secondary infections of *Botrytis cinerea* (Geeson and Robinson, 1975; Geeson and Browne, 1978) or soft-rotting bacteria (Ramsey and Smith, 1961; Nieuwhof, 1969; Sherf, 1972). Poapst *et al.* (1979) suggested that infections by

Alternaria spp. actually predispose cabbages to attack by other pathogens during storage and transit by stimulating the production of the plant senescence hormone, ethylene.

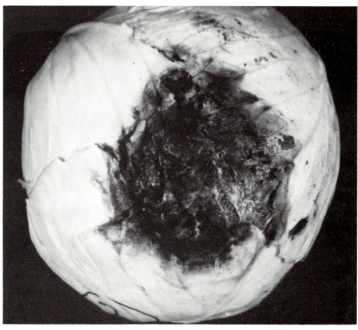

Fig. 6.1 Winter white cabbage showing lesion caused by *Alternaria brassicicola*.

The development and spread of infection by *Alternaria* spp. is favoured by the conditions of high r.h. used for cabbage storage (Ramsey and Smith, 1961), but reports on the effects of temperature are conflicting. Changsri and Weber (1963) observed no growth of *A. brassicae* or *A. brassicicola* at 4°C, and Adair (1971) concluded that *A. alternata* could not develop on cabbage stored at 1–5°C. Although Ryall and Lipton (1972) and Sherf (1972) stated that the development of *Alternaria* diseases is restricted at storage temperatures of 0–1°C, both *A. alternata* and *A. brassicicola* are common and frequently serious pathogens of white cabbage stored for 6–9 months at these temperatures.

C. *Mycosphaerella brassicicola*

Mycosphaerella brassicicola is widely known as the cause of a field leaf disease, "ring spot", on brassicas (Nieuwhof, 1969; Ogilvie, 1969; Channon and Maude, 1971), including winter white cabbage (Anon, 1977), but the fungus has also been reported to be responsible for post-harvest diseases of stored cabbage in the U.K. (Geeson and Robinson, 1975; Geeson and Browne, 1978) and in the United States (Ramsey and Smith, 1961; Ryall and Lipton, 1972).

The lesions of *M. brassicicola* are dark grey-brown to black in colour, and the infected tissue becomes dry and leathery or corky in texture (cf. lesions caused by *Alternaria* spp.). There is a limited growth of mycelium on the surface, but the lesions are best distinguished from those caused by *Alternaria* spp. by the random development over the leaf surface of small brown pycnidia (Fig. 6.2) (Geeson and Browne, 1978). The mature pycnidia of the *Asteromella* conidial state of the fungus (see Punithalingam and Holliday, 1975) exude pale pink or white droplets of pycnospores (Geeson and Browne, 1978). The fungus frequently infects bruised or damaged outer leaves or via the cut stem and petiole bases, where the lesions may penetrate through several layers of leaves or cause a black discolouration of the vascular tissue of the stem, thus necessitating extensive trimming (Geeson and Browne, 1978). *Mycosphaerella brassicicola* rarely spreads to adjacent heads in store (Geeson and Robinson, 1975).

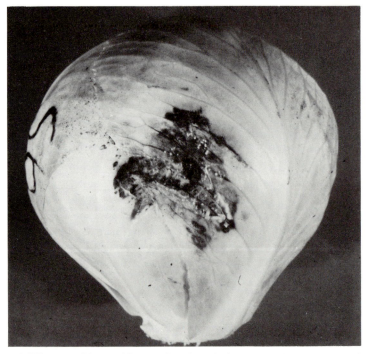

Fig. 6.2 Winter white cabbage showing lesion caused by *Mycosphaerella brassicicola*.

The fungus survives in the field on infected plant debris or on neighbouring brassica crops (Channon and Maude, 1971), and as a post-harvest pathogen, *M. brassicicola* appears to be associated only with certain growing sites. However, since "ring spot" disease is favoured by cool, moist conditions (Ogilvie, 1969; Channon and Maude, 1971), the conditions (*c.* 0–2°C, 95% r.h.) used for commercial storage of cabbage would be expected to encourage infection.

D. *Phytophthora porri*

A rot of stored cabbage caused by a *Phytophthora* sp. was first described in Norway by Semb (1971), and one of the isolates from this outbreak was subsequently identified as *P. porri* (D. J. Stamps, personal communication). In England, storage rots of winter white cabbage caused by *P. porri* were first recorded in crops from Lincolnshire by Geeson (1976), and have also occurred in other parts of the country (Anon, 1977, 1978). Although the disease is infrequent in white cabbage crops stored in the U.K., most infected heads are completely lost after long-term storage, and serious wastage occurs in heavily infected crops (Geeson, 1976; Kear *et al.*, 1977). The disease also occurs in Holland (J. D. Geeson, unpublished observation).

The symptoms of the disease have been described in detail by both Semb (1971) and Geeson (1976). Infected heads are characterized by a dark brown or grey-brown discolouration spreading upwards from the base of the stem into the leaves. Some lesions are restricted to the stem and leaves on only one side of the cabbage, although in some cases, the entire head may become affected. Unlike the spreading, brown soft rots caused by *B. cinerea*, the infected tissue remains quite firm, and has a distinctive acidic or "vinegar-like" odour. The absence of any superficial mycelial growth, even on completely rotted heads, may have led to previous confusion with the symptoms of bacterial rotting or frost injury (Semb, 1971), and severe rotting by *Phytophthora* is frequently followed by secondary infections of soft-rotting bacteria (Geeson and Browne, 1978). However, limited growth of white hyphae is visible in the spaces between the wrinkled heart leaves, and within the cavities in the stem medulla, which are also characteristics of this disease (see Geeson, 1976, plate II). No sporangia or sexual spores have been observed on rotted tissue.

Both Semb (1971) and Geeson (1976) concluded that the fungus, surviving in the soil as oospores or chlamydospores, enters the cut stem of the cabbages during harvesting operations by soil splash or mud on the cutting knife. This is consistent with the prevalence of the disease when cabbages are cut in very wet field conditions (Derbyshire and Shipway,

1978; Geeson and Browne, 1978), and the fact that spread to neighbouring cabbages during storage is very rare (Semb, 1971; Geeson, 1976). Geeson also reported that losses due to *P. porri* were increased by mid-storage trimming, presumably due to re-inoculation of fresh tissue by the trimming knife.

Phytophthora porri is more commonly known as the causal organism of "white tip" in leeks (*Allium porrum*) and salad onions (*A. cepa*). Although there is only one documented case of *P. porri* infection of winter white cabbage on a field on which leeks had been grown previously (Anon, 1977), soils can remain infective for at least 3 years after a diseased crop of *Allium* spp. (Ogilvie, 1969).

E. *Fusarium avenaceum*

Geeson and Robinson (1975) described a storage rot of winter white cabbage caused by *Fusarium avenaceum*, although the disease is very infrequent in cabbage crops stored in the U.K. The fungus causes a rapidly spreading, brown soft rot similar to that caused by *B. cinerea*, but distinguished from the latter by a dense woolly overgrowth of pink or pinkish-white mycelium. *Fusarium avenaceum* was classified by Booth (1971) in the section of the genus *Fusarium* previously grouped under the name *F. roseum* by Snyder and Hansen (1945). *Fusarium roseum* (*sensu* Snyder and Hansen) has been isolated by Adair (1971) from small dry lesions up to 1 cm diameter on the leaves of stored cabbage in the United States, and *Fusarium* species have been a minor cause of wastage in stored cabbage in Canada (Lockhart, 1976).

F. Other Fungal Diseases

Several other fungi, which have not been recorded as post-harvest pathogens in this country, are associated with market, storage or transit diseases in Holland or in North America. "Watery soft rot" caused by *Sclerotinia sclerotiorum* is responsible for serious wastage in stored cabbage crops in the United States (Ramsey and Smith, 1961; Sherf, 1972) and in the U.S.S.R. (Tsupkova, 1978). From initial infections in the field, the fungus causes a watery collapse of the whole cabbage and spreads rapidly to affect entire lots, which are covered by a dense growth of cottony white mycelium, bearing, in the later stages, black sclerotia (Ramsey and Smith, 1961; Sherf, 1972). Although the disease cannot be completely prevented by low temperatures, its spread is greatly restricted by storage close to 0°C (Ryall and Lipton, 1972; Sherf, 1972).

Infection by *Peronospora parasitica*, the causal organism of "downy mildew" in seedlings and mature plants of cruciferous crops, is characterized by the appearance on the leaves of pale yellowish or brown angular spots, later covered by fine, white, downy mycelium (Ramsey and Smith, 1961; Nieuwhof, 1969; Ryall and Lipton, 1972). The disease can spread after harvest, particularly at temperatures of 4°C and above (Ryall and Lipton, 1972), and where the outer leaves of cabbages are badly spotted, this can detract from the market value (Ramsey and Smith, 1961). In the most severely affected plants, the fungus causes a black discolouration in the stem, and can extend as far as the growing tip and infect inner leaves (Ramsey 1935; Ramsey and Smith, 1961; Nieuwhof, 1969).

When cabbages from crops infected with "black-leg" disease caused by *Phoma lingam* are stored at suitable temperatures, the fungus can spread to produce a black rot of the stem and leaf bases (Nieuwhof, 1969; Sherf, 1972). *Rhizoctonia solani*, another fungus usually associated with root and stem infections in the field (Wellman, 1932), can also continue development after harvest to produce head rots (Ramsey and Smith, 1961; Nieuwhof, 1969).

Although not found in cool-stored cabbage crops, *Rhizopus stolonifer* (≡ *R. nigricans*; von Arx, 1970) readily infects wounded tissue under warm, moist conditions to produce a watery soft rot (Ramsey and Smith, 1961; Nieuwhof, 1969; Ryall and Lipton, 1972).

G. Bacterial Diseases

Pseudomonas marginalis can cause a wet, slimy soft rot, characterized by an unpleasant sour smell in cabbages in refrigerated storage. Even at temperatures of 0°C, the disease spreads rapidly by contact and by dripping onto the heads below, and is encouraged by inadequate air circulation within the store (Geeson and Browne, 1978). Spoilage by soft-rotting pseudomonads has also been reported in barn-stored cabbage (Kear and Symons, 1973). Cabbages that have been subjected to severe frosting in the field, and especially those cut while still frozen, are highly susceptible to infection by these soft-rotting organisms (Geeson and Browne, 1978; Anon, 1979).

Soft rots of stored cabbage have also been attributed to *Erwinia carotovora* (Ramsey and Smith, 1961; Nieuwhof, 1969; Ryall and Lipton, 1972; Anon, 1979).

H. Virus Diseases

Large necrotic leaf spots up to 0.5 or even 1 cm in diameter and frequently coalescing to form irregular, sunken, papery black or dark brown areas in the leaves of stored winter white cabbage have been described by Geeson and Browne (1978) and by Walkey and Webb (1978). Walkey and Webb (1978)

demonstrated that this necrosis was due to field infection by the aphid-borne turnip mosaic virus. Glaeser (1970) isolated turnip mosaic virus from stored white cabbage showing much smaller necrotic lesions a few millimetres in diameter. Cauliflower mosaic virus has also been reported to cause internal necrosis in stored white cabbage in Holland (van Hoof, 1952) and in the United States (Natti, 1960).

I. Use of Fungicides to Control Fungal Diseases

Few attempts to control post-harvest fungal spoilage of cabbage by pre-harvest fungicide treatments have been described. Successful control of spoilage by *B. cinerea* following sprays of benomyl or thiophanate methyl applied 1 or 2 weeks prior to harvesting have been reported from the University of Weihenstephan (Anon, 1972) and by Tahvonen (1981), respectively. However, Seidel and Baresel (1978) found that pre-harvest spraying with benomyl gave no detectable advantages, and attributed this failure to poor fungicide coverage of the cabbage heads which were stored. The problems of poor coverage by field sprays, and the rubbing off of any fungicide residues on the cabbage heads during harvesting operations have also been emphasized by Geeson and Kear (1978).

In common with other long-term storage crops, such as apples and carrots, many studies have been made to evaluate different post-harvest fungicides and methods of application for cabbage. The earliest reported success with chemical control was the use of pentachloronitrobenze (PCNB) (Apeland, 1964; Vogel and Neubert, 1964). Although PCNB dusting was used commercially in Germany (Struck and Grutz, 1972) and in Holland (Whitwell, 1971), this compound has never been granted clearance for post-harvest use in the U.K. Brown *et al.* (1975) reported that dicloran applied as a dust controlled spoilage by *B. cinerea* and increased marketable yields after 5–7 months storage, but both Davies (1976) and Shipway (1976) concluded that dicloran dusts had no beneficial effect on fungal spoilage and trimming losses.

The post-harvest treatments that have been studied in most detail and which have achieved most in terms of practical control of *B. cinerea* and other spoilage fungi of cabbage are applications of the benzimidazole fungicides, benomyl or thiabendazole (TBZ) or of the chemically unrelated iprodione.

Brown *et al.* (1975) and Shipway (1977c) reported that dusting with TBZ reduced losses in stored white cabbage caused by fungal spoilage, and in a later series of experiments, Kear and his co-workers (Kear *et al.*, 1977; Geeson and Kear, 1978) made similar observations for both TBZ (30% active ingredient (a.i.) applied at 110 g a.i. t^{-1}) and iprodione (5% a.i.

applied at $30\,g\,a.i.\,t^{-1}$) dust treatments. Both treatments considerably reduced the incidence of rots caused by *B. cinerea* and increased the recovery of trimmed marketable cabbage after 18–26 weeks storage; iprodione dust also controlled rotting by *Alternaria brassicicola*. In the same series of experiments, Kear *et al.* (1977) also evaluated high volume and ultra-low volume spray applications of iprodione, which in most cases were as effective as the dust treatment in their control of *B. cinerea* and reduction of trimming losses. High volume sprays of benomyl and of TBZ have been used successfully to control *B. cinerea* on a commercial scale in East Germany (Seidel and Baresel, 1978) and in West Germany (Crüger, 1978), respectively.

Although Seidel and Baresel (1978) described the design of a conveyor and spray tunnel for post-harvest treatment of cabbage, most of the spray and dust applications described have entailed treating the cabbages layer by layer as they are loaded into store. These methods of application are not only time-consuming and therefore increase labour costs, but all additional handling steps inevitably lead to more mechanical damage, which in itself makes the cabbages more susceptible to infection, especially by spoilage fungi and bacteria which are insensitive to these compounds. Since most growers in the U.K. harvest winter white cabbage directly into the pallet bins used for storage, a system of fungicide treatment of filled bins of cabbage such as dipping or overhead drenching is most practicable on a commercial scale (Geeson and Browne, 1979a).

In experiments carried out at Stockbridge House Experimental Horticulture Station (Anon, 1973) and by Kear and Symons (1973), post-harvest dips of benomyl controlled *Botrytis*, although Kear and Symons (1973) reported that similar dip treatments of dichlofluanid or dichlorophen were not effective. Wale and Epton (1981) also found that benomyl dipping ($0.5\,g\,litre^{-1}\,a.i.$) controlled *B. cinerea* and increased by 2–25% the recovery of trimmed cabbage after 6 months cool storage. Similarly, Shipway (1979, 1981) reported good control of spoilage by *B. cinerea* and *A. brassicicola* following dip treatments of iprodione ($0.5\,g\,litre^{-1}\,a.i.$). Lockhart (1976) reported successful control of *B. cinerea* and *Alternaria* spp. by drenching with a mixture of benomyl and dicloran, and in commercial-scale trials, Geeson and Browne (1979a) demonstrated that post-harvest drenches of benomyl, iprodione ($0.5\,g\,litre^{-1}\,a.i.$) or TBZ ($0.6\,g\,litre^{-1}\,a.i.$) gave equally effective control of *B. cinerea* and *Mycosphaerella brassicicola*; iprodione drenches also controlled rots caused by *Alternaria* spp. (see Fig. 6.3). These drench treatments also increased, by up to 11%, the recovery of trimmed cabbage suitable for coleslaw manufacture after 30–36 weeks. Storage of cabbages with large amounts of water on the surface or trapped beneath outer wrapper leaves following dipping or drenching can encourage bacterial

Fig. 6.3 Winter white cabbage (cv. Langedijk 4) stored for 28 weeks at 0–1°C, 95% r.h. (a) Bin drenched in benomyl (0.5 g litre^{-1} a.i.) prior to storage; (b) undrenched control.

soft rotting (*Pseudomonas marginalis*). Shipway (1981) stressed the need for thorough draining of the treated cabbage before storage, and Geeson and Browne (1979a) recommended deliberate drying out of drenched cabbage during the first few weeks of storage to offset the risk of bacterial spoilage.

In a comparative experiment, Geeson *et al.* (1982) found that thermal fog applications of iprodione (8 g a.i. t^{-1}) or TBZ (33 g a.i. t^{-1}) were almost as effective as the corresponding drench treatments for the control of *B. cinerea*. If the problems of distribution of such fogs within the store could be overcome, this could provide a rapid method of fungicide treatment for entire storage chambers of cabbage.

In addition to their effects on spoilage fungi, several post-harvest fungicide treatments have demonstrated physiological effects on the cabbages. In the earliest experiments, fungicide dips were reported to cause an increase in the incidence of leaf necrosis symptoms described as "black specking" (Anon, 1973; Kear and Symons, 1973), although Wale and Epton (1981) were unable to detect any effect of benomyl dips on the incidence of "black speck". Geeson and Browne (1979a) found that fungicide drenching sometimes accelerated the early development of "pepper spot", but concluded that the occurrence of pepper spotting made little or no difference to the extent of final trimming necessary for drenched and undrenched cabbages. Two post-harvest fungicide treatments used experimentally have proved to be highly phytotoxic. Very severe leaf scorching and subsequent colonization of the damaged tissue by *B. cinerea* was observed following treatment of white cabbage with drenches of pimaricin (0.2 g litre^{-1} a.i.) (J. D. Geeson and K. M. Browne, unpublished) or a smoke formulation of iprodione (J. D. Geeson, R. C. Robinson and D. J. Williams, unpublished). The delay of leaf senescence and improved retention of green colour following benzimidazole dip (Wale and Epton, 1981) or drench treatments (J. D. Geeson and K. M. Browne, unpublished) can be attributed to the known cytokinin-like activity of these compounds (Thomas, 1974). However, J. D. Geeson and K. M. Browne (unpublished) also observed this effect on cabbages drenched with iprodione, which is not a chemical analogue of cytokinins.

Post-harvest fungicide treatments should not be considered as a substitute for careful harvesting and handling of cabbage for storage, since damaged heads are more susceptible to infection by spoilage bacteria and fungi, particularly those which are insensitive to the compound(s) used (Geeson and Browne, 1979a). However, in addition to the direct gain of reduced trimming losses, these treatments may have other indirect benefits. Delaying the development of spoilage fungi, particularly *B. cinerea*, may allow the storage season to be extended, thus improving the continuity of supply (Geeson and Browne, 1979a). Secondly, cabbages with fewer rots are quicker and more pleasant to handle, hence saving time and labour costs involved in preparing the cabbage for marketing or processing (Geeson and Browne, 1979a; Shipway, 1981). Post-harvest drench or dip treatments with benomyl or iprodione are now used on many commercial crops of winter white cabbage stored in the U.K.

III. Cauliflowers

In the U.K. cauliflowers are harvested throughout the spring (from over-wintered crops), summer and autumn from March to October, although the continuity of production is frequently interrupted by changes in the weather, particularly hot, dry periods. Short-term storage is used commercially to fill these gaps or to accommodate fluctuations in supply and demand. Although cauliflowers can be stored for up to 6 weeks under optimum conditions of c. 0°C, 95% r.h. (Robinson et al., 1975), both Shipway (1978) and Anon (1979) have recommended that, in practice, they should not be stored for more than 2–3 weeks, even in an ice-bank cooled store. The flower bud or "curd" tissue of cauliflowers is very susceptible to mechanical damage and to spoilage by fungi and bacteria, and any blemishing or discolouration of the curd can result in downgrading or rejection. Therefore, spotting or decay of cauliflower curds caused by fungal or bacterial diseases can cause serious losses, not only during cool storage (Lutz and Hardenburg, 1968; Robinson et al., 1975; Derbyshire and Shipway, 1978), but even during the normal time scale of distribution and marketing. In addition, the now widely used practice of film over-wrapping or enclosing cauliflower heads in polythene bags to protect them from damage, desiccation and dirt can aggravate or even cause its own disease problems, because of the creation of very humid conditions around the curd tissue.

A. *Alternaria* spp.

Leaf spots caused by the *Alternaria* spp., which attack other brassica crops (see above) and which may develop either before or after harvest on the wrapper leaves of cauliflower heads, may cause downgrading or necessitate extra trimming. However, it is the *Alternaria* infections of the white curds that cause much more serious wastage of cauliflowers during storage and marketing.

J. D. Geeson and B. M. Lund (unpublished) isolated both *A. alternata* and *A. brassicicola* from small black or dark grey spots which developed on over-wrapped cauliflower heads of several cultivars maintained for 7 or 8 days at 20°C. The spots, some of which showed limited surface growth of dark mycelium, were slightly sunken, typically <0.5 cm in diameter, but spread and coalesced to form larger lesions. Similar symptoms were reproduced on cauliflower heads back-inoculated with isolates of either species and incubated for 7 days under comparable shelf-life conditions.

In the U.S.A., brown rot of cauliflower curds during storage, transit and marketing caused by *A. brassicae* has been described by Weimer (1924), Ramsey and Smith (1961) and by Ryall and Lipton (1972), and the disease has also been reported from Eastern Europe (Beranek, 1969; Amariutei *et al.*, 1977). The fungus causes a brown discolouration of buds or groups of buds in the curd, becoming greenish-black with age (see Ramsey and Smith, 1961, plate 6A). As in the case of the spots caused by other *Alternaria* spp. (see above), the lesions, which may be firm or soft in texture, frequently coalesce to form large discoloured areas on the curd. The disease is favoured by warm, moist conditions, under which superficial mycelial growth becomes visible (Ramsey and Smith, 1961), and Beranek (1969) found that attack by *A. brassicae* was particularly severe in cauliflower heads that were enclosed in non-perforated rather than perforated polythene bags. Ramsey and Smith (1961) and Ryall and Lipton (1972) recommended rapid pre-cooling and transportation or holding of cauliflowers at 4°C or lower to retard the development of brown rot. Böttcher (1967) found that trimming away the outer leaves of cauliflowers resulted in increased spoilage by *Alternaria* sp., which in turn reduced the practical storage life of the crop.

Both *A. brassicicola* and *A. brassicae* are seed-borne fungi (Channon and Maude, 1971), but can also survive in debris in the soil. It seems most probable that the spores that infect cauliflower curds either before or after harvest arise either from infected debris or from sporulating lesions on older leaves of the growing crop (Ramsey and Smith, 1961).

B. *Peronospora parasitica* (Downy Mildew)

Downy mildew is a widespread and frequent disease of brassicas, particularly young seedlings (Nieuwhof, 1969), but *P. parasitica* can also cause a disease of cauliflower curds, both in the field (Jenkins, 1964) and after harvest (Ramsey and Smith, 1961; Ryall and Lipton, 1972; Lund and Wyatt, 1978). Infection of cauliflower heads is characterized by a pale grey or brown discolouration of the surface of the curd, which when cut open reveals grey or black spotting or streaking of the thickened flower stems, often extending several centimetres from the surface (see Ramsey and Smith, 1961, plate 4A; Lund and Wyatt, 1978, plate II). As with the storage rot of white cabbage caused by another Oomycete fungus *Phytophthora porri*, the infected tissue remains quite firm and very little surface mycelial growth of the causal organism is visible. However, Lund and Wyatt (1978) observed that the discoloured tissue was frequently decayed by soft-rotting bacteria before any sporangiophores were produced.

Development and spread of the disease is favoured by warm, moist conditions. Ramsey and Smith (1961) reported that spores, probably produced from sporulating lesions on infected wrapper leaves, germinated readily at temperatures between 8 and 18°C and could cause new infections, which were visible after only 3 days at 20°C.

Similarly, Lund and Wyatt (1978) observed spoilage by *P. parasitica* on curds, which were apparently healthy at harvest, after 3–7 days under simulated supermarket shelf-life conditions (20°C, 70% r.h.). Apart from careful selection to exclude infected heads or trimming off infected wrapper leaves before storage, further development, and particularly the spread of infection to other cauliflower heads, can be most effectively controlled by low storage temperatures. Ramsey and Smith (1961) reported that disease development could be retarded at temperatures below 4°C, and Ryall and Lipton (1972) concluded that storage near 0°C was the only useful control when infection was known to be present.

C. Other Fungal Diseases

Mycosphaerella brassicicola, the cause of field and post-harvest diseases (ring spot) of brassica crops in several countries (see above), has also been associated with post-harvest spoilage of cauliflowers in the U.S.A. (Ramsey and Smith, 1961) and in Roumania (Amariutei *et al.*, 1977). Although all parts of the cauliflower are susceptible to infection, leaf spots are seen most frequently. These lesions are small, dark-centred, surrounded by a greenish-yellow, water-soaked zone, and typically develop blackish pycnidia or perithecia either concentrically or randomly on their surface with age (Ramsey and Smith, 1961; Ryall and Lipton, 1972). Development of ring spot is favoured by relatively cool, moist conditions, and whereas lesions do develop on white cabbage during several months storage at 0–2°C (Geeson and Robinson, 1975), Ramsey and Smith (1961) reported that development of existing or new lesions caused by *Mycosphaerella* rarely occurs on cauliflowers that are stored at 0°C for a relatively short time.

The grey mould fungus, *Botrytis cinerea*, causes a light brown soft rot of cauliflower curd stored at 0°C (Ramsey and Smith, 1961; Ryall and Lipton, 1972) or held under simulated shelf-life conditions at 20°C (J. D. Geeson, unpublished), and spoilage of cauliflowers during transit by *Cladosporium* spp. has also been recorded (Khristov, 1968).

D. Bacterial Diseases

Although leaf spotting of cauliflowers caused by *Pseudomonas maculicola* has been described by Ramsey and Smith (1961) and by Ryall and Lipton (1972), it is those bacterial diseases which cause soft rots of the stem or curd which result in the most serious losses in cauliflowers after harvest. *Erwinia carotovora* frequently causes soft-rotting of both curds (Ramsey and Smith, 1961; Ryall and Lipton, 1972; Lund, 1981) and of the cut stem or butt (Lund, 1981). Lund (1981) also isolated *Pseudomonas marginalis* from soft rots of cauliflower curds, and *Pseudomonas* spp. have been reported as post-harvest spoilage organisms of cauliflowers by Lipton and Harris (1976) and by Amariutei *et al*. (1977).

These organisms frequently invade damaged or bruised areas of the curd, and some physical damage is almost inevitable during the harvesting, packing and transportation of such a delicate crop. In addition, contamination of the knives used for harvesting or trimming cauliflower heads can be a major factor in the spread of bacterial soft rots from one head to another (Ryall and Lipton, 1972; Lund, 1981). Soft-rotting bacteria may also develop as secondary infections of lesions caused by other pathogens such as *Alternaria* spp. (Ramsey and Smith, 1961). Bacterial infections of the curd typically begin as small yellow or brown water-soaked spots (see Ramsey and Smith, plate 6B) but spread rapidly, especially in bruised tissue, under favourable conditions to form extensive slimy, discoloured patches on the curd. The importance of moisture on the curd in the initial infection and decay by bacteria was highlighted by Lund (1981), who found an increased incidence of rotting when heads harvested in wet field conditions were over-wrapped; this problem can also be caused by condensation on pre-cooled cauliflower heads which are then over-wrapped. Bacterial soft rots develop most rapidly at temperatures above 10°C, and their development can be arrested by storage at *c*. 0°C (Ryall and Lipton, 1972).

IV. Brussels Sprouts

Brussels sprouts are an autumn and winter crop in the U.K. The cultivation of a succession of different cultivars, now almost exclusively F_1 hybrids, can provide a continuous supply for the fresh market or for processing (especially freezing) from September until February or March, although crops of late-maturing cultivars may be destroyed by severe winter conditions. After removal from the stem, sprouts, either pre-packed or in nets, are usually only stored for a few days to accommodate fluctuations in demand or over weekends. For this duration, yellowing or browning of the cut stems are usually the factors limiting acceptability. However, when stored on the stem, Brussels sprouts can be kept for several weeks at 1–2°C, c. 95% r.h., when fungal spoilage may be one of the major problems (Robinson et al., 1975). In recent years, ice-bank cooled stores have proved to be the most successful technique for short-term holding of sprouts (Shipway, 1977b).

Several fungal diseases, essentially similar to those of stored cabbage, have been described on Brussels sprouts during storage and transit. K. M. Browne and J. E. Robinson (unpublished) considered that grey mould, B. cinerea, was the major cause of spoilage of sprouts stored on the stem at 2°C. In addition to the buds actually infected by the fungus, they also observed yellowing and senescence of adjacent, apparently healthy, sprouts, which could be due to ethylene production by the fungus or the infected host tissue (see Poapst et al., 1979).

Another spreading soft rot of stored sprouts caused by Rhizopus stolonifer ($\equiv$ R. nigricans; von Arx, 1970) has been recorded in the U.S.A. (Ramsey and Smith, 1961; Ryall and Lipton, 1972). The rotted tissue is brown, very soft and watery, and infected buds are frequently covered by a surface growth of characteristic coarse, stringy mycelium and black sporangia (see Ramsey and Smith, 1961, plate 5). Damaged tissue is particularly susceptible to infection, and the disease is favoured by warm, moist conditions. Ryall and Lipton (1972) considered that the disease was inevitable at 13°C but was slower to develop at 7°C and did not occur at 4°C, as is consistent with published data on the temperature requirements for this organism (Dennis and Cohen, 1976).

Leaf spots of Brussels sprouts caused by Alternaria spp. and by the ring spot fungus Mycosphaerella brassicicola have been described by Ramsey and Smith (1961) and by Ryall and Lipton (1972). Ring spot, which also occurs as a field disease on Brussels sprouts (Ogilvie, 1969; Channon and Maude, 1971), causes small, dark spots surrounded by a greenish-grey zone and bearing black fruiting bodies in the central zone (see Ramsey and Smith, 1961, plate 5). Development of ring spot is virtually prevented at storage temperatures of 0°C (Ryall and Lipton, 1972).

Other spoilage organisms recorded post-harvest on Brussels sprouts are Cladosporium sp. (Stoll, 1974), and the soft-rotting bacteria Erwinia carotovora and Pseudomonas spp. (Ramsey and Smith, 1961; Anon, 1979).

V. Other Brassicas

The immature inflorescences of calabrese (usually referred to as broccoli in North America) are harvested from July to October, and sprouting broccoli (either purple or white) from over-wintered crops in March to April in the U.K. Both crops have a short post-harvest life, and in practice are usually marketed or processed within a few days of harvesting. However, successful storage for 2–3 weeks under ideal conditions (0°C, *c.* 95% r.h.) has been reported by Lutz and Hardenburg (1968), Ryall and Lipton (1972) and by Robinson *et al.* (1975). Storage is normally limited by loss of quality due to wilting and softening, yellowing or opening and abscission of flower buds, rather than by pathological problems. Smith (1940) found that cool storage at excessively high (98–100%) r.h. encouraged surface mould growth, and similarly Stork (1981) observed that over-wrapping of broccoli heads, although reducing weight loss and improving colour maintenance, also increased the risk of rotting.

The post-harvest diseases of broccoli are similar to those of cauliflower (Ramsey and Smith, 1961). Infection by *Botrytis cinerea* causes grey or light brown, moist lesions of the inflorescences, but the infected tissue is not appreciably softened (Ryall and Lipton, 1972; see also Ramsey and Smith, 1961, plate 3). Other fungal diseases that have been reported during storage or marketing include: *Cladosporium* and *Mucor* spp. (Lipton and Harris, 1974); *Alternaria brassicae*, *Mycosphaerella brassicicola* and *Peronospora parasitica* (Gorini, 1978); and watery soft rot due to *Sclerotinia sclerotiorum* (Ramsey and Smith, 1961), which has also been associated with a field infection of broccoli heads (Farmer *et al.*, 1971). Both Ramsey and Smith (1961) and Gorini (1978) have also reported bacterial diseases: soft rotting by *Erwinia carotovora* and leaf spotting caused by *Pseudomonas maculicola*.

Chinese cabbage is still a minor crop in the U.K., grown outdoors for harvest in June to September, or in tunnels for a longer cropping season from March to early December (Stokes, 1981). Although Chinese cabbage is not stored commercially in this country, storage trials indicate that the post-harvest life is limited to only a few weeks at 0–1°C (van den Berg and Lentz, 1974; Jones, 1979). The occurrence of leaf spots caused by *Alternaria* spp. has been reported in Chinese cabbage stored at 0–1°C, 95–97% r.h. (Sozzi *et al.*, 1980) or maintained under shelf-life conditions of 18°C, 65% r.h. (Jones, 1979).

Hansen and Bohling (1980) found that bacterial soft rots could cause serious losses, particularly where ventilation of stores was inadequate, and soft-rotting caused by *Erwinia carotovora* was considered by Ramsey and Smith (1961) to be a serious transit and storage disease of Chinese cabbage. Bacterial soft rot of the cut stem, particularly in over-wrapped heads, is a major cause of wastage during supermarket shelf-life in the U.K. (J. R. Geary, personal communication). Black discolouration spreading up through the leaf venation has been described by Ryall and Lipton (1972) and by Sozzi *et al.* (1980), and the former attributed this problem to infection by *Xanthomonas campestris*.

Both savoys and kale are hardy over-wintering crops in the U.K., and are normally only harvested as required for marketing. Nevertheless, some disease problems have occurred during short-term storage and marketing. The storage and market diseases affecting savoys and the symptoms are similar to those described for cabbage (Gorini, 1979), although Stoll (1974) specifically mentioned spoilage by *Botrytis cinerea*. Both Ramsey and Smith (1961) and Ryall and Lipton (1972) stated that the diseases of kale are similar to those of other leafy brassicas, and Ramsey and Smith considered that bacterial soft rot (*Erwinia carotovora*) and the yellow discolouration of the foliage caused by *Fusarium conglutinans* were the most important market diseases of kale. Hruschka (1971) recorded soft rotting by *E. carotovora* and by *Sclerotinia sclerotiorium* as serious problems, particularly in washed, pre-packed kale, and concluded that the spread of infection by these decay organisms was largely via the washing water.

VI. Effects of Controlled and Modified Atmospheres

A. Cabbage

Long-term refrigerated storage of white cabbage in controlled atmospheres (CA) containing 2–6% carbon dioxide and 1–5% oxygen with the remainder nitrogen, rather than in air, has resulted in considerable benefits in terms of lower storage and trimming losses, better retention of fresh colour, flavour and texture, and reduced fungal spoilage (Isenberg, 1968; Isenberg and Sayles, 1969; van den Berg and Lentz, 1973; Bohling and Hansen, 1977; Geeson and Browne, 1979b, 1980). However, few of these reports have provided information on the effects of CA conditions on specific spoilage organisms. Bohling and Hansen (1977) and Henze (1977) reported reduced spoilage by *Botrytis cinerea* in CA storage, and Geeson and Browne (1979b, 1980) found that storage in a CA containing 5–6% CO_2 and 3% O_2 consistently reduced the incidence and severity of rots caused by *B. cinerea* (see Fig. 6.4). In subsequent experiments, combined treatments of CA storage and post-harvest drenches of the fungicides benomyl or iprodione more or less eliminated spoilage by *B. cinerea* during 28–35 weeks storage (Geeson and Browne, 1981) (see Table 6.2). Wells and Uota (1970) found that, although conidia of *B. cinerea* were able to germinate even in 1% oxygen, the rate of mycelial growth of the fungus was reduced by *c.* 50% in 4% oxygen and thereafter decreased linearly with further decreases in oxygen concentration. Similarly, inhibition of mycelial growth of *B. cinerea* has been reported in oxygen concentrations below 1.7% and below 1% by Adair (1971) and Follstad (1966), respectively. However, it seems most probable that the controlling effect of these storage atmo-

Fig. 6.4 Winter white cabbage (cv. Bartolo) stored for 34 weeks at 0–1°C in (a) air or (b) a controlled atmosphere containing 5% CO_2, 3% O_2 and the remainder N_2.

J. D. GEESON

Table 6.2 Effect of controlled atmosphere (CA: 5% CO_2, 3% O_2, 92% N_2) and post-harvest fungicide drenches on the incidence of rots by Botrytis cinerea in winter white cabbage stored at 0–1°C

Season	Cultivar	Storage Period (wks)	Incidence of B. cinerea (%)			
			Air	CA	CA + Benomyl Drench	CA + Iprodione Drench
1979–80	Ladena	35	81	13	0	0
	Langedijk 4-Decema Extra	34	97	45	16	1
1980–81	Bartolo	33	63	36	2	0
	Hidena	34	70	28	8	1

spheres on spoilage by *B. cinerea* is a consequence of delayed leaf senescence, which reduces the ability of the fungus to colonize the tissue (see above).

Geeson and Browne (1979b, 1980) found that the incidence of several other cabbage spoilage fungi—*Alternaria* spp. and *Mycosphaerella brassicicola*—was not consistently controlled by CA storage, and in some cases, these fungi were more evident on CA-stored cabbages, where they were not masked by the more extensive lesions caused by *Botrytis*.

As with post-harvest fungicide treatments which effectively control *B. cinerea*, CA storage also provides the indirect benefits of reducing time and labour costs in preparing stored cabbage for marketing or processing, and enables the storage season to be extended. Winter white cabbage is now stored commercially for up to 10 months under CA conditions in the U.K., and this has substantially improved the continuity of supply of home-grown cabbage for coleslaw manufacture. Similar CA storage is also carried out commercially in New York State and in West Germany (Isenberg, 1979).

B. Cauliflower and Broccoli

Many studies have been made of the effects of controlled and modified atmosphere on cauliflower and broccoli (see reviews by Stoll, 1974 and Isenberg, 1979), which can be very effective in preventing loss of green colour in the leaves and delaying further development of flower buds. However, the limiting factors are typically changes in texture or the development of off-odours and flavours rather than microbial spoilage. Amariutei *et al.* (1977) reported that storage of cauliflowers in atmospheres containing 6% CO_2 reduced spotting by *Alternaria brassicae*, but that these advantages were offset by increased bacterial spoilage. In their studies of the effects of different oxygen concentrations on cauliflower, Lipton and Harris (1976) found that although bacterial soft-rotting was not affected by O_2 concentrations between 2 and 6%, more soft rots occurred at 1% O_2, which they attributed to physiological damage of the curd tissue. In a similar study of CA storage of broccoli, Lipton and Harris (1974) observed that whereas *Alternaria* spp. and *Cladosporium* were the major spoilage organisms in the air, *Mucor* spp. predominated in low-oxygen atmospheres.

References

Adair, C. N. (1971). Influence of controlled atmosphere storage conditions on cabbage postharvest decay fungi. *Pl. Dis. Reptr* **55**, 864–868.
Amariutei, A., Tataru, D. P. and Tasca, G. (1977). Research results on refrigerated

152 · J. D. GEESON

and CA storage of cauliflower *Brassica oleracea* L. var. *botrytis*. *Acta Hort.* **62**, 31–39.

Anon (1972). *A. Rep. Tech. Univ., Weihenstephan, Munich 1971*, 98 pp.

Anon (1973). Cabbage Dutch white: control of *Botrytis*. *A. Rep. Stockbridge House Exp. hort. Stn*, 20–22.

Anon (1977). *A. Rep. A.D.A.S. agric. Sci. Serv. 1976*, 171.

Anon (1978). *A. Rep. A.D.A.S. agric. Sci. Serv. 1977*, 161.

Anon (1979). "Refrigerated Storage of Fruit and Vegetables", Ministry of Agriculture, Fisheries and Food Reference Book 324. H.M. S.O. London.

Anon (1981). "Output and Utilization of Farm Produce in the UK 1974–1980", Ministry of Agriculture, Fisheries and Food. H.M.S.O., London.

Apeland, J. (1964). Eldre og nyare forsøksresultat frå lagringsforsøk med hovudkål. *Gartneryket* **54**, 2117–1218; 1228–1229.

Beranek, R. (1969). Long-term cold storage of cauliflower in polyethylene packages. *Prum. Potravin* **20**, 77–79.

Bohling, H. and Hansen, H. (1977). Storage of white cabbage (*Brassica oleracea var. capitata*) in controlled atmospheres. *Acta Hort.* **62**, 49–54.

Booth, C. (1971). "The genus *Fusarium*." Commonwealth Mycological Institute, Kew.

Böttcher, H. (1967). Zur Lagerfähigkeit von Blumenkohl. *Dte Gartenb.* **14**, 300–304.

Brouk, B. (1975). "Plants Consumed by Man." Academic Press, London.

Brown, A. C. Kear, R. W. and Symons, J. P. (1975). Fungicidal control of *Botrytis* on cold-stored white cabbage. *Proc. 8th Br. Insectic. Fungic. Conf., Brighton, 1975*, 339–346.

Brumshtejn, V. D. and Metlickij, L. V. (1963). Biochemical methods of plant selection in breeding for resistance to micro-organisms. *Dokl. Akad. Nauk. S.S.S.R.* **149**, 1197–1199.

Bunnemann, G. and Hansen, H. (1973). "Frucht und Gemüselagerung. Eine Anleitung für den Lagerwart." Verlag Eugen Ulmer, Stuttgart.

Changsri, W. and Weber, G. F. (1963). Three *Alternaria* species pathogenic on certain cultivated crucifers. *Phytopathology* **53**, 643–648.

Channon, A. G. and Maude, R. B. (1971). Vegetable diseases. *In* "Diseases of Crop Plants" (J. H. Western, Ed.), 323–363. MacMillan, London.

Crüger, G. (1978). Chemischer Vorratsschutz von Gemüse, Obst und Kartoffeln. *Ernähr.–Umsch.* **25**, 170–174.

Davies, A. C. W. (1976). Cabbage, Dutch white—control of diseases during long-term storage 1974–75. *A. Rep. Luddington Exp. hort. Stn 1975*, 97.

Dennis, C. and Cohen, E. (1976). The effect of temperature on strains of soft fruit spoilage fungi. *Ann. appl. Biol.* **82**, 51–56.

Dennis, C. and Geeson, J. D. (1980). Physiology of spoilage fungi in relation to survival. *Bienn. Rep. agric. Res. Coun. Fd Res. Inst. 1977 & 1978*, 47.

Derbyshire, D. M. (1973). Post-harvest deterioration of vegetables. *Chemy. Ind.*, No. 22, 1952–1954.

Derbyshire, D. M. and Shipway, M. R. (1978). Control of post-harvest deterioration in vegetables in the UK. *Outl. Agric.* **9**, 246–252.

Ellis, M. B. (1971). "Dematiaceous Hyphomycetes." Commonwealth Mycological Institute, Kew.

Farmer, L. J., Firstman, A. and Betz, J. (1971). A firm head rot of broccoli. *Pl. Dis. Reptr* **55**, 1136.

Follstad, M. N. (1966). Mycelial growth rate and sporulation of *Alternaria tenuis*,

Botrytis cinerea, *Cladosporium herbarum* and *Rhizopus stolonifer* in low-oxygen atmospheres. *Phytopathology* **56**, 1098–1099.

Furry, R. B., Isenberg, F. M. and Jorgensen, M. C. (1977). Post-harvest storage response of cabbage subjected to various diurnal freeze–thaw regimes. *Acta Hort.* **62**, 217–228.

Geeson, J. D. (1976). Storage rot of white cabbage caused by *Phytophthora porri*. *Pl. Path.* **25**, 115–116.

Geeson, J. D. (1980). Storage disorders of winter white cabbage. *Bienn. Rep. agric. Res. Coun. Fd Res. Inst. 1977 & 1978*, 46.

Geeson, J. D. and Browne, K. M. (1978). Careful harvest is vital for white cabbage storage success. *Grower* **89**(1), 27–31.

Geeson, J. D. and Browne, K. M. (1979a). Effect of post-harvest fungicide drenches on stored winter white cabbage. *Pl. Path.* **28**, 161–168.

Geeson, J. D. and Browne, K. M. (1979b). CA keeps coleslaw crop greener. *Grower* **92**(14), 36–38.

Geeson, J. D. and Browne, K. M. (1980). Controlled atmosphere storage of winter white cabbage. *Ann. appl. Biol.* **95**, 267–272.

Geeson, J. D. and Browne, K. M. (1981). Storage disorders of winter white cabbage. *Bienn. Rep. agric. Res. Coun. Fd Res. Inst. 1979 & 1980*, 45.

Geeson, J. D. and Kear, R. W. (1978). Fungicide dips for cabbage storage. *Hort. Ind.*, October, 38–39.

Geeson, J. D. and Robinson, J. E. (1975). Damage will mean trouble in store. *Comml Grower*, No. 4147, 1245–1246.

Geeson, J. D. and Wallwork, L. J. (1981). Physiology of *Botrytis cinerea* in relation to survival in stores. *Bienn. Rep. agric. Res. Coun. Fd Res. Inst. 1979 & 1980*, 44–45.

Geeson, J. D., Robinson, R. C. and Williams, D. J. (1982). Evaluation of fungicide fogs for the control of *Botrytis cinerea* in stored winter white cabbage. *Tests Agrochem. Cultivars* (*Ann. appl. Biol.* **100**, Suppl.), No. 3, 36–37.

Glaeser, G. (1970). Wodurch enstenen schwarze stippen an lagerkraut? *Pflanzenarzt* **23**, 122–123.

Gorini, F. (1978). Schede orticole. 6. Ortaggi da infiorescenza. 6.3. Cavolo broccolo. *Inf. Ortofruttic.* **19**, 3–6.

Gorini, F. (1979). Schede orticole. 2. Ortaggi da foglia. 2. 24 Cavolo verza. *Inf. Ortofruttic.* **20**, 3–6.

Hansen, H. and Bohling, H. (1980). Long-term storage of Chinese cabbage. *Acta Hort.* **116**, 31–34.

Harnack, H., Baumann, J. and Meinke, G. (1973). Richtige Lüftungsmassnahmen in Kopfkohllagern senken Verluste. *Gartenbau* **20**, 227–229.

Harrison, S. G., Masefield, G. B., Wallis, M. and Nicholson, B. E. (1969). "The Oxford Book of Food Plants." Oxford University Press, London.

Henze, J. (1977). Influence of CA-storage on fermentation of white cabbage (*Brassica oleracea* L.) *Acta Hort.* **62**, 71–78.

Hruschka, H. W. (1971). "Storage and Shelf-life of Packaged Kale", Marketing Research Report No. 923. United States Department of Agriculture, Washington, D.C.

Isenberg, F. M. (1968). Fresh green cabbage after seven months of storage. *Am. Veg. Grow.* **17**, 24–49.

Isenberg, F. M. (1979). Controlled atmosphere storage of vegetables. *Hort. Rev.* 337–394.

Isenberg, F. M. and Sayles, R. M. (1969). Modified atmosphere storage of Danish cabbage. *J. Am. Soc. hort. Sci.* **94**, 447–449.

Jarvis, W. R. (1977). *Botryotinia* and *Botrytis* species: taxonomy, physiology and pathogenicity. *Monogr. Can. Dep. Agric.*, No. 15, 195 pp.

Jenkins, J. E. E. (1964). Downy mildew on cauliflower curds. *Pl. Path.* **13**, 46.

Jones, S. R. (1979). Chinese cabbage: storage potential in an ice-bank store. *A. Rep. Kirton Exp. hort. Stn 1978*, 74–75.

Kear, R. W. and Symons, J. P. (1973). Post-harvest deterioration of stored Dutch white cabbage. *Meded. Facult. Landbouwwetensch., Rijksuniv., Gent* **38**, 1549–1560.

Kear, R. W., Williams, D. J. and Stevens, C. C. (1977). The effect of iprodione on the fungal deterioration of stored white cabbage. *Proc. Br. Crop Protect. Conf., Brighton, 1977*, 189–195.

Khristov, A. (1968). Protection of cauliflower from rotting during transport and its preservation. *Gradinarstvo* **10**, 45–48.

Lipton, W. J. and Harris, C. M. (1974). Controlled atmosphere effects on the market quality of stored broccoli (*Brassica oleracea* L. Italica group). *J. Am. Soc. hort. Sci.* **99**, 200–205.

Lipton, W. J. and Harris, C. M. (1976). Responses of stored cauliflower (*Brassica oleracea* L. *botrytis*) to low-oxygen atmospheres. *J. Am. Soc. hort. Sci.* **101**, 208–211.

Lockhart, C.' L. (1976). Reducing disease losses in stored cabbage. *Can. Agric.* **21**, 9–10.

Lund, B. M. (1981). The effect of bacteria on post-harvest quality of vegetables. *In* "Quality in Stored and Processed Vegetables and Fruit" (P. W. Goodenough and R. K. Atkin, Eds), 287–300. Academic Press, London.

Lund, B. M. and Wyatt, G. M. (1978). Post-harvest spoilage of cauliflower by downy mildew *Peronospora parasitica*. *Pl. Path.* **27**, 143–144.

Lutz, J. M. and Hardenburg, R. E. (1968). "The Commercial Storage of Fruits, Vegetables and Florist and Nursery Stocks", Agriculture Handbook No. 66, 94 pp. United States Department of Agriculture, Washington, D.C.

Natti, J. J. (1960). Virus diseases causing spotting of cabbage leaves. *Fm Res.*, June, 10–11.

Nieuwhof, M. (1969). "Cole Crops." Leonard Hill, London.

North, C. and Gray, L. S. (1961). Cabbages stored in clamps—a possible new crop for Scotland. *Scott. Agric.* **41**, 145–147.

Ogilvie, L. (1969). "Diseases of Vegetables", Ministry of Agriculture, Fisheries and Food Bulletin 123. H.M.S.O., London.

Paliliov, N. A. and Frolov, A. M. (1974). Storability and the physiological and biochemical processes in head cabbage in relation to resistance to pathogenic micro-organisms. *Referat. Zh.* **8**, 467.

Pendergrass, A. and Isenberg, F. M. (1974). The effect of relative humidity on the quality of stored cabbage. *Hortscience* **9**, 226–227.

Poapst, P. A., Ramsoomair, B. A. and Gourley, C. O. (1979). On the promotion of senescence in *Brassica oleracea* var. *capitata* by *Alternaria brassicicola* and by *Botrytis cinerea*. *Can. J. Bot.* **57**, 2378–2386.

Polegaev, V. I. and Pastukhov, V. M. (1976). Effect of temperature and relative humidity of storage environment on infection of cabbages by grey mould rot and bacteriosis. *Dokl. sel'. sk. Akad. Imeni K. A. Timiryazeva* **221**, 131–135.

Polegaev, V. I., Chzhao, A. E. and Avilova, S. V. (1979). Effect of different fertilizer combinations on the resistance of white cabbage to grey mould and its storage quality. *Izv. timiryazev. sel'.—khoz. Akad.*, No. 2, 193–195.

Punithalingam, E. and Holliday, P. (1975). "Descriptions of Pathogenic Fungi and Bacteria", No. 468. Commonwealth Mycological Institute, Kew.

Ramsey, G. B. (1935). *Peronospora* in storage cabbage. *Phytopathology* **25**, 955–957.

Ramsey, G. B. and Smith, M. A. (1961). "Market Diseases of Cabbage, Cauliflower, Turnips, Cucumbers, Melons and Related Crops", Agriculture Handbook, No. 184, 49 pp. United States Department of Agriculture, Washington, D.C.

Rangel, J. F. (1945). Two *Alternaria* diseases of cruciferous plants. *Phytopathology* **35**, 1002–1007.

Robinson, J. E., Browne, K. M. and Burton, W. G. (1975). Storage characteristics of some vegetables and soft fruits. *Ann. appl. Biol.* **81**, 399–408.

Rodin, M. N. (1963). The influence of forced ventilation on the temperature regime and keeping quality of cabbages in clamps. *Dokl. mosk. sel'.—khoz. Akad. K. A. Timiryazeva* **83**, 300–305.

Roosenboom, J. W. (1978). Caged cabbages for store. *Hort. Ind.*, April, 12–14.

Rubin, B. A. and Ivanova, T. M. (1960). Dynamics of phenols in the tissues of cabbage infected with *Botrytis cinerea*. *Dokl. Akad. Nauk. S.S.S.R.* **131**, 445–448.

Ryall, A. L. and Lipton, W. J. (1972). "Handling, Transportation and Storage of Fruits and Vegetables", Vol. 1, "Vegetables and Melons." AVI Publishing, Westport, Connecticutt.

Seidel, M. and Baresel, F. (1978). Der Einsatz eines Spritztunnels zur Behandlung von Kopfkohl gegen Lagerfäulen in der LPG "Ann Meer des Friedens" Elmenhorst. *NachrBl. PflSchutz D.D.R.* **32**, 55–56.

Semb, L. (1971). A rot of stored cabbage caused by a *Phytophthora* sp. *Acta Hort.* **20**, 32–35.

Sherf, A. F. (1972). Storage diseases of cabbage and carrots. *Ext. Inf. Bull. N.Y. St. Coll. Agric. Life Sci.* **35**, 8 pp.

Shipway, M. R. (1976). Winter white cabbage: use of fungicidal dusts and polythene lining of storage bins. *A. Rep. Kirton Exp. hort. Stn 1975*, 35–36.

Shipway, M. R. (1977a). Winter white cabbage: mechanical harvesting. *A. Rep. Kirton Exp. hort. Stn 1976*, 44–46.

Shipway, M. R. (1977b). Ice bank cooling system: investigation of its value for a range of vegetable crops. *A. Rep. Kirton Exp. hort. Stn 1976*, 35–38.

Shipway, M. R. (1977c). Winter white cabbage: use of fungicidal dust TBZ (thiabendazole). *A. Rep. Kirton Exp. hort. Stn 1976*, 46.

Shipway, M. R. (1978). Ice bank cooling system: investigation of its value for a range of vegetable crops. *A. Rep. Kirton Exp. hort. Stn 1977*, 61–63.

Shipway, M. R. (1979). Winter white cabbage: use of iprodione (Rovral) for the control of disease in store. *A. Rep. Kirton Exp. hort. Stn 1978*, 65–66.

Shipway, M. R. (1981). Refrigerated storage of winter white cabbage. *Ann. Rev. Kirton Exp. hort. Stn 1980*, 44–50.

Simmons, E. G. (1967). Typification of *Alternaria*, *Stemphylium* and *Ulocladium*. *Mycologia* **59**, 67–92.

Smith, W. H. (1940). The storage of broccoli and cauliflower. *J. hort. Sci.* **18**, 287–293.

Snyder, W. C. and Hansen, H. N. (1945). The species concept in *Fusarium* with reference to Discolor and other sections. *Am. J. Bot.* **32**, 657–666.

Sofronov, A. M. (1959). Study of a storage system for white cabbage. *Nauch. Trudy Ukr. n.-i. Inst. Ovoschevod. Kartofelya* **5**, 287–300.

Sozzi, A., Gorini, F. L. and Uncini, L. (1980). Storage suitability of the Chinese cabbage as affected by lining. *Acta Hort.* **116**, 157–163.

Stokes, D. A. (1981). Chinese cabbage. *Rev. Lea Valley Exp. hort. Stn 1980*, 1–2.

Stoll, K. (1974). Storage of vegetables in controlled atmospheres. *Bull. int. Inst. Refrig.* **54**, 1302–1324.

Stork, H. W. (1981). Kleurbehoud broccoli. *Groenten Fruit* **36**, 37.

Struck, P. and Grutz, E. M. (1972). Erfolgreiche Lagerung von Kopfkohl. *Gartenbau* **19**, 201–202.

Suhonen, I. (1969). On the storage life of white cabbage in refrigerated stores. *Acta Agric. scand.* **19**, 19–32.

Tahvonen, R. (1981). Storage fungi of cabbage and their control. *J. scient. agric. Soc., Finland* **53**, 211–227.

Thomas, T. H. (1974). Investigations into the cytokinin-like properties of benzimidazole-derived fungicides. *Ann. appl. Biol.* **76**, 237–241.

Tsupkova, N. A. (1978). Diseases of vegetables in storage. *Zasch. Rast.* **12**, 44–45.

van den Berg, L. and Lentz, C. P. (1968). The effect of relative humidity and temperature on survival and growth of *Botrytis cinerea* and *Sclerotinia sclerotiorum*. *Can. J. Bot.* **46**, 1477–1481.

van den Berg, L. and Lentz, C. P. (1973). High humidity storage of carrots, parsnips, rutabagas and cabbage. *J. Am. Soc. hort. Sci.* **98**, 129–132.

van den Berg, L. and Lentz, C. P. (1974). High humidity storage of some vegetables. *Can. Inst. Fd Sci. Technol. J.* **7**, 260–262.

van Hoof, H. A. (1952). Stip in kool, een virusziekte. *Meded. Dir. Tuinb.* **15**, 727–742.

Vogel, G. and Neubert, P. (1964). Ergebnisse bei der Anwendung von Phomasan (Pentachloronitrobenzol) zur Kopfkohlagerung. *Dte Gartenb.* **11**, 205–208.

von Arx, J. A. (1970). "The Genera of Fungi Sporulating in Pure Culture." Verlag von J. Cramer, Lehre.

Wale, S. J. and Epton, H. A. S. (1981). The effect of field nitrogen and postharvest fungicide dips on storage of Dutch white cabbage. *In* "Quality in Stored and Processed Vegetables and Fruit" (P. W. Goodenough and R. K. Atkin, Eds), 301–312. Academic Press, London.

Walkey, D. G. A. and Webb, M. J. W. (1978). Internal necrosis of stored white cabbage caused by turnip mosaic virus. *Ann. appl. Biol.* **89**, 435–441.

Weimer, J. L. (1924). *Alternaria* leafspot and brown rot of cauliflower. *J. agric. Res.* **29**, 421–441.

Wellman, F. L. (1932). *Rhizoctonia* bottom rot and head rot of cabbage. *J. agric. Res.* **45**, 461–469.

Wells, J. W. and Uota, M. (1970). Germination and growth of five fungi in low-oxygen and high-carbon dioxide atmospheres. *Phytopathology* **60**, 50–53.

Whitwell, J. D. (1970). Kirton compares keeping qualities in winter cabbage variety trials. *Grower* **74**, 1008–1010.

Whitwell, J. D. (1971). Winter storage cabbage. *A.D.A.S. Q. Rev.* **1**, 38–48.

Wiltshire, S. P. (1947). "Species of *Alternaria* on *Brassicae*", Mycological Paper No. 20. Commonwealth Mycological Institute, Kew.

Yoder, O. C. (1977). Development of methods for long-term cabbage storage. *Acta Hort.* **62**, 301–310.

Yoder, O. C. and Whalen, M. L. (1975a). Factors affecting post-harvest infection of stored cabbage tissue by *Botrytis cinerea*. *Can. J. Bot.* **53**, 691–699.

Yoder, O. C. and Whalen, M. L. (1975b). Variation in susceptibility of stored cabbage tissues to infection by *Botrytis cinerea*. *Can J. Bot.* **53**, 1972–1977.

7
Salad Crops

C. DENNIS

I. Introduction

Produce from salad crops includes fruits, such as tomatoes, cucumbers and peppers, and leafy tissues, such as lettuce and celery. Fungal and bacterial diseases are often the main type of deterioration which reduce the acceptability and cause spoilage of these crops during post-harvest storage and retail distribution. The relatively high pH (>4.5) of the fruits and of the leafy tissues often allows soft-rotting bacteria to cause spoilage, especially at high distribution temperatures. The bacterial diseases will be discussed in a later chapter and the present chapter will therefore be confined to fungal diseases.

Botrytis cinerea (grey mould) is an important post-harvest pathogen of all the salad crops mentioned and in certain instances may be the cause of the only commercially important disease encountered. In common with other fruits (see Chs 1 and 2), infection by *B. cinerea* often occurs at the flowering stage and may form latent or quiescent infections (see Ch. 1). The other important post-harvest fungal pathogens are specific to certain crops, except for *Alternaria* spp. which under certain conditions infect both tomatoes and peppers (see Table 7.1).

II. Tomatoes

Botrytis cinerea is the major cause of post-harvest fruit rotting of tomatoes (Chastagner and Ogawa, 1979; Dennis and Davis, 1980), although on out-door-grown fruit *Alternaria* spp. are often of similar or greater importance (Bartz, 1972; Pearson and Hall, 1975; Dennis *et al.*, 1979). Other fruit-rotting fungi are *Stemphylium* spp., *Fusarium* spp., *Cladosporium* spp. and *Rhizopus stolonifer*.

The relative importance of these fungi on stored fruit, apart from being

158   C. DENNIS

Table 7.1 *Fungi responsible for the major post-harvest diseases of salad crops*

Crop	Fungi
Cucumber	*Botrytis cinerea, Didymella bryoniae*
Peppers	*B. cinerea, Alternaria* spp.
Tomatoes	*B. cinerea, Alternaria* spp.
Celery	*B. cinerea, Mycocentrospora acerina, Sclerotinia sclerotiorum*
Lettuce	*B. cinerea, S. sclerotiorum, S. minor*

affected by growing conditions, is influenced by post-harvest treatments such as fungicide applications (Dennis *et al.*, 1979; Chastagner and Ogawa, 1979; Spalding, 1980), as well as the storage conditions, temperature, relative humidity and composition of the gaseous environment (Kidd and West, 1933; Eaves and Lockhart, 1961; Tomkins, 1963; Parsons *et al.*, 1970; Dennis *et al.*, 1979). Due to the sensitivity of tomatoes to low temperatures (see below), they cannot be stored for prolonged periods below 12°C (Rhodes, 1980), and if the relative humidity exceeds 93–94%, rapid fungal rotting usually occurs at non-chilling temperatures.

A. *Botrytis cinerea* (Grey Mould)

Most of the rots caused by *B. cinerea* (except where chilling injury occurs) develop in one of two ways: either from a quiescent infection in the calyx (see Ch. 1) or from an infected fruit by contact. The infected areas around the calyx may have a water-soaked appearance and may eventually become covered with greyish-white mycelium. The fungus frequently sporulates on the calyx but does not generally sporulate on the fruit surface, the only sign of infection being the water-soaked appearance. The infected areas are extremely soft, and the skin of the fruit often cracks and exudes liquid at the site of infection. Tomkins (1963) demonstrated the advantageous effect of removal of the calyx on the rotting of the fruit. A second advantage is the reduction in loss of water vapour from fruit during storage.

Tomatoes, in common with other fruits of tropical or sub-tropical origin, suffer chilling injury when held for prolonged periods at low temperatures (Moline, 1976; Rhodes, 1980). Chilling injury is a physiological disorder which results from the exposure of sensitive plant species to temperatures in the range 0–12°C (Lyons, 1973). It is generally associated with the necrosis of groups of cells situated either externally, leading to the formation of depressed areas, pitting and external discolouration, or internally, leading to internal browning. In the case of tomatoes, chilling injury also results in the inability of unripe fruits to ripen.

Lyons and Raison (1970) suggested that chilling injury which occurs below critical temperatures involves a phase change in the lipids of sensitive crops at this critical temperature. This has been postulated as the primary response of sensitive species to chilling temperatures and is thought to lead to changes in membrane permeability and in the activity of membrane-bound enzymes with the accumulation of toxic intermediates and hence damage or even death of the cells (Lyons, 1973). A number of cellular functions of the tomato fruit associated with membranes show marked changes at about 12°C (Lyons and Raison, 1970; Shneyour *et al.*, 1973; Nobel, 1974). Ultrastructural changes also occur in the mitochondria and plastids of tomatoes during chilling (Moline, 1976). Rhodes and Wooltorton (1977) reported increased metabolism of chlorogenic acid in chilled tomatoes, which has also been reported to occur in chilled peppers (Kozube and Ogata, 1971).

In addition to these physiological changes (pitting, discolouration, failure of unripe fruit to ripen), chill-injured tomatoes suffer enhanced microbial spoilage, and hence the onset of chilling injury is associated with major wastage of the stored commodity. Frequently, these symptoms of chilling injury do not become apparent during the period of cold storage and only develop if the tomatoes are returned to a warmer temperature. Tomatoes can indeed be exposed for short periods to temperatures below their threshold temperature without injury, but beyond a certain minimum period of exposure, symptoms of injury become apparent and become increasingly severe as the period of chilling temperature is extended (Tomkins, 1963; Dennis *et al.*, 1979; Dennis and Davis, 1980).

Thus in Fig. 7.1 the rate of infection of fruits by *B. cinerea* increases with increasing exposure time at 2°C when transferred to 20°C. There is little change in the rate of rotting after 4 days exposure at 2°C, compared with continuous storage at 20°C, whereas after 7 days at 2°C the rate of infection markedly increases. After storage at low temperature, surface rots tend to be established more quickly than those spreading from the calyx, and the storage life is, to an increasing extent, terminated by rots originating at sites other than the calyx, as shown in Fig. 7.2 (see also Tomkins, 1963). Surface rots are usually circular, blister-like and have a very water-soaked appearance with no surface mycelium or conidiophores. Exposure to low temperature, although delaying the spread of infection from the calyx, clearly predisposes the surface to infection by spores of *B. cinerea*.

R. P. Davis and C. Dennis (unpublished) observed that the frequency of rots caused by *B. cinerea* increased relative to those caused by other fungi with increasing exposure to chill injury and that ripe tomatoes were more susceptible than quarter-ripe fruit to increased rotting after storage at 2°C. The rate and extent of spoilage is also affected by the level of inoculum of *B. cinerea* on the fruit. For example, the number of fruits infected at 20°C after 7 days initial storage at 2°C was markedly higher for fruit sprayed with a high

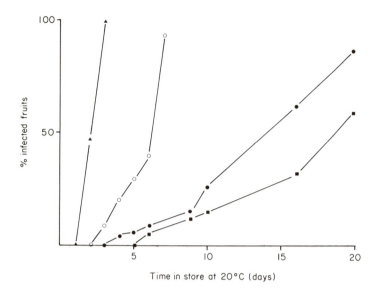

Fig. 7.1 Effect of chilling at 2°C on the rate of infection of stored tomatoes by *B. cinerea* (cv. Sonato). ▲——▲, 11 days at 2°C; ○——○, 7 days at 2°C; ●——●, 4 days at 2°CC; ■——■, 20°C only.

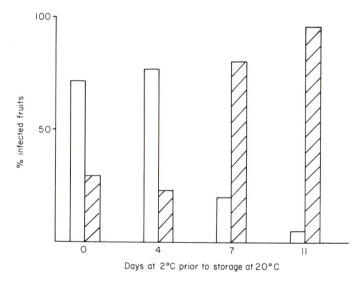

Fig. 7.2 Effect of chilling at 2°C on the site of infection of stored tomatoes by *B. cinerea* (cv. Sonato). ☐, initiated at calyx; ▨, initiated on fruit surface.

concentration (10^4 spores ml^{-1}) of spores of *B. cinerea*, compared with uninoculated fruit. This difference was due to an increase in the number of rots originating on the surface of the fruit: 70% as opposed to 30%. The difference was much less with fruit stored continuously at 20°C, with the majority of the lesions developing from the calyx (see Fig. 7.3). Thus commercially, fruit that is heavily contaminated with *B. cinerea* will rapidly spoil if exposed to chilling temperatures sufficient to cause injury.

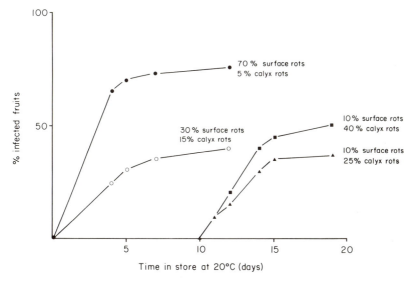

Fig. 7.3 Effect of inoculum on the infection of stored tomatoes after chilling at 2°C (cv. Winterbrid). ○——○, 7 days at 2°C, natural inoculum only; ●——●, 7 days at 2°C, sprayed with conidial suspension of *B. cinerea* (10^4 conidia ml^{-1}); ■——■, 20°C only, natural inoculum; ▲——▲, 20°C only, sprayed with conidial suspension of *B. cinerea* (10^4 conidia ml^{-1}).

Further observations (Dennis and Davis, 1980) on the behaviour of conidia of *B. cinerea* on the surface of tomatoes indicated that the rate of germination and formation of appressoria at 20°C increased after initial storage at 2°C and that both the percentage and rate of germ tubes forming appressoria was higher on the surface of ripe fruit than on quarter-ripe, thus accounting for the greater susceptibility of the former.

The increased susceptibility to fungal infection was found to vary between varieties, in that Winterbrid showed a greater increase in spoilage after chilling than did Sonato or Eurocross BB (C. Dennis and R. P. Davis, unpublished). These workers also confirmed the early report by Kidd and West (1933) that fruit harvested in the summer was more resistant to chilling injury than that harvested from the same plants at the end of the fruiting period in the autumn. No explanation for this change in behaviour of fruit during the harvesting period is available at present.

Botrytis cinerea produces a range of pectolytic enzymes (Jarvis, 1953; Verhoeff and Warren, 1972) which undoubtedly are involved in lesion development. In addition, Shishiyama *et al.* (1970) demonstrated that *B. cinerea* produced a cutinase enzyme which attacked tomato cutin and they suggested that such an enzyme may play a role in cuticular invasion by the fungus, by causing a reduction in the mechanical strength of the cutin; although they further suggested that there seems to be little possibility for complete decomposition of cutin by the enzyme.

Apart from chill-injured fruit, *B. cinerea* also infects the surface of immature fruit to cause the ghost-spotting symptom that occurs during growth of the fruit. The conidia germinate and penetrate the cuticle of immature fruits although no further growth occurs, despite the fungus remaining alive (Verhoeff, 1970). Verhoeff and Liem (1975) suggested further growth of *B. cinerea* could be prevented by the toxic action of the glycoalkaloid, tomatine. However, this does not explain why renewed growth of the fungus does not occur in ripe fruits where no tomatine can be detected. The author has never observed further development of *B. cinerea* from ghost spot symptoms during storage at temperatures from 2 to 20°C and, unlike Verhoeff (1970), has often failed to re-isolate *B. cinerea* from ghost-spot lesions. The lack of further growth may be due to the presence of phenolic compounds. Tomato fruits, unlike many fruits, show an increase in concentration of phenolic compounds during ripening (Walker, 1962).

B. *Alternaria* and *Stemphylium*

Rotting of tomato fruits by *Alternaria* spp. has been reported wherever tomatoes are grown (McColloch *et al.*, 1968), although on glasshouse-grown fruit spoilage by these species rarely occurs (Dennis *et al.*, 1979). *Alternaria* and *Stemphylium* are generally considered to be weak pathogens of tomato fruits, requiring injured or weakened tissue in which to germinate and develop (McColloch and Worthington, 1952; Bartz, 1972; Pearson and Hall, 1975).

Lesions caused by both *Alternaria* and *Stemphylium* are firm, flattened or slightly sunken, usually extending into the carpel wall and sometimes into the locule of the fruit, often with a dense, velvety, olive green or black spore mass on the surface. Lesions occur at any point on the fruit surface, but are often associated with the calyx scar (McColloch and Worthington, 1952; Dennis *et al.*, 1979). The fungi are considered to become established on the corky tissue around the stem scar of tomato fruits while in the field. Thus at the margin of the stem scar, fruits often show incipient infections that usually remain quiescent unless the fruits are subjected to adverse and weakening conditions such as low temperature (McColloch and Worthington, 1952). During storage under controlled atmosphere conditions using artificially high carbon dioxide and low oxygen atmospheres, Dennis *et al.* (1979) observed dark brown or black lesions developing around the calyx scar. Such lesions were only a few millimetres in diameter and did not develop into progressive rots. Similarly, Pearson and Hall (1975) reported incipient or quiescent infections caused by *Alternaria* on the surface of green fruit in the field which failed to become aggressive on ripening. This is analogous to the ghost-spot symptom caused by *B. cinerea* on tomatoes and where again no explanation for the lack of further development is known.

According to McColloch and Worthington (1952), tomatoes that were not chilled failed to become actively infected by *Alternaria*, whereas those that were chilled developed rots caused by *Alternaria* in proportion to the length of exposure to 0°C or to the decrease in temperature below 10°C. However, progressive lesions did occur on field-grown fruit stored under controlled atmospheres at 12.5°C, which is considered to be a non-chilling temperature for tomatoes (Dennis *et al.*, 1979). Thus it would appear that although chilling injury markedly increases the susceptibility of tomato fruits to infection by *Alternaria* (and *Stemphylium*) these species occasionally cause rots in the absence of such injury. Mechanical wounding is also known to increase susceptibility of fruits to both *Alternaria* and *Stemphylium* species.

C. Other Diseases

Cladosporium spp., *Fusarium* spp. and *Rhizopus stolonifer* are occasionally encountered as the cause of post-harvest rotting of tomatoes in the U.K., although they have been reported to cause significant losses in the U.S.A. (McColloch *et al.*, 1968; Ceponis and Butterfield, 1979).

Infection by *Cladosporium* spp. results in a firm lesion similar to that caused by *Alternaria* and *Stemphylium*, being slightly sunken and covered in a green to black mass of conidia. The lesions are usually confined to the superficial layers of the carpel wall but may extend into the locule.

Rots caused by *Fusarium* spp. are softer and extend into the centre of the fruit. Externally, the rotted tissue is usually water-soaked and becomes covered by white, yellow or pinkish mycelium according to the species of *Fusarium* involved. Internally, the infected tissue is discoloured and appears pale brown.

Infection by *R. stolonifer* results in characteristic lesions which are usually large and at first somewhat distended (blister-like). There is no discolouration and the infected area appears water-soaked through the

distended skin. The rot develops rapidly throughout the fruit tissue, resulting in total decay. The skin of infected fruit frequently ruptures, resulting
in exudation of abundant liquid and total collapse of the fruit. The tissue is
held together by the coarse mycelium of the fungus which rapidly
sporulates on the surface of the infected fruit and spreads to adjacent fruit,
thus resulting in rapid spread throughout a container of fruit.

III. Cucumbers

A. *Didymella bryoniae* (Black Rot)

Didymella bryoniae causes a variety of symptoms on cucumber plants;
leaves, stems, growing tips and fruits can all be attacked, and many other
cucurbits can also be infected (Chupp and Sherf, 1960; Luepschen, 1961;
Ramsey and Smith, 1961; Van Steekelenburg, 1978). The black fruit rot
fungus is known by many synonyms in the phytopathological literature.
Those most frequently used are *Mycosphaerella citrullina* (Wiant, 1945)
and *M. melonis* (Chiu and Walker, 1949a,b). The correct names are
Didymella bryoniae for the perfect state and *Phoma cucurbitacearium* for
the imperfect state (Boerema and Van Kesteren, 1972), although the
imperfect form is referred to as *Ascochyta cucumis* by Svedelius and Unestam (1978).

Infection of cucumber fruits by *D. bryoniae* causes economic losses and is
often the most important cause of disease during storage and handling of
the fruits after harvest (Veenman, 1972; Van Steekelenburg, 1982). The
fungus causes two types of rot. First, when infection occurs only at the
blossom end, *D. bryoniae* produces a brown heart rot (internal rot) with no
discolouration except for the dried decaying blossom remains and
blackening at the extreme tip, which forms a conspicuous symptom of this
type of infection. Alternatively, lesions may occur at any point on the
surface of the fruit (external rot). Initially, such lesions appear as greasy or
water-soaked spots on which grey-white mycelium may develop. The somewhat irregularly circular spots are first yellow to light brown in colour, but
soon darken, finally becoming black with abundant production of pycnidia
and eventually perithecia. However, the lesions may become considerably
advanced without turning black, although such discolouration is the characteristic symptom resulting from infection by the fungus on the surface of
the fruit. Blackening of the lesion is accompanied by a drying out of the
diseased tissues with shrivelling and wrinkling of the cucumber. Beneath
the dark outer lesions an extensive rot is found. Internal symptoms,
however, are much less characteristic than the external ones. The affected
tissues are spongy-soft, ranging from rather dry to fairly moist, and are

frequently somewhat vacuolated. The softening is brought about by pecto-lytic and cellulolytic enzymes produced by the fungus (Curren, 1969). Brownish discolouration is usually apparent in the older areas of the lesions.

Black rot is particularly troublesome in apparently healthy fruits that are stored under warm, humid conditions, when up to 25% of the fruits, de-pending on the fungicides sprayed on the crop, have been reported to be caused by surface infections of *D. bryoniae* (Van Steekelenburg, 1978). Van Steekelenburg and Van de Vooren (1981) showed that the percentage of fruits with rots initiated at the blossom end varied according to harvest date, with 14% as the highest recorded. These reports suggest that the infection after harvest usually occurs at points on the surface of the fruit.

Van Steekelenburg (1982) subsequently confirmed an earlier report (Wiant, 1945) that *D. bryoniae* is a wound parasite. Inoculated, non-woun-ded and slightly wounded fruits did not rot, suggesting that the peel of the fruit forms a mechanical and/or physiological barrier to fungal infection. Similarly, Svedelius and Unestam (1978) demonstrated that mechanical injury facilitates invasion by the fungus of cucumber leaves and that the release of nutrients from damaged cells rather than the rupture of the protective outer layer accounted for the increased invasion by the fungus. Van Steekelenburg (1982) reported increased rotting of wounded fruits with higher fertilization rates and suggested that this is a consequence of the nutrient availability to the fungus. He also observed increased resistance to rotting in the dark (both of fruit and leaves) and suggested the only possible explanation to be a light-induced change in the biochemistry of resistance or susceptibility of the fruit.

Temperature also has a pronounced effect on rotting. Van Steekelenburg (1982) reported that 10°C was the minimum at which growth occurred on inoculated fruits but that exposure to 6°C for 1 day predisposed the fruits to more severe rotting on transfer to 20°C. One is tempted to suggest that this was due to chilling injury caused by exposure to the low temperature, although Wiant (1945) claimed that there is little danger of chilling injury occurring at 5–7°C unless the period of exposure exceeds 1 week. Conver-sely, an 8 h exposure to 20°C followed by storage at 11°C was enough to stimulate the rotting process. Thus cucumbers should be rapidly cooled to 10–12°C and held at this temperature to prevent rotting by *D. bryoniae*. Prolonged exposure to lower temperatures will undoubtedly cause chilling injury, and predispose fruits to attack by this fungus.

In common with other post-harvest diseases, it is important that the fungus is controlled in the growing crop. In the case of *D. bryoniae*, the fungus is seed-borne and therefore seed treatment followed by appropriate fungicide treatment during production of the crop is necessary (Van Steekelenburg, 1978). After harvest it is important to avoid damaging and wounding the fruit as damage to the cuticle will enhance infection by the fungus.

B. *Botrytis cinerea* (Grey Mould)

Botrytis cinerea is the cause of the only other important post-harvest fungal disease of cucumbers, and in common with other fruit diseases (see Chs 1 and 2) the fungus often becomes established in the flower parts and after harvest may infect the fruit at the blossom end. Infected areas appear water-soaked, are soft and have a yellowish-brown appearance, and often become covered with a dense mass of grey conidia. In some cases, typical black sclerotia are formed on the surface of infected cucumbers. Infection via other parts of the fruit may occur if wounding occurs (infection some-times occurs by the cut stem end) or if flower remains adhere to the surface, as sometimes occurs in shrink-wrapped cucumbers. In the latter case the fungus uses the infected flower parts as a food base to infect the fruit surface. In order to reduce infection by *B. cinerea*, the fruit should be rapidly cooled to a safe storage temperature as mentioned above, but unlike *D. bryoniae*, *B. cinerea* is able to cause rots at temperatures between 0 and 10°C.

IV. Peppers

The main post-harvest diseases of peppers are those caused by *Botrytis cinerea* and *Alternaria* spp., with bacterial soft-rotting also being important under certain conditions (Risse *et al.*, 1979; see Ch. 9).

Botrytis cinerea is the most important fungal disease of glasshouse-grown peppers, for example, as in the U.K., whereas *Alternaria* spp. are of greater importance on outdoor-grown crops, as in the U.S.A. (McColloch, 1963; Quebral and Shertleff, 1965). Both fungi are, however, weak path-ogens, requiring physical wounding to allow penetration or, alternatively, as in the case of tomatoes, exposure of the fruits to periods of low tem-perature. However, the critical threshold temperature is slightly lower than for tomatoes. McColloch (1966) reports 10°C to be critical for infection by *B. cinerea*, whereas storage at 8–9°C is recommended in order to avoid predisposition to rotting by *Alternaria* spp. (McColloch, 1963).

If the fruits are injured through excessive chilling, lesions may develop anywhere on their surface without the necessity for wounding (McColloch *et al.*, 1968). McColloch (1966) demonstrated that although conidia of *B.*

cinerea germinated on the surface of peppers when incubated in a humid atmosphere no infection occurred unless the fruits had been exposed to chilling temperatures, and the extent of rotting increased proportionally with increased exposure to 0°C. Such observations indicate a marked change in the susceptibility of the fruit surface to infection after chilling similar to that described for tomatoes, and the underlying biochemical changes are known to be similar (Kozube and Ogata, 1971; Rhodes and Wooltorton, 1977; Dennis, 1981).

In the absence of chilling injury, infection of peppers by *B. cinerea* and *Alternaria* spp. occurs via wounds in the skin surface or through the stem end (Quebral and Shertleff, 1965; Shawkat *et al.*, 1978). Infection by *B. cinerea* results in round, water-soaked lesions which become grey to greyish-brown in colour as they develop, although the infected tissue remains relatively firm.

In the early stages, lesions caused by *Alternaria* appear as small, circular, slightly depressed, water-soaked spots with a greyish-green colour. Infected areas gradually become enlarged and sharply sunken. Under conditions of high humidity, grey mycelium develops on the surface and when sporulation occurs the lesions become dark green to black in colour (see McColloch *et al.*, 1968).

V. Celery

A. *Mycocentrospora acerina* (Liquorice Rot)

Storey and Wilcox (1953) first reported the occurrence of this fungus in Great Britain when they observed several crops of celery in Bedfordshire to be infected and showing typical black rot or liquorice rot symptoms. It was already known to be a serious storage rot of celery in the United States and Canada (Newhall, 1944; Truscott, 1944), although Truscott (1944) regarded the disease as affecting senescent plants only. Lowings (1956) subsequently reported the fungus to cause substantial losses of celery during storage under semi-commercial conditions in the U.K. The disease occurred on both early (self-blanching) and main-crop varieties harvested from fields in Cambridgeshire and Lincolnshire. The introduction of cool storage as an extension to vegetable growing in East Anglia in the 1960s confirmed the importance of *Mycocentrospora acerina* as a storage pathogen of celery (Derbyshire and Crisp, 1971) as well as of carrots (Derbyshire and Crisp, 1971; see Ch. 5). In fact, Derbyshire and Crisp (1971) stated that this disease caused such heavy commercial losses that it was difficult to justify the use of cool storage in the absence of control measures. Losses due to *M. acerina* reached 30% after 8 weeks storage and losses of 100% were recorded.

Thus very similar disease patterns have been observed in Britain (Lowings, 1956) and North America (Newhall, 1944; Truscott, 1944), with symptoms suddenly appearing after several weeks storage at 2°C. Infection usually occurs near the base of the outermost petioles, at the junction of the petioles with the stem or "butt". This position of the rot is most characteristic of the disease together with the development of rapidly spreading dark-coloured lesions (see Day *et al.*, 1972), which makes the plants unsaleable.

Attempts to locate the fungus on stored plants prior to lesion development were initially unsuccessful (Truscott, 1944), until Day *et al.* (1972) carried out a detailed study of the infection processes. Observations by these workers on plants grown in a field with a previous history of infection showed that 70% of the plants were infected and that the infection was confined to the underground parts of senescent petioles, many of which became detached and remained in the soil when the plants were lifted. Further observations indicated that chlamydospores were the most likely source of inoculum due to their abundance in lesions and common occurrence in dead remains of the petiole around the leaf scar and in adjacent soil particles. Conidia, on the other hand, were rarely observed and are known to be rapidly killed (Neergaard and Newhall, 1951).

Day *et al.* (1972) also showed that germ tubes from chlamydospores grew randomly over the petiole surface and penetrated via appressoria through either intact cuticle or damaged areas but seldom through stomata. Once penetration had occurred, the fungus grew both intercellularly and intracellularly, with chlamydospores forming abundantly in tissue several millimetres behind the advancing edge of the lesion. The fungus caused maceration of the cells, usually up to 1 mm in advance of the hyphae, and eventual collapse of infected tissue. Lesions are initially cinnamon in colour, gradually darkening and finally becoming black as chlamydospores are formed. The lesions develop rapidly in a longitudinal manner due to impedance of lateral spread by collenchyma cell walls.

Day *et al.* (1972) also analysed the occurrence of the position of the lesions of naturally infected plants during storage and showed that over 90% of the lesions occurred on the lower regions of the petiole with the remainder on the cut stem ("butt" or "root knot").

The infected plants were randomly distributed in samples in storage experiments, thus confirming the earlier suggestion by Truscott (1944) that spread from plant to plant in storage was rare.

The lag phase in infection, which has been reported to be of similar duration by a number of workers (Newhall, 1944; Truscott, 1944; Lowings,

1956; Day *et al.*, 1972), appears to coincide with a phase of high plant resistance. Day *et al.* (1972) detected three forms of resistance, namely failure of chlamydospore germination, failure of appressoria to penetrate the cuticle and failure of hyphae to grow through tissue after minor cuticle damage. They went on to suggest that diffusion of fungistatic substances from the cells of the celery tissue into the chlamydospores was the most likely explanation and would also account for the suppression of hyphal development in surface tissues following minor damage to the cuticle. Only if severe damage enabled the fungus to penetrate the underlying tissues did infection occur. Similarly, resistance of the cuticle during early storage may involve fungistatic toxins since appressoria that fail to penetrate do not develop further, as they do in late storage. However, mechanical impedance of the cuticle, as has been postulated for several other diseases (Martin, 1964), may also contribute to resistance.

Disappearance of the inhibitory effects was found to be coincident with rapid senescence of the petioles, although no studies have been carried out in an attempt to isolate and characterize possible inhibitory substances as was subsequently done for carrot tissue (Garrod *et al.*, 1978; Garrod and Lewis, 1979). The correlation between senescence and disease development confirmed this suggestion by Truscott (1944) and is in agreement with observations by Lowings (1956) and Derbyshire and Crisp (1971). Trimming plants prior to storage removes the oldest, outermost petioles and therefore together with storage at low temperature (*c.* 2°C) would explain the considerable delay in infection.

The trimming process, although removing contaminated and possibly infected outermost petioles, will not remove all sources of inoculum (e.g. chlamydospores in soil) and, as Day *et al.* (1972) claimed that microscopic amounts of inoculum can cause infection, it is unlikely that an assessment of inoculum levels after trimming would give a reliable estimate of storage potential.

However, an understanding of the timing of infection and the processes involved indicates that even when plants are contaminated with inoculum of the fungus the consistency in length of the lag phase means that plants could be stored free of rotting by this fungus for up to 5 or 6 weeks at 2°C. Storage for longer periods necessitates treatment to either prolong host resistance or to inactivate the inoculum. Since the pathogen does not penetrate the plant until late in storage, post-harvest treatment with effective fungicides such as Benlate (50% active ingredient benomyl) reduces the development of lesions (Derbyshire and Crisp, 1971). In the same trials, high concentrations of Iodophor appeared to accelerate senescence and consequently disease incidence. The effective reduction in lesion development by Benlate was no doubt partly due to its direct effect on the fungus, but may also have resulted from a delay in senescence, as benomyl is well known to have cytokinin-like effects on plant tissue (Tomlinson and Rich, 1973). Thus other treatments that delay senescence should delay the onset of rotting of celery by *M. acerina*. One aspect that warrants further investigation in this context is the use of modified or controlled atmospheres. Low oxygen and high carbon dioxide atmospheres are known to delay senescence in crops such as Dutch white cabbage (Geeson and Browne, 1980) and indeed have also been reported by Canadian workers (van den Berg and Lentz, 1974) to maintain colour and reduce decay of celery, although they did not report the cause of the decay.

Although damage to the petiole surface clearly enhances infection and should be avoided as far as possible, it is inevitable that a great deal of minor damage occurs during harvesting and handling. In practice, the different types of lesion are not equally deleterious. Whereas petiole lesions can be easily removed by further trimming of the plants after storage if the disease is noticed in time, lesions that spread from the stem ("butt" or "root knot") cause loss of the plant.

B. *Botrytis cinerea* (Grey Mould)

Botrytis cinerea is generally more widespread than *M. acerina* or *Sclerotinia sclerotiorum*, although its effect is not usually as devastating as that of the last two species.

Infection by *B. cinerea* occurs at any point along the length of the petiole and also may originate from the leaf blade. Damaged areas are often sites of infection. Mechanical damage on the petiole and leaf blade, which may be inflicted during harvesting and trimming, or lesions caused by *Septoria apii-graveolentis*, predispose the tissues to infection by *B. cinerea*.

Initially, the infected areas appear as water-soaked, gradually becoming brownish and covered in the typical grey mycelium and conidia. Prior to sporulation and aerial development of the fungus, the symptoms resemble those of bacterial soft rots (see Ch. 9), with the tissue becoming soft and slippery. The lesions often develop in an irregular pattern, but in common with *M. acerina*, progress along the length of the petioles rather than across them.

Smith *et al.* (1966) considered *B. cinerea* to be the principal cause.of loss of celery stored for longer than 4 weeks and it was the most common cause of rotting in small-scale trials with celery grown in Cambridgeshire stored in an ice-bank store in the late 1970s (C. Dennis, unpublished data).

No detailed study of the sequence of events leading to infection and rotting of celery by *B. cinerea* has been carried out, although the innoculum appears to originate from soil particles and leaf debris in common with diseases caused by *B. cinerea* of other vegetables and fruits (see Chs 2, 4 and 5). Once petioles become infected, further spread of the fungus occurs by contact between healthy and infected petioles and in this way it spreads from plant to plant. Derbyshire and Crisp (1971) reported *B. cinerea* to occur only during the later stages of storage, especially where leaf and soil debris were present.

Control of *B. cinerea* on stored celery can be achieved by careful handling and trimming prior to storage, refrigerated storage at 0–2°C and the use of post-harvest dips with fungicides such as Benlate, providing a high proportion of benomyl-resistant strains of the fungus are not present.

C. *Sclerotinia sclerotiorum* (Watery Soft Rot)

Sclerotinia sclerotiorum does not appear to be a common cause of post-harvest rotting of celery in the U.K., the only published account being that of Lowings (1956), whereas in the U.S.A. it appears to be a more serious cause of post-harvest loss (Poole, 1922; Purdy, 1979). Lowings (1956) reported that the fungus was frequently found growing over the roots and stem (crown), although these were rarely severely rotted. However, where plants were stored "head" to "tail" in alternate rows, the fungus had often spread from the infected plants to the foliage and petioles of adjacent layers, causing a very destructive wet rot similar to that described by Ramsey and Wiant (1941) and Smith *et al.* (1966). Infected leaves and petioles appear water-soaked at first, sometimes with a faint pink discolouration, followed by rapid collapse of the infected tissue with liberation of much fluid. Abundant aerial mycelium and occasional sclerotia are also produced.

Watery rot of celery is considered to be a soil-borne disease, as observations by Lowings (1956) showed the inoculum to be confined to the roots and stem. The long-term survival of sclerotia of the fungus (Adams and Ayers, 1979) is no doubt important in maintaining adequate inoculum levels in the soil. There has been no detailed study of the infection of celery by this fungus, although much work has been done with other hosts and many general features have emerged (Lumsden, 1979). Infection is considered to occur via mycelia rather than from direct infection by germinated ascospores and unless infection is via stomata, appressora are formed, which are usually large, complex, multi-celled structures (Lumsden, 1979; see Ch. 5).

In common with other hosts, the fungus can penetrate celery in the absence of wounds, although wounding favours immediate infection and may result in increased rotting (Smith *et al.*, 1966). After infection of hosts, *S. sclerotiorum* generally produces a range of cell wall and middle lamella

degrading enzymes, and toxins and enzymes to degrade host tissue and
defence substances, which together result in a rapid infection of many hosts
(Lumsden, 1979). The defence mechanisms of host tissue against *S. scle-*
rotiorum are not well understood, but it would appear with celery that the
leaves and petiole are more susceptible than the roots and stem (Lowings,
1956). There would also appear to be differences in varietal susceptibility,
as Lowings (1956) reported that approximately 80% of plants in one batch
of early celery were severely attacked after 12 weeks in store, whereas
losses of main-crop celery were less severe. Although the level of inoculum
of these two varieties was reported to be similar, further work is required to
substantiate variation in resistance of different celery tissue and varieties as
well as the effect of growing conditions on this resistance.

Recent work carried out with strains of *S. sclerotiorum* from a different
host has indicated that the fungus grows extremely slowly, if at all, at 0–1°C
on culture media or carrot tissue (D. Bibby, C. Dennis and B. G. Lewis,
unpublished) and thus storage at this temperature in an ice-bank store (see
Dennis, 1981; Lindsay and Neale, 1981) or in a jacketed store (van den Berg
and Lentz, 1974) should provide a storage system that would prevent infection by this fungus.

VI. Lettuce

Fungal rots caused by *Botrytis cinerea* and *Sclerotinia* spp. can cause post-
harvest losses of lettuce, although bacterial soft-rotting is often of equal or
greater importance (Ramsey *et al.*, 1967; Ceponis, 1970; Ceponis *et al.*,
1970).

A. *Botrytis cinerea* (Grey Mould)

Lesions caused by *B. cinerea* are characteristically water-soaked areas,
being greyish-green or brown in colour. The infected tissue becomes soft,
slimy and covered with the characteristic grey conidia and conidiophores.
Spread of infection by contact with diseased tissues rapidly occurs under
favourable conditions.

Delon *et al.* (1977) showed that the fungus penetrates the cuticle and
rapidly spreads in the foliar parenchyma, destroying the cell walls and cell
contents in advance of the hyphae. The host cell ultrastucture is rapidly
altered: dictyosomes and mitochondria disappear first and although the
chloroplasts resist longer, they are markedly modified with inflating and
stacking of thylakoids, an increase in the number of osmiophilic globules
and the disintegration of the plastidial membrane. Delon *et al.* (1977) also

demonstrated that bacteria are present in the interaction between *B. cinerea* and lettuce tissue, although their role in the development of the disease is uncertain. It is suggested by Delon *et al.* (1977) that such bacteria may only attack substrates previously liberated by the fungus, despite the fact that soft rot coliforms and pseudomonads are known to be able to invade lettuce tissue (Lund, 1971; see Ch. 9). The observation of hyphae filled with bacteria indicates the latter play a role in decompositon of the invading hyphae (Delon *et al.*, 1977).

Rots of lettuce caused by *B. cinerea* can be effectively controlled by ensuring that old wrapper leaves, debris and soil are removed, that there is prompt and rapid cooling and that a low distribution temperature (<5°C) is maintained.

B. *Sclerotinia* spp. (Watery Soft Rot)

Watery soft rot caused by *S. sclerotiorum* and *S. minor* mostly occurs on the lower parts of the heads but may be found on any part. The infected tissues appear water-soaked and light or pinkish-brown in colour. A white cottony mycelium develops as the infection spreads and the infected heads eventually become a watery mass due to the activity of the cell wall degrading enzymes (see Lumsden, 1979). In very advanced stages the irregular black sclerotia are produced on the infected tissue (Ramsey *et al.*, 1967). Spread of infection from head to head readily occurs under favourable conditions and infection of plant parts before harvest often results in post-harvest disease by spread of the fungi from diseased to healthy tissue in storage and shipping containers (Purdy, 1979). In order to minimize this spread of disease during storage and distribution, prompt and rapid cooling together with the removal of sources of inoculum such as outer wrapper leaves, debris and soil (Lumsden, 1979; Purdy, 1979) is essential.

The field disease of lettuce caused by *S. sclerotiorum* and *S. minor* is known as "lettuce drop" or "wilt" (Abawi and Grogan, 1979; Lumsden, 1979; Purdy, 1979). According to Abawi and Grogan (1979) infection of lettuce by *S. sclerotiorum* almost always occurs at ground level, because it usually originates from ascosporic infection of senescent lower leaves. Lumsden (1979), however, concluded that for *S. sclerotiorum* mycelium infection from sclerotia rather than infection directly from germinated ascospores appears to be the primary means of host penetration. Infection by *S. minor* can occur either at the soil line through senescent lower leaves or below ground as deep as 10 cm through root or stem tissues. The role of ascospores in "lettuce drop" epidemics caused by *S. minor* appears to be of importance and if involved they probably would infect senescent lower leaves. In addition, a source of organic matter for inoculum nutrition is usually a prerequisite for penetration, although Adams and Tate (1976) reported direct penetration of lettuce plants by *S. minor* in the absence of available organic matter. This was considered to be due to the fact that sclerotia of this species germinate by producing a mass or "plug" of mycelium which apparently has sufficient nutrient reserves to allow direct penetration. The epidemiology, physiology and symptoms of "lettuce drop disease" have been reviewed in more detail by Abawi and Grogan (1979), Lumsden (1979) and Purdy (1979).

References

Abawi, G. S. and Grogan, R. G. (1979). Epidemiology of diseases caused by *Sclerotinia* species. *Phytopathology* **69**, 889–904.

Adams, P. B. and Ayers, W. A. (1979). Ecology of *Sclerotinia* species. *Phytopathology* **69**, 896–899.

Adams, P. B. and Tate, C. J. (1976). Mycelial germination of sclerotia of *Sclerotinia sclerotiorum* on soil. *Pl. Dis. Reptr* **60**, 515–518.

Bartz, J. A. (1972). Studies on the causal agent of black fungal lesions on stored tomato fruit. *Proc. Fla St. hort. Soc.* **84**, 117–119.

Boerema, G. H. and Van Kesteren, H. A. (1972). Enkele bijzondere Schimmelaantastingen IV (Mycologische Waarnemingen No. 16). *Gewasbescherming* **3**, 65–69.

Ceponis, M. J. (1970). Diseases of California head lettuce on the New York market during the spring and summer months. *Pl. Dis. Reptr* **54**, 964–966.

Ceponis, M. J. and Butterfield, J. E. (1979). Losses in fresh tomatoes at the retail and consumer levels in the greater New York area. *J. Am. Soc. hort. Sci.* **104**, 751–754.

Ceponis, M. J., Kaufman, J. and Butterfield, J. E. (1970). Relative importance of grey mould rot and bacterial soft rot of Western lettuce on the New York market. *Pl. Dis. Reptr* **54**, 263–265.

Chastagner, G. A. and Ogawa, J. M. (1979). A fungicide–wax treatment to suppress *Botrytis cinerea* and protect fresh market tomatoes. *Phytopathology* **69**, 59–63.

Chiu, W. F. and Walker, J. C. (1949a). Morphology and variability of the cucurbit black rot fungus. *J. agric. Res.* **78**, 81–102.

Chiu, W. F. and Walker, J. C. (1949b). Physiology and pathogenicity of the cucurbit black rot fungus. *J. agric. Res.* **78**, 589–615.

Chupp, C. and Sherf, A. F. (1960). "Vegetable Diseases and their Control", 314–317. Ronald Press, New York.

Curren, T. (1969). Pectic and cellulolytic enzymes produced by *Mycosphaerella citrullina* and their relation to black rot of squash. *Can. J. Bot.* **47**, 791–794.

Day, J. R., Lewis, B. G. and Martin, S. (1972). Infection of stored celery by *Centrospora acerina*. *Ann. appl. Biol.* **71**, 201–210.

Delon, R., Kiffer, E. and Mangenot, F. (1977). Ultrastructural study of host–parasite interactions. II. Decay of lettuce caused by *Botrytis cinerea* and phyllosphere bacteria. *Can. J. Bot.* **55**, 2463–2470.

Dennis, C. (1981). The effect of storage conditions on the quality of vegetables and salad crops. *In* "Quality in Stored and Processed Vegetables and Fruit" (P. W. Goodenough and R. K. Atkin, Eds). Academic Press, London.

Dennis, C. and Davis, R. P. (1980). Fungal spoilage of stored tomatoes. *Bienn. Rep. Fd Res. Inst. 1977 & 1978*, 44–45.

Dennis, C., Browne, K. M. and Adamicki, F. (1979). Controlled atmosphere storage of tomatoes. *Acta Hort.* **93**, 75–83.

Derbyshire, D. M. and Crisp, A. F. (1971). Vegetable storage diseases in East Anglia. *Proc. 6th Insectic. Fungic. Conf.*, 167–172.

Eaves, C. A. and Lockhart, C. L. (1961). Storage of tomatoes in artificial atmospheres using calcium hydroxide absorption method. *J. hort. Sci.* **36**, 85–95.

Garrod, B. and Lewis, B. G. (1979). Location of the antifungal compound falcarindiol in carrot root tissue. *Trans. Br. mycol. Soc.* **75**, 166–169.

Garrod, B., Lewis, B. G. and Coxon, D. T. (1978). Cis-heptadeca-1,9-diene-4, 6-diyne-3,8-diol, an antifungal polyacetylene from carrot root tissue. *Physiol. Pl. Path.* **13**, 241–246.

Geeson, J. D. and Browne, K. M. (1980). Controlled atmosphere storage of winter white cabbage. *Ann. appl. Biol.* **95**, 267–272.

Jarvis, W. R. (1953). Comparative studies of the pectic enzymes of *Botrytis cinerea* Pers. and *Erwinia aroideae* (Townsend) Stapp. Ph.D.Thesis, University of London.

Kidd, F. and West, C. (1933). Gas storage of tomatoes. *Rep. Fd Invest. Bd, Lond.*, 209–211.

Kozube, N. and Ogata, K. (1971). Physiological and chemical studies of chilling injury in pepper fruits. Part 3. Changes of phenylaline ammonia-lyase and tyrosine ammonia-lyase activities in pepper seeds stored at 6°C and 20°C. *J. Jap. Soc. hort. sci.* **40**, 416–420.

Lindsay, R. T. and Neale, M. A. (1981). Cold stores for fresh vegetables. *In* "Quality in Stored and Processed Vegetables and Fruits" (P. W. Goodenough and R. K. Atkin, Eds). Academic Press, London.

Lowings, P. H. (1956). Some storage rots of celery caused by *Centrospora acerina* and other fungi. *Pl. Path.* **4**, 106–107.

Luepschen, N. S. (1961). The development of *Mycosphaerella* black rot and *Pellicularia rolfsii* rot of watermelons at various temperatures. *Pl. Dis. Reptr* **45**, 557–559.

Lumsden, R. D. (1979). Histology and physiology of pathogenesis in plant diseases caused by *Sclerotinia* species. *Phytopathology* **69**, 890–896.

Lund, B. M. (1971). Bacterial spoilage of vegetables and certain fruits. *J. appl. Bact.* **34**, 9–20.

Lyons, J. M. (1973). Chilling injury in plants. *A. Rev. Pl. Physiol.* **24**, 445–466.

Lyons, J. M. and Raison, J. K. (1970). Oxidative activity of mitochondria isolated from plant tissues sensitive and resistant to chilling injury. *Pl. Physiol.* **45**, 386–389.

Martin, J. T. (1964). Role of the plant cuticle in the defense against disease. *A. Rev. Phytopath.* **2**, 81–100.

McColloch, L. P. (1963). "Chilling Injury and *Alternaria* Rot of Bell Peppers", Marketing Research Report No. 536. United States Department of Agriculture, Washington, D.C.

McColloch, L. P. (1966). "*Botrytis* Rot of Bell Peppers", Marketing Research Report No. 754. United States Department of Agriculture, Washington, D.C.

McColloch, L. P. and Worthington, J. T. (1952). Low temperature as a factor in the susceptibility of mature-green tomatoes to *Alternaria* rot. *Phytopathology* **42**, 425–427.

McColloch, L. P., Cook, H. T. and Wright, W. R. (1968). "Market Diseases of Tomatoes, Peppers and Eggplants", Agriculture Handbook No. 28, 51–61. United States Department of Agriculture, Washington, D.C.

Moline, H. E. (1976). Ultrastructural changes associated with chilling of tomato fruit. *Phytopathology* **66**, 617–624.

Neergard, P. and Newhall, A. G. (1951). Notes on the physiology and pathogenicity of *Centrospora acerina* (Hartig) Newhall. *Phytopathology* **41**, 1021–1033.

Newhall, A. G. (1944). A serious storage rot of celery caused by the fungus *Ansatospora macrospora* n. gen. *Phytopathology* **34**, 92–105.

Nobel, P. S. (1974). Temperature dependence of the permeability of chloroplasts from chilling-sensitive and chilling-resistant plants. *Planta* **115**, 369–372.

Parsons, C. S., Anderson, R. E. and Penney, R. W. (1970). Storage of mature-green tomatoes in controlled atmospheres. *J. Am. Soc. hort. Sci.* **95**, 791–794.

Pearson, R. C. and Hall, D. H. (1975). Factors affecting the occurrence and severity of blackmould on ripe tomato fruit caused by *Alternaria alternata*. *Phytopathology* **6**, 1352–1359.

Poole, R. F. (1922). The sclerotinia rot of celery. *Bull. New Jers. agric. Exp. Stn* **359**, 27 pp.

Purdy, L. H. (1979). *Sclerotinia sclerotiorum:* history, diseases and symptomology host range, geographic distribution and impact. *Phytopathology* **69**, 875–880.

Quebral, F. C. and Shertleff, H. C. (1965). *Alternaria* rot, a serious disease of bell peppers in Illinois. *Phytopathology* **55**, 1072.

Ramsey, G. B. and Smith, M. A. (1961). "Market Diseases of Cabbage, Cauliflower, Turnips, Cucumbers, Melons and Related Crops," Agriculture Handbook No. 184, 21–25. United States Department of Agriculture, Washington, D.C., No. 184, 21–25.

Ramsey, G. B. and Wiant, J. S. (1941). "Market Diseases of Fruit and Vegetables", Miscellaneous Publications No. 440, 56–58. United States Department of Agriculture, Washington, D.C.

Ramsey, G. B., Friedman, B. A. and Smith, M. A. (1967). "Market Diseases of Beets, Chickory, Endive, Escarole, Globe Artichokes, Lettuce, Rhubarb, Spinach and Sweet Potatoes", Agriculture Handbook No. 155, 6–15. United States Department of Agriculture, Washington, D.C.

Rhodes, M. J. C. (1980). Chilling injury—some underlying biochemical mechanisms. *In* "Opportunities for Increasing Crop Yields" (R. G. Hurd, P. V. Biscoe and C. Dennis, Eds), 377–394. Pitman, London.

Rhodes, M. J. C. and Wooltorton, L. S. C. (1977). Changes in the activity of enzymes of phenylpropanoid metabolism in tomato fruit stored at low temperatures. *Phytochemistry* **16**, 655–659.

Risse, L. A., Smoot, J. J., Dow, A. T., Moffitt, T. and Cubbedge, R. (1979). Harvest conditions, packing house treatments and shipping temperatures for export of Florida bell peppers. *Proc. Fla. St. hort. Soc.* **92**, 192–194.

Shawkat, A. L. B., Michail, S. H., Tarabeih, A. M. and Al-Zarari, A. J. (1978). *Alternaria* fruit rot of pepper. *Acta Phytopath. Acad. Sci. Hung.* **13**, 349–355.

Shishiyama, J., Araki, F. and Akai, S. (1970). Studies on cutin-esterase II. Characteristics of cutin esterase from *Botrytis cinerea* and its activity on tomato cutin. *Pl. Cell Physiol.* **11**, 937–945.

Shneyour, A., Raison, J. K. and Smillie, R. M. (1973). The effect of temperature on the rate of photosynthetic electron transport in chloroplasts of chilling sensitive and chilling resistant plants. *Biochem. Biophys. Acta* **292**, 152–161.

Smith, M. A., McColloch, L. P. and Friedman, B. A. (1966). Market diseases of asparagus, onions, beans, peas, carrots, celery and related vegetables. Agriculture Handbook No. 303, 41–53. United States Department of Agriculture, Washington, D.C.

Spalding, D. H. (1980). Control of *Alternaria* rot of tomatoes by postharvest application of Imazalil. *Pl. Dis.* **64**, 169–171.

Storey, I. F. and Wilcox, H. J. (1953). *Centrospora acerina* on celery. *Pl. Path.* **2**, 72.

Svedelius, G. and Unestam, T. (1978). Experimental factors favouring infection of attacked cucumber leaves by *Didymella bryoniae. Trans. Br. mycol. Soc.* **71**, 89–97.

Tomkins, R. G. (1963). The effects of temperature, extent of evaporation and restriction of ventilation on the storage life of tomatoes. *J. hort. Sci.* **38**, 335–347.

Tomlinson, H. and Rich, S. (1973). Anti-senescent compounds reduce injury and steroid changes in ozonated leaves and their chloroplasts. *Phytopathology* **63**, 903–906.

Truscott, J. H. L. (1944). A storage rot of celery caused by *Ansatospora macrospora* (Osterw.) Newhall. *Can. J. Res.* **22**, 290–304.

van den Berg, L. and Lentz, C. P. (1974). High humidity storage of some vegetables. *Can. Inst. Fd Sci. Technol. J.* **7**, 260–262.

Van Steekelenburg, N. A. M. (1978). Chemical control of *Didymella bryoniae* in cucumbers. *Neth. J. Pl. Path.* **84**, 27–34.

Van Steekelenburg, N. A. M. (1982). Factors influencing external fruit rot of cucumber caused by *Didymella bryoniae. Neth. J. Pl. Path.* **88**, 47–56.

Van Steekelenburg, N. A. M. and Van de Vooren, J. (1981). Influence of the glasshouse climate on development of diseases in a cucumber crop with special reference to stem and fruit rot caused by *Didymella bryoniae. Acta Hort.* **118**, 45–56.

Veenman, A. F. (1972). *Mycosphaerella* in kan kommers. *Tuiriderji* **12**, 24–27.

Verhoeff, K. (1970). Spotting of tomato fruits caused by *Botrytis cinerea. Neth. J. Pl. Path.* **76**, 219–226.

Verhoeff, K. and Liem, J. I. (1975). Toxicity of tomatine to *Botrytis cinerea* in relation to latency. *Phytopath. Z.* **82**, 333–338.

Verhoeff, K. and Warren, J. M. (1972). In vitro and in vivo production of cell wall degrading enzymes by *Botrytis cinerea* from tomato. *Neth. J. Pl. Path.* **78**, 179–185.

Walker, J. R. L. (1962). Phenolic acids in cloud and normal tomato fruit wall tissue. *J. Sci. Fd Agric.* **12**, 362–367.

Wiant, J. S. (1945). *Mycosphaerella* black rot of cucurbits. *J. agric. Res.* **71**, 193–213.

8

Potatoes

C. LOGAN

I. Introduction

After harvest, potatoes are stored under various systems and environmental conditions for different periods of time, depending on whether they are required for human consumption, processing or seed. Few crops are now kept in outdoor clamps, the advantages and convenience of indoor storage in bulk, pallets or in bins having been realized increasingly by today's specialist potato grower. It is now possible to keep potatoes in store with correct management and sprout suppression for periods as long as 10–12 months without excessive weight or quality loss (Hesen, 1981). However, the storage life of a particular potato crop is determined prior to storage by a combination of factors that were prevailing during its growth and especially the interaction of the cultivar (early or main-crop) with fertilizer treatment, cultural practices, herbicide treatment, damage, weather conditions and its physiological state at harvest. Essentially the same factors influence the level of contamination and infection (progressive or latent) by the pathogens which cause post-harvest disease, and improper conditions of harvesting, transport and storage favour the development of epidemics.

Although Boyd (1972) stated that, for all practical purposes, infection by almost all potato storage diseases occurs prior to harvest, it is more correct to assume that at harvest the soil on the tuber surface is contaminated by bacterial and fungal pathogens and that infection and subsequent rot development are largely influenced by damage and by the storage environment. Potato tubers are susceptible to infection by all pathogens that attack the stem and roots of the growing plant, as they are simply the swollen tips of stolons which are underground lateral shoots (Cutter, 1978), and thus differ basically from seeds and fruits which are usually beset with disease problems different from those of the parent plant. A later chapter deals with bacterial diseases so this chapter will be confined to diseases caused by fungi.

In the British Isles the main fungal storage diseases of ware and seed potatoes are late blight, gangrene and dry rot. Although tuber blight is still a serious disease in certain years and has been responsible for large losses in the field and in store as recently as 1981, there has been a notable change in prominence of the true storage diseases. Dry rot has become of relatively little significance during the last three decades (Brenchley and Wilcox, 1979), whereas gangrene, first recognized in Scotland as a storage mycosis of potato tubers by Alcock and Foister (1936), has since spread to all parts of the British Isles and most of the world's temperate potato-growing regions. It was estimated that gangrene was present in 94% of seed stocks in England (Hirst *et al.*, 1970), and Logan (1967a) described it as being the commonest and most troublesome disease to the seed potato trade because it not only causes direct losses to the grower but may also lead to forfeiture of confidence in his seed if extensive gangrene develops after transfer to the customer's store.

Control of storage diseases has made considerable advances in recent years as the various host–parasite–environment interactions have been elucidated. The control measures fall into the main categories of sanitation, chemical treatment, resistant varieties, the propagation of disease-free basic seed stocks and storage management with adequate control of environmental conditions. Dumps of discarded potatoes have long been known to be a source of infection for blight (Boyd, 1973a), and recently it has been established that pycnidia of the gangrene pathogen form on moribund stems (Khan and Logan, 1968). Fungicidal spray programmes are now standard practice for protection against foliar blight and the disinfection of potato tubers has progressed from dipping them in highly toxic organo-mercury compounds (Boyd, 1960a) to misting them with the relatively safe thiabendazole fungicide (Logan *et al.*, 1975).

With regard to the exploitation of genetical variability, plant breeders, following the rapid appearance of new virulent races of *Phytophthora infestans*, have turned away from the use of major gene resistance to a search for field resistance controlled by a series or complex of minor genes. The use of stem cuttings to initiate basic seed stocks developed from the realization (Hirst *et al.*, 1970) that the major storage diseases are tuber-borne and that once a stock of potatoes becomes infected by any disease it is impossible to eliminate it. The importance of starting with absolutely disease-free stocks is self-evident. However, the relative importance of the various sources of recontamination has yet to be elucidated and a knowledge of the time and method of infection is essential, particularly in relation to chemical and storage management techniques which may involve considerable expense. It is clear that storage disease control requires the coordination and the integration of various practices to assure that the recommendations of one discipline do not nullify or contradict those of another.

II. Important Storage Diseases

The more important fungal diseases arranged in order of time of symptom appearance are:

(1) Late blight: *Phytophthora infestans* (Mont.) de Bary
(2) Pink rot: *Phytophthora erythroseptica* Pethybr.
(3) Rubbery rot: *Oospora lactis* (Fres.) Sacc.
(4) Watery wound rot: *Pythium ultimum* Trow
(5) Gangrene: *Phoma exigua* Desm. var. *foveata* (Foister) Boerema
(6) Dry rot: *Fusarium solani* var. *coeruleum* (Sacc.) Booth
(7) Skin spot: *Polyscytalum pustulans* (Owen and Wakefield)

The reader is referred to the booklet "Potato Diseases", published jointly by the National Institute of Botany, Cambridge, and the Potato Marketing Board, London, for clear, coloured photographs of these diseases.

A. Late Blight

The disease caused by *Phytophthora infestans* (Mont.) de Bary is usually referred to as late blight of potato to distinguish it from early blight (*Alternaria solani* Sorauer) which, although widely distributed in the British Isles, occurs sporadically and causes little damage (Moore, 1959). On the other hand, late blight is the most important disease of potato and, during wet, cool seasons, is responsible for very serious crop loss in the field and in the store; it is one of the few plant diseases that have played a major part in the social history of the British Isles.

This old, widely known disease, which first attracted attention when it devastated the potato fields in Europe and the U.S.A. during the early 40s of the last century, apparently reached the apex of its virulence in the years 1845–47 in Ireland, where the loss of the staple crop in two consecutive seasons brought famine and the emigration of over 1 million people by 1851 (Beaumont, 1959). It has thus been the subject of much research for well over a century but, despite the introduction of increasingly more efficient fungicides and spraying equipment and the breeding of resistant cultivars, control measures are still inadequate in seasons when the weather favours the spread and development of infection. Baker (1972) estimated that in the 20 years between 1950 and 1970, the incidence of blight was sufficiently prominent on a national scale for about half the years to be declared blight years, which, if taken with the recorded occurrence in Great Britain of some 70 physiological races of *P. infestans* (Malcolmson, 1969), serves to highlight the problems facing plant pathologists and plant breeders.

Losses due to blight occur in two distinct ways: (1) when the disease prematurely destroys all or part of the aerial parts of the plant and (2) when it affects the tubers. Further, tuber blight predisposes potatoes to infection by secondary organisms which rapidly invade the tissues that have been killed and subsequently cause wet rots to develop.

1. *The Causal Organism*

The pathogen *P. infestans* occurs on a number of solanaceous hosts apart from the potato, but the only other of importance is the tomato. It exists in the form of many physiological races for which a standard nomenclature has been adopted, based on the international system proposed by Black *et al.* (1953). In the British Isles race 0 was initially the most common, but in recent years it has been replaced by race 4, which was first recorded in 1947 coincident with the identification of genotype R_4 (Black, 1957). The fungus is not readily cultivated on a chemically defined medium (Hall, 1959; Scheepens and Fehrmann, 1978), but can be successfully grown axenically on an agar medium supplemented with various substances such as lima beans (Clinton, 1908) or peas (Keay, 1953), and can be produced in large quantities on autoclaved peas (Cruikshank, 1953).

2. *Disease Symptoms*

The earliest symptom of tuber blight is a slight brownish, irregular area which spreads under the skin of the tuber, initially producing a brown speckled, marbled discolouration. These patches become sunken as the tissues dry out and, when the tubers are cut, rusty brick-red markings are visible just below the skin and extend inwards for a variable distance as rusty-brown spots or streaks with a characteristic granular appearance.

Tubers with blight lesions frequently sprout in advance of healthy tubers, but certain other rots or mechanical damage may cause the same effect (Boyd, 1972). Under dry storage conditions cavities are formed internally and the tuber becomes shrunken, or may be invaded by fungi of the dry rot type, usually *Fusarium* spp.

Under wet conditions in store, secondary organisms, usually species of the bacteria *Pseudomonas*, *Erwinia* and *Bacillus*, gain entrance readily into blighted tubers and cause a wet rot to develop, turning the diseased tuber into a wet, slimy mess. The blight fungus does not usually spread from tuber to tuber during storage (Murphy, 1921), but bacteria may continue to spread, producing pockets of soft rot in the potato pile. However, whether or not other fungi or bacteria participate in the production of either a dry or a wet rot, *P. infestans* can usually be identified as the original invader by the rusty-brown, granular tissue which tends to persist even in thoroughly decayed tuber tissue (McKay, 1955).

3. *Inoculum and Infection*

The inoculum for tuber infection comes directly from infected haulm, but may also occur as a result of the fungus sporulating underground on the eyes and lenticels of blighted tubers, particularly in the tubers of Up-to-Date and King Edward (Lapwood, 1966) and of Home Guard (C. Logan, 1981, unpublished). Lacey (1967) found that, when the soil water content was greater than 20% and sporulation on the inoculated tubers was maximal, young developing tubers of Ulster Ensign and King Edward, inoculated with *P. infestans*, infected healthy tubers up to 13 mm away but similarly inoculated Up-to-Date and Majestic tubers failed to infect healthy tubers.

The incidence of tuber infection is influenced greatly by rainfall and may often be independent of the severity of haulm infection, though it is encouraged by irrigation (Lapwood, 1965). Lapwood (1964, 1977) observed that cultivar growth habit and stolon length also affect the chances of tubers becoming infected and Lacey (1966) found that infection tends to occur at the apical end of shallow set tubers, whereas in deeper set ones stolon-end infection is more prevalent.

Although it is generally accepted that tuber infection commonly occurs when sporangia produced on infected haulms are washed down by rain, there is no surer way of getting heavy infection than by harvesting when the haulm is green and partially blighted. In these circumstances tubers become infected through contact with the foliage or from sporangia shaken or blown onto the soil, the disease developing during storage usually within 3 weeks of harvesting (McKay, 1955).

4. *Predisposing Factors*

Resistance to blight in tubers is reputed to increase with maturity, but this was not confirmed by Malcolmson (1981) who found that maturity had no influence on the incidence of blight recorded following tuber inoculation with suspensions of sporangia. She found, however, that inoculation on the day of harvest resulted in 36% of the tubers becoming infected whereas only 6% were blighted when inoculation was delayed by 1 week.

The physiological basis for tissue resistance of tubers has been the object of extensive study. It is generally acknowledged that tissue carrying R genes conferring resistance to *P. infestans* responds hypersensitively when inoculated with incompatible races of the pathogen. However, Varns *et al.* (1971) found that the tissues of all cultivars tested, even those without R genes, reacted hypersensitively when treated with cell wall components of the fungus. The hypersensitive reaction involves the rapid death of a limited number of host cells, tissue browning and the accumulation of terpenoid compounds, particularly rishitin and phytuberin (Varns *et al.*, 1971). In the susceptible reaction caused after infection by compatible races of *P. infestans*, the fungus develops in the host tissue for at least 3 days without causing cell collapse (Tomiyama, 1967) or significant terpenoid accumulation, presumably due to the suppression of the hypersensitive reaction. The latter hypothesis has been substantiated by Garas *et al.* (1979), who found that a water-soluble fraction from the hyphal walls of a compatible race of *P. infestans* suppressed the hypersensitive reaction in potato tissue either inoculated with an incompatible race of the fungus or treated with elicitors of terpenoid accumulation.

Recently, Doke and Tomiyama (1980) summarized the knowledge on this host–pathogen interaction by proposing the following concepts: (1) elicitation of the hypersensitive response of protoplasts by hyphal wall components of *P. infestans* may involve an initial reaction triggering a series of resistance reactions including the accumulation of phytoalexins; (2) the reactivity of tissue cells to hyphal wall components may reveal the intensity of resistance to each type of host; (3) specific suppression of the reactivity of host cell to hyphal wall components by glucan from compatible races may contribute to the establishment of compatible interactions as well as host–parasite specificity.

5. *Control*

Measures to control blight by sanitary and chemical methods have recently been covered adequately by Brenchley and Wilcox (1979) and therefore need only to be brought up to date. Despite their contention that the chances of avoiding blight by the choice of a resistant variety are limited, breeding has moved to field resistance (polygenic or non-specific resistance) following the cessation of the use of R genes mainly from *Solanum demissum*. Holden (1977) concluded that the levels of resistance in both foliage and tubers of clones emerging from the Scottish Plant Breeding Station were very encouraging and cultivars capable of surviving blight years without spraying seemed to be a distinct prospect. However, these cultivars have yet to pass the test of the permanence of field resistance and it is to be hoped that their effective life will be much longer than was the case with cultivars such as Pentland Dell, which carries major genes R_1, R_2 and R_3.

The performance of recently introduced systemic fungicides seems to parallel the breeding story. The mode of action of the well-tried protective fungicides such as dithiocarbamates, captafol and tin-based compounds is generally non-specific, interfering with many vital functions of fungi, whereas that of systemic fungicides appears to be more selective, affecting perhaps only one vital function. Such is the case with metalaxyl, which interferes primarily with nucleic acid synthesis (Kerkenaar, 1981) and which was first used commercially in 1978 in the Republic of Ireland under the trade name, Ridomil. Its reputation as an efficient blighticide grew rapidly, so that by 1980 it was used there extensively but, unfortunately, in that year, the Republic had its worst outbreak of blight for 10 years. Tests showed that a strain of *P. infestans* resistant to metalaxyl had developed and similar strains have since been isolated from crops sprayed with Fubol (10% w/w metalaxyl and 48% w/w mancozeb) in the U.K. (Cooke, 1981). The occurrence of metalaxyl-resistant strains of *P. infestans* has also been reported in Dutch potato crops in 1980 (Davidse *et al.*, 1981).

The future strategy for blight control should be based, as suggested by Fry (1977), on the integrated effects of polygenic resistance combined with improved forecasting facilities to provide more effective fungicide use, these being the factors which can reduce the rate of epidemic development. The rapid build-up of strains of *P. infestans* resistant to systemic fungicides may be reduced by applying the fungicides to crops only when necessary, by rotating fungicides with different modes of action within each season and by using mixtures of fungicides with different modes of action.

B. Pink Rot

This disease was first recorded by Pethybridge (1913) in the west of Ireland, where it was called "water rot" or "water slain". A few years later it was observed in Scotland in 1919 (Cotton, 1921) and in England in 1921 (Cotton, 1922). It is now world-wide in distribution, occurring in several European countries, Australia, Canada and the United States of America (O'Brien and Rich, 1976). In the British Isles, although its incidence varies from one year to the next, past records show that pink rot has been particularly in evidence in years when late blight has been absent (Baker, 1972) and the reverse was also true in 1981 in Northern Ireland when pink rot was virtually absent. It is a disease that can often result in serious losses in particular crops, but is so markedly seasonal and so often very local that its general importance in England is comparatively small (Brenchley and Wilcox, 1979). However, in Scotland, Boyd (1973b) considered that the disease may have become more common and the factors influencing its development were assessed by Lennard (1980). It is a disease that is usually observed at harvest, may be an important rot in storage, but is rarely found on the market.

1. *The Causal Organism*

The main cause of pink rot is *Phytophthora erythroseptica* Pethybr., but other species of *Phytophthora* have been recorded as causing similar symptoms—*P. megasperma* Dreschl., under natural conditions in the British Isles, and *P. cactorum* (Leb. and Cohn) Schroet. and *P. cryptogea* Pethybr. and Laff, experimentally (Boyd, 1972).

2. *Symptoms*

Affected tubers are soft and rubbery, exuding juice when pressed and may often be confused with tubers exposed to frost or prolonged chilling. These symptoms usually occur first at the stolon end, the skin over the diseased part becoming discoloured and darkened around the lenticels. In coloured tubers the first symptom is a fading of the colour, whereas in white tubers the affected area turns brown. On cutting a partly affected tuber, the diseased area is found to be watery in texture and slightly off-white, but on exposure to air it turns pink; after about half an hour the colour intensifies and eventually becomes purplish-brown or black due to the oxidation of phenolic compounds by tyrosinase (White, 1946). The pink discolouration is, however, not completely diagnostic of pink rot, as it occasionally occurs in tubers soft-rotted by *Erwinia carotovora* Jones, and almost invariably in rubbery rot caused by *Geotrichum candidum* Link ex Pers. (*Oospora lactis*).

The rot develops rapidly under high temperatures (15–20°C) and moist conditions in the store, tubers becoming mummified unless invaded by secondary bacteria, when typical soft rot develops. Thus, poor ventilation in storage usually leads to the development of pockets of wet, rotting tubers in the potato pile. The disease spreads into new tubers from the stem through the stolon, or the pathogen may penetrate directly through eyes or lenticels, especially in very wet soils (Cairns and Muskett, 1933). As the plant tissues decay, oospores of *P. erythroseptica* are produced abundantly in stem bases, stolons and roots, as well as occasionally under the skin of rotted tubers (Pethybridge, 1913; Cairns and Muskett, 1939). Tubers only slightly affected when lifted will rot in storage, the disease spreading from tuber to tuber, causing heavy loss under conditions of high humidity and poor aeration.

Cunliffe *et al.* (1977) found that during storage at 9°C under high humidity *P. erythroseptica* grew from rotting, wound-inoculated tubers to the surface tissues of adjacent tubers in which oospores were formed, and reported that these oospores may lead to pink rot in daughter tubers in crops grown in previously uninfected land. It is, of course, possible that the disease is also spread in contaminated manure (Cotton, 1922; Blodgett, 1945).

3. *Predisposing Factors*

Lennard (1980) found in the Lothian region of Scotland that earlier lifting usually resulted in the development of more pink rot in storage because tubers carrying inoculum or latent infection could not be detected and discarded. He also observed that application of fertilizer in the drill tended to give more disease than broadcast application and suggested the possibility of chemical damage to roots and stolons facilitating entry of the pathogen. He suggested that dry seasons may aggravate this effect and confirmed the previous findings of Van Haeringen (1938) and Boyd (1960b) that pink rot is most frequently observed in the largest tubers.

Infection may also occur through damage caused at harvest, but rot development requires warm, moist conditions which are usually provided by the rise of temperature during early storage. Pink rot also develops very rapidly when aeration of the store is poor, causing the temperature and humidity to rise, and under these conditions Cairns and Muskett (1933) recorded up to 50% loss.

No cultivars are immune from pink rot, but some, such as Arran Pilot, Home Guard, Arran Victory and King Edward, are more susceptible than Record or Stormont Enterprise (Lennard, 1980).

4. *Control*

Inoculum i   soil should be reduced by a combination of collecting haulm debr after harvest and burning immediately. A 4 or 5 year rotation and good, deep drainage, especially of known wet patches or fields, should also help to reduce the incidence of infection. Finally, tubers from crops known to be affected with pink rot should be ventilated in the store to cause diseased tubers to mummify and thus reduce the chances of pockets of soft rot developing.

It is clear that more information on the epidemiology of pink rot is required to elucidate the importance of climatic and soil factors in affecting disease incidence. However, as long as the disease continues at its present low level it is doubtful if the cost of such research could be justified.

C. Rubbery Rot

This disease was first reported in the British Isles in 1948 when it was found in clamps and in the field in the East Midlands, Essex and Glamorganshire (Moore, 1959). It was observed in the 1957 potato crop in Yorkshire and Lancashire (Baker, 1972) and later in 1968 in an early ware crop in Pembrokeshire (Humphreys-Jones, 1969). The disease is associated with the fungus *Oospora lactis* (Fres.) Sacc. (syn. *Geotrichum candidum* Link ex Pers.), which has also been reported as the cause of a watery breakdown in ripe tomato fruit (Baker, 1972). During the past decade the disease has been observed in stored potatoes in Northern Ireland, affecting mainly the early cultivar Ulster Sceptre and more recently Estima; its incidence is increasing.

1. *Symptoms*

These may be noticed at harvest or after a time in storage and are somewhat similar to those of pink rot or frost damage. Restricted lesions appear as irregular brown patches with clearly visible black margins. Usually, the whole tuber is affected, the tuber tissues becoming rubbery in texture, leaking water if the skin is broken and exuding a distinct fishy odour. The small, greyish pustules of *O. lactis* that grow out from lenticels, and scab pustules or breaks in the skin, especially after tubers are incubated at 20°C for 48 h, are almost completely diagnostic (Humphreys-Jones, 1969).

    If affected tubers are cut open, the flesh turns a dirty pink colour in 3–4 h, later turning black. Affected tubers eventually become invaded by secondary soft rot bacteria.

2. *Predisposing Factors*

Much has yet to be elucidated concerning the aetiology and epidemiology of rubbery rot. Previous outbreaks (Humphreys-Jones, 1969) occurred following exceptionally warm weather and were associated with either heavy irrigation of crops within 3 weeks of harvest or with panned, naturally wet, or poorly drained land. In Northern Ireland outbreaks have occurred in crops from land that has been waterlogged for several days shortly before harvest. Preliminary observations (C. Logan, unpublished) have substantiated that the tuber skin must be immature for infection to occur unless it has been damaged and that the disease develops more rapidly at 90% relative humidity than in completely saturated conditions.

3. *Control*

Measures are similar to those recommended for pink rot, with the omission of the destruction of haulm debris as, so far, there is no evidence that the pathogen attacks the haulm. The disease should be prevented from causing pockets of soft rot to develop in storage by adequate ventilation shortly after lifting.

D. Watery Wound Rot

Watery wound rot is a rapid rot of newly lifted potato tubers and is known very descriptively as "leak" or "shell rot" in the U.S.A. It may occur wherever potatoes are grown, but serious loss occurs only in certain years when potatoes are harvested early and moved during warm weather (Pethybridge and Smith, 1930; O'Brien and Rich, 1976). According to Pethybridge and Smith (1930), the disease was first described and studied in Germany in 1874, but nothing further was written about it until Hawkins (1916) described *Pythium debaryanum* as one of the causes of potato "leak", a disease of considerable importance in potatoes shipped from the delta region of the San Joaquin River, California. Since a serious outbreak in Lincolnshire and Cambridgeshire, in 1929, further reports of serious losses have been few, e.g. in Idaho, U.S.A. in 1944 (Blodgett and Ray, 1945), in Salop, in 1958, and in the West Midlands, in 1959 and 1961 (Baker, 1972). Boyd (1972) suggested that in the U.K. watery wound rot seldom presents a serious storage problem because temperatures at the time of normal harvest are not sufficiently high. The disease has also been reported in two localities in East Germany (Pett and Hahn, 1974) and in Egypt as a new disease (El-Helaly *et al.*, 1971).

Watery wound rot differs in certain important respects from more common rots such as those caused by *Phytophthora erythroseptica* (pink rot), *Phytophthora infestans* (blight) and *Erwinia carotovora* var. *atroseptica* (black leg). It is not found in the tubers while they are still attached to the plant, although it does become evident shortly after lifting. Also the rot does not start, as pink rot and black leg rots usually do, from the heel or stolon-end of the tuber.

1. *The Causal Organism*

In the British Isles the main cause of watery wound rot is *Pythium ultimum* Trow (Boyd, 1972), although in the U.S.A. the closely allied

P. debaryanum is also considered responsible, as are other species of *Pythium* (O'Brien and Rich, 1976).

2. *Symptoms*

The rot begins at a wound on the tuber surface as a slight discolouration which rapidly becomes moist with a dark boundary line separating it from the healthy area. The most characteristic symptom is the rapid breakdown of the internal tissues into a greyish, soft decay bordered by a dark zone, leaving in most cases a shell of healthy vascular and cortical tissue. On exposure to air the grey, pulpy rot turns black. At high temperatures during the early part of storage, liquid drips from affected tubers in great quantities, so that later the diseased tissue dries up, forming cavities and finally producing a hollow tuber.

3. *Infection Cycle and Predisposing Factors*

Tubers become contaminated in the field, infection taking place at lifting through wounds or bruises which are often inconspicuous. Infection of tubers occurs during warm spells of weather and frequently occurs in tubers damaged by sunscald, especially if the tubers remain on warm soils after being dug (O'Brien and Rich, 1976). In the U.K. tubers are predisposed to infection if dug early during dry, warm weather when the tuber skin is not fully mature and is at considerable risk of abrasion and wounding by harvesting machinery. At temperatures above 16°C, rots become visible within 36 h (O'Brien and Rich, 1976), progressing rapidly under humid conditions (Brien, 1940), the mycelium invading the tissues by both intra- and intercellular hyphae (Pethybridge and Smith, 1930) so that a good-sized tuber may be completely rotted within 2–3 days.

Under natural conditions, no reproductive bodies have been found in diseased potato tissue, but oospores are readily formed in culture on various agar media and in the soil when the moisture content is high and when oogonial production can take place in water-filled pores (Bainsbridge, 1970).

4. *Control*

Losses from watery wound rot in storage or in transit can be kept to a minimum by ensuring that potatoes to be harvested early are mature and that care is taken to prevent damage during lifting and handling. Tubers should be kept as cool and as dry as possible.

E. Gangrene

Since gangrene was first recognized, described and differentiated from dry rot by Alcock and Foister (1936), it has become the major storage disease of seed potatoes in the British Isles. This was not the first description of a *Phoma* potato rot, as Prillieux and Delacroix (1890) had described a similar condition in France which they attributed to *P. solanicola* (Prill. and Delac.) sp. nov., and Melhus (1914) reported a storage rot of Irish potatoes in Maine, U.S.A. which closely resembled Alcock and Foister's description of gangrene, the causal organism being named *P. tuberosa* sp. nov. (Melhus *et al.*, 1916).

Although gangrene was not recognized in its own right until 1936, there are reports of *Phoma* rots on Scottish seed potatoes for planting in Yorkshire and Caernarvonshire in 1923, Berkshire in 1924, Lancashire in 1928 and in Lincolnshire in 1930 (Pethybridge, 1926; Pethybridge *et al.*, 1934). Moore (1948) recorded that the disease had spread throughout the British Isles during the 1940s, but the first record of it affecting English-grown seed is in early 1960 in Salop (Baker, 1972). It is now recognized world-wide, occurring in Australia (Harrison, 1959), Germany (Kranz, 1958), the Netherlands (Boerema, 1967), Russia (Nikolaeva, 1970) and Sweden (Bang, 1972). Seed potato exports, especially, are economically sensitive to gangrene as symptom expression occurs long after infection and thus tubers which are apparently healthy on despatch may arrive in a rotting condition.

1. *The Causal Organism*

When Foister (1940) described the gangrene pathogen, he named it *Phoma foveata* Foister sp. nov. and stated that it was morphologically indistinguishable from *P. solanicola* and *P. tuberosa*, but different in its growth characteristics in culture. Both Dennis (1946) and Malcolmson (1958a) examined the morphological, cultural and pathological characteristics of these *Phoma* species, and whereas Dennis divided them into two groups, with *P. foveata* the sole member of one group, Malcolmson argued that as there was no decisive distinguishing morphological characteristic they were all special forms of the same species and that they should be called by the first established name, *P. solanicola*. She found that, of some 400 isolates from tuber and stem rots from Britain and the U.S.A., those which were pathogenic when inoculated into potato tubers were of two types, irrespective of their origin. One type, like the *P. foveata* described by Foister (1940), was characterized by the production of a brown, diffusible pigment in the culture medium and was named *P. solanicola* f. *foveata* (Malcolmson,

1958b). The other type, named *P. solanicola*, was morphologically identical but did not pigment the medium.

Maas (1965), who worked with the flax root-rot fungus *Aschochyta linicola*, advocated its inclusion with *P. solanicola* into the older species *P. exigua* Desm. and described two varieties, *P. exigua* var. *linicola* (syn. *A. linicola*) and *P. exigua* var. *exigua* (syn. *P. solanicola*). Subsequently, Boerema (1969) considered that the pathogenicity and cultural differences were sufficient to distinguish between the anthraquinone pigment and non-pigment producing isolates by more than the term *forma specialis* and classed them as varieties. Thus the generally accepted name of the causal organism of gangrene is now *Phoma exigua* Desm. var. *foveata* (Foister) Boerema.

Further work has shown that var. *foveata* is by far the most prevalent gangrene pathogen (Logan, 1967a; Todd and Adam, 1967; Griffith, 1970), being responsible for about 97% of the gangrene in the U.K. (Hide, 1981). Boerema and Howeler (1967) and Logan and O'Neill (1970) further subdivided var. *foveata* into strains on the basis of the production of substance "E", an antibiotic and fungicidal compound. Strains that do not produce this chemical produce higher concentrations of anthraquinone pigments and more aerial hyphae than "E"-positive strains. Pathogenicity differences were found to exist between "E"-positive and "E"-negative strains (Logan and Woodward, 1971) in the colour and type of rot produced, and this was confirmed by Wastie and Stewart (1977). The most common isolate from gangrene rots in the U.K. is now the "E"-negative strain.

Compared with var. *foveata*, var. *exigua* is a more ubiquitous soil-borne fungus which may be isolated from the moribund stems of many herbaceous plants (Todd and Adam, 1967; Paulson and Schoeneweiss, 1971) and which may occasionally cause gangrene. Logan and Khan (1969) found that isolates of var. *exigua* were morphologically similar to those of var. *foveata* in Northern Ireland, but differed markedly in appearance in culture, in maximal and optimal temperatures for spore germination and growth in culture, and in degree of pathogenicity to potato tubers; isolates of var. *foveata* caused more extensive tuber rotting and produced gangrene symptoms in unwounded tubers. However, Walker and Wade (1976) state that in Tasmania about 90% of gangrene rots are caused by var. *exigua*.

2. Symptoms of the Disease

The initial symptoms are small, rounded, black or dark brown depressions in the skin situated around a lenticel, eye or in relation to some form of injury which may often be too slight to be detected. The rot may remain small, often reaching only the size of a thumb print, but in susceptible cultivars, e.g. Ulster Sceptre, may extend to over 40 mm in diameter. The size of the external lesion gives little indication of the severity of internal rotting which, depending on cultivar, age and storage conditions of the tuber, may remain shallow, the affected tissue being easily peeled from the healthy tissue, or may extend deep into the centre of the tuber, forming cavities.

It is characteristic of the disease that the necrotic tissue is differentiated from the healthy tissue by a definite margin, black or dark brown in colour. Cavities are lined with greyish mycelium overlying a salmon-pink or dark purple necrotic tissue in which black ostiolate pycnidia are embedded. Pycnidia may also be found embedded in the necrotic skin on the surface of a lesion. According to Malcolmson and Gray (1968a), deep lesions tend to be associated with warm storage conditions, and shallow lesions, particularly of the skin necrosis type, with low temperatures. The condition called "skin necrosis" is the reaction of certain cultivars, particularly Arran Banner, to var. *foveata* or var. *exigua*, where completely superficial, extensive, dark and irregularly shaped lesions spread over the surface of the tuber, usually under conditions of high atmospheric humidity.

3. *Inoculum and Infection*

Gangrene is both a soil- and tuber-borne disease. The pathogen can remain infective in soil over winter, and potato crops, planted in land in which a contaminated crop was grown in the previous year, can become infected (Logan, 1970a). It has also been found to survive in the soil in the absence of a potato crop for at least 5 years in Northern Ireland (Khan and Logan, 1968) and for 7 years in Scotland, where the distribution of the disease is related to soil moisture (Malcolmson and Gray, 1968b), being more prevalent where the moisture content remains high throughout the growing season. However, in English soils Adams (1979) found that the survival of the pathogen was prolonged by low temperatures in dry soils where the water-holding capacity was between 20 and 50%. He concluded that soil is unlikely to be an important reservoir of inoculum for potato crops, except in the production of high-grade seed stocks, where even small populations could result in introducing the disease.

In contrast to var. *exigua*, which colonizes many herbaceous plants, var. *foveata* was found only in association with potato plants (Todd and Adam, 1967). However, Fox and Dashwood (1972) reported the presence of var. *foveata* in a number of self-sown weed seedlings growing in contaminated soil, as well as on several species of weeds growing in an infected potato crop. They later found that barley seedlings may be symptomlessly affected

(Fox and Dashwood, 1978). Thus, if the host range of var. *foveata* is more diverse than was originally thought, this could explain its survival in soil, in the absence of a potato crop, for at least 7 years.

There is no doubt about the importance of tuber-borne infection, which, either as contaminated (lesion-free) or diseased (gangrene-rotted) tubers, increases the amount of inoculum on progeny tubers, seed with rots being normally, but not invariably, a better source of inoculum than contaminated seed (Khan and Logan, 1968; Griffith, 1969; Logan, 1974).

Within 3 weeks of planting, gangrene spreads either from infested soil or the diseased parent tuber into the base of stems, colonizing them without symptom expression (Todd and Adam, 1967; Khan and Logan, 1968), and remains latent during the active growth of the plant. Its distribution within the plant appears to be discontinuous and, although difficult to isolate from stem tissue, seems to be more easily isolated from leaves (Bannon, 1978; Croke, 1980). Little is known of the structure of var. *foveata* during its period of symptomless colonization of leaves and stems, though Fox *et al.* (1970) stated that its mycelium can spread inter- and intracellularly and that its pycnidiospores can be transported upwards as far as 60 cm in 24 h in actively transpiring stems.

The pycnidiospores from stem pycnidia are an important source of inoculum for infection of progeny tubers (Logan, 1967b, 1970b) and for spread within and between potato crops. Logan (1976) concluded that rain splash was the main method of spread of var. *foveata* within the crop, as the most effective spread occurred within a 1.2 m radius of a central focus of infection. Entwistle (1972) found evidence of aerial spread of the pathogen using filter-paper bait strips placed at various sites and heights within and outside a potato crop. This has recently been confirmed by Carnegie (1980), who consistently recovered var. *foveata* from the open air during rainfall at distances of at least 800 m down-wind of the nearest potato crop, thus supporting the general conclusion of Faulkner and Colhoun (1976) that rainfall is an important generator of aerosols of sphaeropsidaceous fungi.

However, the parent seed tuber may also provide a direct source of inoculum underground, as Todd and Adam (1967), Malcolmson and Gray (1968b) and Fox and Dashwood (1969), all noted a progressive increase in the contamination of tubers before the haulm had become senescent. More recently, Adams (1980) has shown that the inoculum on seed tubers, whether from rots or surface contamination, contributed more to the contamination of progeny tubers at harvest than did the inoculum from pycnidia on stems following desiccation of the haulm. His results substantiated previous observations by Copeland and Logan (1976) that the potential of tubers to develop gangrene in storage was often lowest when tubers were harvested on the day of haulm destruction and almost invariably increased as the interval between haulm destruction and harvest was increased. There was also agreement that, although within years for any cultivar the incidence of var. *foveata* pycnidia was an indicator of relative gangrene potential of progeny tubers, stem infection could not be used generally to predict likely disease levels in store, which are mainly influenced by a complex series of factors which interact during the growing season.

Although tubers become contaminated by var. *foveata* some time before harvest, it is only occasionally that gangrene rots are visible at the time of lifting, usually around eyes or proliferating lenticels (Malcolmson, 1958b). It is generally agreed that when tubers arrive in store, the inoculum is either on the surface or in the epidermal tissues of the tuber and that subsequent rot development is largely influenced by damage and by the storage environment (Boyd, 1972). Although there is some evidence of tuber to tuber spread in store, especially under damp conditions, this is of relatively little importance, at any rate in chitting houses (Brenchley and Wilcox, 1979).

4. *Predisposing Factors*

Virtually all gangrene in commercial stocks of potatoes is associated with damage that occurs at lifting, riddling or during handling, and thus disease development may not reflect accurately the gangrene potential of any particular stock (Boyd, 1972). Recently, Adams (1980) found that inoculum potential and incidence of severe damage both influenced disease development, but that damage was of greater importance.

The nature of the wound is important in determining the predisposition to infection. Griffith (1970) and Henriksen (1976) maintained that ragged crush wounds or bruised tissue with a rupture in the periderm are more likely to cause gangrene development than abrasions of the skin or clean cuts. According to Pietkiewicz and Jellis (1975), shallow wounds in the cortex are less likely to give rise to gangrene lesions than deeper wounds in the medulla, but, in practice, gangrene lesions are by no means always associated with obvious damage and may originate in very small cracks, especially if caused when tubers are bruised. Gangrene incidence is increased by potato harvesting machinery likely to increase tuber damage (Malcolmson and Gray, 1968a), but more important is the damage caused to the tuber surface during riddling, when the knocks and repeated bumping of tubers may also activate the fungus from its latent state (Griffith, 1969; Logan, 1969).

In association with damage, temperature is the other most important factor in the development of gangrene. Because of the inhibition of wound periderm formation at low temperatures (Artschwager, 1927), the healing process in tubers does not function and thus exposure to cold conditions after damage increases the incidence of the disease (Kranz, 1958). Thus, the influence of temperature on tuber infection and development of gangrene is largely mediated through its effect on tuber resistance, because the alternative, i.e. mediation through germination and growth of the pathogen, has been shown not to be possible. The optimum and maximum temperatures for *in vitro* germination and growth of var. *foveata* are 22 and 24°C, respectively, much higher than either the optimum (5°C) or the maximum (20°C) temperatures for the development of gangrene (Logan and Khan, 1969).

Because the tuber responses that are involved in resistance are temperature dependent, attempts have been made to relate phenolic compound and phytoalexin production to gangrene development. Gans (1978) found a relationship between the resistance of cultivars and their ability to accumulate high concentrations of chlorogenic acid and to develop highly fluorescent cell walls adjacent to wounded tissue. Harris and Dennis (1976) tested the activity of the phytoalexins phytuberin, rishitin, anhydro-β-rotunol and solavetivone against var. *foveata* and other fungi and found that all compounds, except phytuberin, showed anti-fungal activity against the majority of the fungi (potato pathogens and non-pathogens) when tested at $100\,\mathrm{g\,ml^{-1}}$. By increasing the concentrations of phytuberin and rishitin, Walker and Wade (1978) demonstrated that germ tube growth was inhibited, so that, if present in infected tissues, these compounds may have a role in containing fungal infection. However, although tissues of the resistant cultivar Arran Consul produce more phytuberin than those of the susceptible Ulster Sceptre, those of the latter produce more rishitin (Croke, 1980), so that the role of phytoalexins in tuber resistance requires further elucidation.

An almost completely saturated atmosphere is necessary for infection by var. *foveata* to occur through unwounded tuber tissue, especially through lenticels and eyes (Malcolmson, 1958b; Khan, 1967). In fact, the official U.K. National List tests of cultivar susceptibility to gangrene are carried out by exposing unwounded tubers, after dipping them in var. *foveata* inoculum, to high humidity conditions by storing them at 5°C in cardboard boxes lined with damp peat (Todd and Adam, 1967). After infection has occurred through wounds, the development of gangrene appears to be little affected by humidity (Malcolmson, 1958b; Khan, 1967). However, Khan (1967) found that humidity encourages rot development in the cortex and low humidity induces deeper penetration of the rot into the pith.

It is generally accepted that as the tuber ages in storage it becomes more susceptible to gangrene. However, during the growing season Khan (1967) and Fox *et al.* (1970) found that tuber susceptibility tends to be high in the early stages of development, decreasing towards maturity about September/October and increasing again during storage until February. Malcolmson and Gray (1968b) suggested that factors which retard the skin setting process as tubers mature influence the level of pre-lifting infection. Thus, although increasing rates of fertilizers reduce tissue susceptibility, the incidence of natural infection can be increased either by greater liability to mechanical damage or by delaying maturity of the tubers (Mackenzie, 1968).

5. *Control*

The control of gangrene during storage may be discussed under four headings: the use of gangrene-free tubers for planting, avoidance of damage, curing, and chemical disinfection.

Because of the tuber-borne nature of the disease, it was hoped that the practice of initiating stocks of seed tubers from stem cuttings (Hirst and Hide, 1967) would reduce the level of gangrene and maintain it at negligible levels in all certified seed potato stocks if associated with a chemical treatment (Graham and Hardie, 1971). However, Hide (1981) has shown in the King Edward and Pentland Crown stocks examined for planting in England in 1975 and 1976 that the average amount of diseased tubers was similar whether or not the stocks were derived from stem cuttings. He had found earlier (Hide, 1978) that on Scottish seed tubers derived from stem cuttings gangrene became the major cause of rots in fourth and fifth year stocks. More recently, Carnegie *et al.* (1981) found that contamination by both skin spot and gangrene pathogens was much greater on commercial VTSC (virus-tested stem cuttings) farms than at the Department of Agriculture for Scotland farm at Ingraston, near Edinburgh, and that both fungi were more prevalent on tubers after the third year of multiplication than on those of second year clones. Their results confirm those of Hide (1978) that an annual fungicide treatment is desirable if the health of VTSC tubers is to be maintained during propagation, and also highlight the problem of preventing the recontamination of VTSC stocks.

As mentioned in a previous section, damage is the main predisposing factor affecting the occurrence of gangrene infection but, despite the fact that this was originally emphasized by Alcock and Foister (1936), little seems to have been done by manufacturers to reduce mechanical damage caused by harvesters and graders. Twiss and Jones (1965) showed that almost one-third of potatoes in the national crop suffered severe mechanical damage before they left the farm. The Potato Marketing Board's "National Damage Survey—1973" showed that during harvesting and handling of the crop into store, 21% of tubers were badly damaged and a further 13% were affected by internal damage and bruising. It is necessary, therefore, to take all reasonable precautions to avoid damage at lifting, grading and handling. The risk of damage can be considerably reduced by paying attention to the setting of the machines, by padding dangerous parts and reducing the height of tuber drop to a minimum.

As the direct losses from mechanical damage and bruising are still of the order of 30% or more (Hampson *et al.*, 1980), growers should be prepared to reduce the additional risk of gangrene development by exposing tubers to a temperature of 15–16°C and about 80% relative humidity for 10–14 days to allow wounds to heal as quickly as possible (Malcolmson and Gray, 1968a). Hampson *et al.* (1980) found that a proper curing duration followed by temperature control (5–7°C) during the holding period can substantially reduce wastage losses in store. After grading, the period of curing for effective control of gangrene may be shortened to 7 days or less by raising the temperature to 20°C (Copeland and Logan, 1980). However, the effectiveness of curing may depend on the nature and depth of the wound (Griffith, 1969) and also on the possible additional presence of the dry rot fungus, so that too much reliance cannot be placed on this method as a means of effective control (Boyd, 1972).

One of the first chemicals to be used for tuber disease control in Scotland was tecnazene (1,2,4,5-tetrachloro-3-nitrobenzene), which controls dry rot and to a certain extent gangrene, but as it is a sprout suppressant it should not be used on seed potatoes. During the 1950s and the 1960s substantial quantities of seed tubers were disinfected by dipping in solutions of organo-mercury compounds, especially 2-methoxyethylmercury chloride (Boyd, 1960a), which provided good control of gangrene if carried out immediately after lifting (Boyd and Penna, 1967; Logan, 1967c). Because of the toxicity and persistence problems of mercury, alternatives were sought and 2-aminobutane was introduced commercially as a fumigant for gangrene and skin spot control (Graham *et al.*, 1973), and generally good control of both diseases has been achieved. The systemic fungicides thiabendazole and benomyl both as dip and as dust formulations also gave reasonable disease control if applied immediately after lifting (Hide *et al.*, 1969). The application of thiabendazole as a mist to the surface of tubers was developed commercially as an alternative method of disinfection (Logan *et al.*, 1975) and good control of gangrene, dry rot, skin spot and silver scurf is achieved during storage, particularly if the crop is treated at lifting (Boyd, 1977).

Graham *et al.* (1981) summarized the results of their experiments on the control of potato tuber fungal diseases, mainly with 2-aminobutane and thiabendazole, and concluded that in some cases, where fungicides did not control pathogens, failure was traced to there being too low a residue or poor distribution of the chemical over the tuber surface, suggesting faulty treatment or unsatisfactory methods of application.

6. *Cultivar Reaction to Gangrene*

The tubers of all commercial potato cultivars can develop gangrene when damaged and exposed to low temperature. However, large differences exist between cultivars in both the ease with which tubers become infected and the extent of the resultant rot. For instance, Home Guard is relatively susceptible to infection through wounds and especially through eyes, but the resultant rots are small compared with those that develop in tubers of Ulster Sceptre.

Commercial cultivars available in the U.K. may be classified into four groups using a combination of the results from laboratory experiments and practical experience:

Group I   Very susceptible   Blanka, Gracia, Ulster Premier, Ulster Sceptre, Vanessa

Group II   Susceptible   Cara, Estima, Arran Peak, Craigs Royal, Désirée, King Edward, Majestic, Ulster Chieftain, Ulster Prince, Pentland Crown, Pentland Dell, Pentland Squire

Group III   Less susceptible   Arran Banner, Arran Pilot, Dunbar Standard, Home Guard, Kerrs Pink, Maris Peer, Pentland Javelin, Pentland Meteor, Wilja

Group IV   Fairly resistant   Arran Consul, Golden Wonder, Maris Piper

F. Dry Rot

Dry rot is a common disease of the potato tuber that develops only after lifting and was, during the first half of this century, one of the most important storage diseases of potato in the British Isles. It is now world-wide in distribution. Pethybridge and Bowers (1908) first identified this disease in Ireland, although according to Moore (1959) it was probably known in England long before that time. The losses caused by this disease have been second only to those caused by late blight in the British Isles but, whereas blight develops within a month of harvest, dry rot does not usually appear until December and becomes progressively prevalent as winter advances (McKay, 1955). Dry rot is caused by infection of the tubers with soil-borne *Fusarium* species, the species responsible varying with country and locality and, although the disease has declined considerably in the British Isles since 1950, *Fusarium* tuber rots still cause, particularly in late cultivars, greater storage and transit losses in the U.S.A. than any other post-harvest disease of potatoes (Smith and Wilson, 1978).

The reason for *Phoma exigua* var. *foveata* becoming the dominant storage rot organism in the British Isles at the expense of the dry rot pathogens has not yet been fully explained. It is probable that the use of increased rates of fertilizers as suggested by Boyd (1972), the demise of the dry rot susceptible cultivars Catriona, Arran Pilot and Doon Star coincident with the introduction of the gangrene-susceptible cultivars Ulster Sceptre, Ulster Premier and Pentland Crown and the marked competitiveness of *P. exigua* var. *foveata* on the tuber surface (C. Logan, unpublished) have all interacted to the disadvantage of the *Fusarium* pathogens.

1. *The Causal Organisms and Symptoms*

There are three main species of *Fusarium* associated with potato dry rot and the symptom expression varies with each. The most common cause of dry rot in Europe and the northern potato-growing areas of the U.S.A. is *F. solani* var. *coeruleum* (Sacc.) Booth = *F. coeruleum* (Lib.) Sacc., which causes dark brown external lesions, usually without a distinct edge. As the tuber loses water, the affected area shrinks and the skin develops a series of concentric rings or wrinkles. Internally, the affected tissue is fawn or light brown, sometimes with dark streaks, and the margin of the rot is diffuse, usually merging with the healthy tissue. Cavities form within the rotting tissue as it continues to lose water and these become lined with greyish-white mycelium. Sporodochial pustules, which frequently form on the surface of the lesion, are white or pale blue if light is excluded but develop a pinkish tinge when exposed to light (Boyd, 1972). When there is less desiccation of the affected tissue, the rot progresses more rapidly (Moore, 1945) and secondary bacteria hasten the decay of the tuber, often producing wet rots.

Fusarium sulphureum Schlecht. (*F. sambucinum* Fckl f. 6 Wr.) is the main cause of the disease in the northern and central states of Texas and Idaho (Smith and Wilson, 1978), but it has been recorded only occasionally in Scotland (Boyd and Tickle, 1972) and in Northern Ireland (C. Logan, unpublished). The rot caused by *F. sulphureum* is externally and internally very similar to that caused by gangrene (*P. exigua* var. *foveata*); it is a dry, rather mealy, brown rot with a well-defined edge and extensive cavitation of the tissues (Boyd, 1972). As *F. sulphureum* is general in North America and as its temperature range for rot development is 5–20°C, even wider than that of gangrene (C. Logan, unpublished), this fungus is worthy of note as a potentially serious potato storage pathogen in the U.K.

The third species of *Fusarium* that is sometimes involved in dry rot is *F. avenaceum* (Corda) Fr. Sacc., which causes external symptoms similar to those of var. *coeruleum*, but internally the rotted tissue is darker in colour, often black, with a distinct margin between it and the healthy tissue. Moore (1945) compared the role of *F. avenaceum* and var. *coeruleum* in causing wastage in stored tubers, and McKee (1952) showed that *F. anthrosporoides* and *F. tricinctum* may also be involved occasionally, though 90% of dry rot in the U.K. is caused by var. *coeruleum*. Later, McKee (1954) found that with var. *coeruleum* the mycelium is intercellular and the adjacent host cells remain alive for some time, whereas with *F. avenaceum* the mycelium kills and penetrates the cells as it advances; lesion restriction was associated with suberin depositions on the host cell walls and in the intercellular spaces.

Besides causing a rot in storage, *F. sulphureum* and var. *coeruleum* are also responsible for "seed-piece decay" following the cutting of tubers into pieces for planting, which is widely practised in North America and Mediterranean countries. The planting of tubers affected with dry rot or seed-piece decay results in poor stands and subsequent reductions in yield (O'Brien and Rich, 1976).

2. *Inoculum and Infection*

Fusarium solani var. *coeruleum* is present in the soil both in the field and on the tuber surface and damage to the tuber surface brings contaminated soil in contact with the exposed tissues (Small, 1944a; Foister *et al.*, 1945a). Small (1944b) also showed that the pathogen may be viable in soils having a wide range of pH and in fields in which potatoes have not been grown for 5–6 years; other sources of infection are used sacks, seed boxes, diseased tubers and knives used for cutting seed potatoes. Damage may occur at harvesting, grading and during handling, and Foister *et al.* (1952) showed that grading by the commonly used reciprocating riddles, often fitted with bare wire screens, causes more serious damage and subsequently more dry rot than that caused by harvesting or transport or any other form of handling.

After riddling, seed potatoes are bagged and transported, but it may not be for weeks or even months later, depending on environmental conditions, that dry rot symptoms appear. Thus, it is primarily a latent disease of seed rather than ware, the latter being normally utilized immediately, and therefore either re-riddling of seed stocks or re-storing of ware potatoes is potentially to encourage heavy losses (Boyd, 1972).

The planting of seed affected with dry rot, besides resulting in blanking and thus gappy crops (Small, 1946), leads to increased contamination of the soil as assessed at harvest, and also of the soil on the surface of the progeny tubers (Ayers and Robinson, 1956). Thus Boyd (1972) found the disease to be perpetuated mainly by planting diseased or contaminated seed, and he considered tuber-borne infection to be the more important means of carry-over, little being known of the pathogen's activity in soil. However, studies of the epidemiology of var. *coeruleum* and *F. sulphureum* have been carried out by Jeffries and Boyd (1977), using a selective medium (Jeffries, 1978) which enables enumeration of fungal propagules in soil. They found that the application of thiabendazole (2% active ingredient) to the parent tuber consistently reduced the transmission of var. *coeruleum* but not of *F. sulphureum* to the progeny tubers.

3. *Predisposing Factors*

As already detailed above, the main factor that predisposes potato tubers to dry rot infection is damage to the skin. However, several other factors are also important in increasing the incidence of the disease, such as cultivar reaction to the disease, tuber maturity, nutrient availability and environmental conditions.

Among the first to recognize differences in cultivar susceptibility to dry rot were Pethybridge and Bowers (1908), who noted heavy losses in the early cultivar Snowdrop in Ireland. Later, Boyd (1952c) distinguished two factors in the assessment of field susceptibility of a large number of British cultivars: the physiological resistance of the tissue to infection and the mechanical resistance of the periderm to wounding. The cultivars Arran Banner and Arran Victory were the most resistant whereas Catriona and Dunbar Standard were the most susceptible and these are still the standard cultivars used today in the U.K. National List dry rot tests. Moore (1945) suggested that cultivar reaction to various *Fusarium* spp. varies, Doon Star being more susceptible to var. *coeruleum* and King Edward to *F. avenaceum*. In the U.K. the disease is generally more common on early cultivars than on main crop cultivars. Ayers (1956) found amongst North American cultivars that Sebago, Keswick, Chippewa and Pontiac are very susceptible to *F. sambucinum*, whereas Cherokee, Irish Cobbler, Warba and others are resistant; Keswick, Pontiac and Warba are susceptible to var. *coeruleum*, whereas Cherokee, Green Mountain, Ontario and others are resistant.

Another important factor is the fluctuation of tuber resistance over the whole storage period with increasing susceptibility as tubers mature in storage (Pethybridge and Lafferty, 1917; Boyd, 1952b). Boyd (1952b) also found that a high peak of susceptibility to var. *coeruleum* occurs in immature tubers in the post-flowering period during growth, followed by a sharp decrease which reaches a minimum when the haulm is dead. Boyd (1967) maintained that susceptibility in immature tubers is closely related to sucrose content, which increases temporarily in tubers because of rapid translocation from the vigorously growing haulm and that cases where dry rot develops in early harvested seed tubers, with little or no delay after haulm destruction, can be explained on this basis. However, he did not think that these factors are directly related during storage when an increase in the content of mainly reducing sugars is also accompanied by an increase in susceptibility.

Mechanical damage caused at the time of lifting does not usually result in excessive rotting in storage if the tubers are initially given a period of curing but, if damage is followed by low temperatures, rotting may become extensive, although it develops more slowly (Boyd, 1952d). However, the highest incidences of dry rot either follow wounding during grading (Foister et al., 1952), which is normally carried out between November and March while potatoes are gradually becoming more susceptible, or when grading is done shortly before planting, which is then delayed or followed by cold, dry conditions in the soil.

McKee (1954) wounded tubers at intervals throughout storage and found that resistance to infection by F. avenaceum is maintained for a longer period than resistance to var. coeruleum, although eventually tubers become equally susceptible to both fungi. He also showed that resistance to infection increases with increasing time between wounding and inoculation and suggested that this was not simply due to suberization in the wound healing process, but to resistance of the cells near the wound, because lesions resulting from delayed inoculations are smaller than those developing on fresh wounds. He concluded that the initial response of the tuber tissue to infection by Fusarium spp. is similar to that due to Phytophthora infestans, because Muller and Borger (1941) had shown a phytoalexin formed after inoculation with an avirulent strain of P. infestans also restricted growth of var. coeruleum. Despite several investigations involving, for example, chlorogenic acid (Bate-Smith, 1956), solanin (McKee, 1961), sugar content (Boyd, 1967), osmotic pressure (Radtke, 1969), orthodihydric phenols (Griffin, 1964) and phytoalexins (Harris and Dennis, 1976), the nature of the defence mechanism against Fusarium infection is not yet clear.

Boyd (1967) showed that a balanced fertilizer (12:12:18 NPK) applied at 753 kg ha^{-1} significantly reduced the liability to infection in the subsequent crop. However, additional nitrogen (73 kg ha^{-1} ammonium sulphate) raised tuber susceptibility to that in non-fertilized plots, and when sulphate of ammonia was the only treatment given, the tubers were significantly more susceptible than those receiving complete fertilizer. Boyd (1972) suggested that, with the exception of the effect of fertilizer regime, in several respects susceptibility to dry rot appears to be associated with higher dry matter content. For instance, the stolon end of a tuber is more susceptible than the rose end (Boyd, 1952a), and smooth-skinned Olympia tubers have a higher dry matter content than the more susceptible rough-skinned tubers (Schoene, 1967). The results of his experiments (Boyd, 1967) in which he shortened the length of the growth period by either planting late or removing haulm early also showed a reduction in dry matter content and susceptibility to dry rot, although no direct relationship between the two was established.

The optimum temperatures for growth on potato dextrose agar are stated by Moore (1945) as 20–25°C for *F. avenaceum* and 20°C for var. *coeruleum*, the maximum temperatures being >30 and 30°C, respectively. She also found that the cardinal temperatures for infection of wounded tubers are similar to those on agar for *F. avenaceum*, but the optimum for var. *coeruleum* is 15°C and the maximum 25°C. However, at low temperatures var. *coeruleum* causes quicker rotting than does *F. avenaceum* and, in fact, Boyd (1952d) found that seasonal tuber resistance may be overcome by incubating tubers immediately after inoculation at 4°C for some time before placing them at the optimum temperature of 15°C for rot development. Amongst seed tubers chitted in glasshouses, gangrene lesions are sometimes over-run by the dry rot fungus late in storage when temperatures are beginning to rise and tissues becoming more susceptible.

High humidity encourages rotting by *F. avenaceum*, whereas var. *coeruleum* is more tolerant of low humidity (Moore, 1945). The presence of moisture in wounds, and its persistence until infection has occurred, probably is of greater importance in influencing infection than relative humidity alone. Also, the direct effects of temperature and humidity on the pathogens are of less importance in influencing infection than the effects of these factors on wound healing of the tuber (Hooker, 1967).

4. *Control*

As already mentioned, cultural control measures such as avoiding planting diseased seed, reducing the length of the growing season and ensuring an adequate application of a balanced fertilizer may be practised to reduce the susceptibility of the crop. An interval of at least 2 weeks should elapse between burning down and lifting. Control of storage temperatures is also most important. Efficient curing should decrease dry rot incidence and it is also important to ventilate the store with warm air before grading and afterwards to allow the tubers to remain in a temperature in excess of 16°C for 2–3 days to allow wounds to heal. It is essential that damage must be reduced to a minimum at all stages, but particularly at riddling, prior to planting and during planting. The removal of infected tubers should be done only once and as near planting time as possible to minimize the risk of spreading infection.

Dry rot can be effectively controlled by the use of organo-mercury compounds (2-methoxyethylmercury chloride) as disinfectant solutions, with the seed tubers being separated immediately after harvest (Foister, 1940; Boyd, 1960a). Boyd (1960a) found that much of the dry rot inoculum can be removed by washing tubers immediately after lifting but there is a risk of increasing bacterial disease (e.g. *Erwinia carotovora* var. *atroseptica*) if the tubers are washed before disinfection (Logan, 1964). Leach (1971) found that both dusts and dips containing benomyl, thiabendazole or zinc plus maneb effectively controlled dry rot in wounded and inoculated tubers. Smokes containing thiabendazole may also control the disease but this application method is inferior to a mist (ultra-low volume) or to low volume sprays applied to tubers as they are being transported into store (Henriksen, 1975; Logan and Copeland, 1975; Meijers, 1975; Meredith, 1975).

Dusting tubers with thymol (Foister *et al.*, 1945b) or with tecnazene (Foister and Wilson, 1950) at the time of lifting has also provided effective control; the latter acts in addition as a sprout depressant and should therefore not be applied to seed potatoes. Stores and sprouting boxes may be disinfected conveniently with a fog containing 1–3% formaldehyde (Eckert, 1977). However, the use of 2-aminobutane, which as a fumigant effectively controls gangrene and skin spot (Graham *et al.*, 1973), is completely ineffective for control of dry rot. Furthermore, 2-aminobutane can cause rotting or surface necrosis in damp or newly damaged tubers as shown by Hims and Fletcher (1980) and can sometimes stimulate dry rot infection in newly lifted or newly graded tubers (Boyd, 1981).

G. Skin Spot

This is one of the less conspicuous potato diseases which was named skin spot by Pethybridge (1915), and of which the first full account of the disease on the tuber with a detailed description of the causal organism was given by Owen (1919). The ability of *Oospora pustulans* to infect and kill tuber eyes was first recognized by Milburn and Bessey (1915) and its presence on stolon scars was reported by Chamberlain (1935). However, it was much later that Hirst and Salt (1959) discovered that the fungus also attacked and sporulated on roots and stem bases. Skin spot would be of relatively little consequence if it were not associated with blindness caused by eye infection (Boyd, 1957) and Hirst and Salt (1959) maintain that tuber yield is not greatly affected, despite severe damage to the roots.

The disease is restricted to the cool temperate regions of the world and is present in North European countries, Canada and the U.S.A. It is classed as a minor disease in the U.S.A. (Smith and Wilson, 1978), but in the U.K. (Baker, 1972), Norway (Forsund, 1966) and certain parts of Russia it is frequently assessed as of considerable economic importance. Hide (1981) found that during the period 1963–76, 95% of the King Edward samples planted in England and Wales contained tubers with skin spot infection and on average 44% of all tubers were affected. The annual incidence of the disease varied widely, most diseased tubers (78%) being found in 1963 and least (26%) in 1973 and 1974, but there were only small differences in the proportion of stocks affected to different degrees. He found that King Edward and Majestic seed from mid- or southern Scotland had the highest incidence and severity of infections, whereas disease amounts were less from areas further north; by contrast, seed of Pentland Crown from the north and south of Scotland had equal proportions of diseased tubers.

1. *Causal Organism*

The causal fungus, formerly known as *O. pustulans* Owen and Wakefield, has recently been renamed *Polyscytalum pustulans* (Owen and Wakefield) (Ellis, 1976). It has the relatively low optimum temperature of 12°C, growth being practically absent at 0 and at 24°C. It may be cultured on a wide range of agar media and may be isolated directly from soil on an agar medium containing fentin hydroxide and chloramphenicol (Bannon, 1975). *Polyscytalum pustulans* sporulates readily on the affected areas of tubers, stem bases and roots in cool, damp conditions, producing characteristic branched chains of hyaline conidia (Hide *et al.*, 1968). In the soil it produces microsclerotia which are capable of producing conidia after burial in soil for 7 years (Hirst *et al.*, 1970).

2. *Disease Symptoms*

Symptoms on tubers do not become evident until after several months of storage and are more distinctive when tubers have been washed. Individual lesions are superficial, varying in size from barely visible flecks to conspicuous spots about 3–4 mm across and deep. There may be few up to a large number in which case they may coalesce, producing a brownish-purple patch with a clearly marked margin. The eyes of the tuber are often affected and pustules or small areas of necrosis may be formed around them, but a high proportion of affected tubers carry a small area of infection at or near the heel end (Boyd, 1957). The form of the individual spots usually varies with the cultivar (Brenchley and Wilcox, 1979).

A mature pustule when cut across extends to a depth of from 1 to 2 mm; it is black at the surface, shading to brown at the base where the progress of the rot is arrested by the formation of a cork barrier. Generally less common but much more serious is the ability of the skin spot fungus to kill the eyes, especially of susceptible cultivars such as King Edward and, when conditions favour the fungus, all the buds in most of the eyes may be killed, even though there is little skin spotting on the remainder of the tuber. In such cases the presence of the pathogen may be detected only by incubation of the eyes as described by Hide *et al.* (1968).

3. *Inoculum and Infection*

After planting infected tubers, Hirst and Salt (1959) found that the pathogen colonizes the sprouts and parts of the stem bases nearest to the tubers. They stated that severe eye infection may kill sprouts and so cause blanking or uneven emergence, whereas minute infections of bud-scales may infect the sprout without impairing its vigour. Later the pathogen spreads to the stolon and eventually widespread infection of the root systems of most plants in a crop provides ample inoculum for the infection of progeny tubers, which occurs first of all in the eyes (Hide *et al.*, 1968).

According to Allen (1957), tuber eye regions and lenticels are susceptible to infection about lifting time when the pathogen penetrates the tissues to about 12 layers of cells in depth and may cause the death of both central and axillary buds. The pathogen becomes isolated from the healthy tissue by a cork barrier and remains so for at least 2 months when the cork layer to the outside of the infected tissue breaks down and the "spot" becomes evident.

During storage the pathogen continues to penetrate the eye and surface of the tuber, especially under cold, damp conditions, causing death of buds (Edie, 1966; Hide *et al.*, 1969). In damp conditions the fungus sporulates on tuber surfaces and sprouts and may be detected in the air amongst stored tubers (Hide *et al.*, 1969), though the spread of these spores is not regarded as an important source of infection.

In the U.K. it is generally accepted that skin spot is largely tuber-borne (Boyd and Lennard, 1961; Hirst *et al.*, 1963), the degree of infection of the parent seed being reflected in that of the progeny (Boyd *et al.*, 1968). Transmission of the disease from contaminated soil and carry-over on plant debris until the following year appears to be of importance in the U.S.S.R. (Kharkova, 1961).

4. *Predisposing Factors*

There is considerable fluctuation in the seasonal incidence of skin spot, the two most important environmental factors being soil moisture and temperature (Boyd and Lennard, 1962). After studying records in both England and Scotland from 1927 to 1960, they found that above average rainfall over the lifting period in the seed-growing areas and subsequent

below average autumn and early winter temperatures showed a close relationship with above normal incidence of skin spot in the following spring.

Soil type also plays an important part in influencing conditions for infection both of stem bases (Salt, 1964) and tubers; a sandy loam greatly
reduces transmission of infection compared with that from a clay loam
(McGee *et al.*, 1972).

Potato cultivars vary considerably in their reaction to skin spot. Boyd
(1957), Boyd and Lennard (1961) and Nagdy and Boyd (1965) tested
cultivar susceptibility over a range of 30 cultivars and found Golden
Wonder and Dunbar Rover to be the most resistant and Arran Banner and
King Edward the most susceptible; there was a significant correlation between surface and eye infection. However, the reactions of the surface and
eye of tubers may not by themselves indicate field performance: King
Edward and Kerrs Pink are both equally very susceptible in the surface and
eye but, whereas a high degree of seed infection seldom results in blanking
with Kerrs Pink, it is usually extensive with King Edward (Boyd *et al.*,
1968). Sprout vigour is therefore a factor that should be considered when
assessing the field performance of a cultivar (Boyd, 1972).

Ives (1955) described an unusually deep form of skin spot in tubers
treated with the sprout suppressant isopropyl phenylcarbamate (IPC)
which appears to arrest cork barrier formation in tuber tissue. French
(1976) obtained much the same effect with isopropyl 3-chlorophenyl
carbamate (CIPC), which is now more commonly used than IPC. He found
that the depth of penetration of lesions in Désirée was greater at low
temperature (5°C) and significantly greater in the presence of CIPC at 5°C.
However, on raising the temperature to 10–15°C he found no deeper skin
spot penetration in either the presence or absence of CIPC.

5. *Control*

Control of skin spot may be achieved by controlling the storage conditions,
by chemical treatment immediately after harvest or by raising stocks from
stem cuttings. The last method (Hirst and Hide, 1967) has given effective
control in the first two or three generations but, as with gangrene, the
problems associated with recontamination during commercial production
have yet to be solved. There is evidence that VTSC stocks may be kept
relatively healthy by the use of fungicides during multiplication; also, Boyd
(1977) and Copeland *et al.* (1980) found effective control of both skin spot
and silver scurf in progeny tubers by treating the seed tubers, as they were
being prepared for chitting, with the systemic fungicides benomyl or
thiabendazole.

Skin spot can be effectively controlled during storage by treating tubers either soon after lifting with dusts or mists of thiabendazole or benomyl (Hide *et al.*, 1969; Boyd, 1975; Logan *et al.*, 1975) or after an interval of about 2 weeks with the fumigant 2-aminobutane (Graham *et al.*, 1973). These chemicals have replaced disinfection in organo-mercury solutions as originally advocated by Greeves and Muskett (1939).

The incidence of skin spot may be reduced by exposing tubers to dry conditions (e.g. sprouting trays) at the time of lifting. Lennard (1967) found that high temperature (15°C) in the initial holding period is only effective in controlling disease development when coupled with dry conditions (75% relative humidity), but prolonged periods (at least 3 months) of high temperature are necessary when the humidity is high. Thus, Boyd (1972) considered that, although an initial curing period of 10–14 days may in some cases be beneficial, it is unlikely that the relative humidity in bulk storage would usually be low enough to restrict infection satisfactorily.

References

Adams, M. J. (1979). Factors affecting survival of *Phoma exigua* var. *foveata* in soil. *Trans. Br. mycol. Soc.* **73**, 91–97.

Adams, M. J. (1980). The role of seed-tuber and stem inoculum in the development of gangrene in potatoes. *Ann. appl. Biol.* **96**, 17–28.

Alcock, N. L. and Foister, C. E. (1936). A fungus disease of stored potatoes. *Scott. J. Agric.* **19**, 252–257.

Allen, J. D. (1957). The development of potato skin spot disease. *Ann. appl. Biol.* **45**, 293–298.

Artschwager, E. (1927). Wound periderm formation in the potato as affected by temperature and humidity. *J. agric. Res.* **35**, 995–1000.

Ayers, G. W. (1956). The resistance of potato varieties to storage decay caused by *Fusarium sambucinum* f.6 and *Fusarium caeruleum*. *Am. Potato J.* **33**, 249–254.

Ayers, G. W. and Robinson, D. B. (1956). Control of dry rot of potatoes by seed treatment. *Am. Potato J.* **33**, 1–5.

Bainsbridge, A. (1970). Sporulation by *Pythium ultimum* at various soil moisture tensions. *Trans. Br. mycol. Soc.* **55**, 485–488.

Baker, J. J. (1972). "Report on Diseases of Cultivated Plants in England and Wales for the Years 1957–1968", Ministry of Agriculture, Fisheries and Food Technical Bulletin No. 25. H.M.S.O., London.

Bang, H. (1972). Susceptibility of Swedish potato cultivars to *Phoma* rots. *Vaxtskyddsnotiser* **36**, 46–47.

Bannon, E. (1975). A new medium for the isolation of *Oospora pustulans* from potato tubers and soil. *Trans. Br. mycol. Soc.* **64**, 554–556.

Bannon, E. (1978). Detection of species of *Phoma* on stems and leaves of potato. *7th Trienn. Conf. E.A.P.R.*, Warsaw, 215–216 (Abstract).

Bate-Smith, E. C. (1956). Food investigation. *Rep. Fd Invest. Bd Lond. 1955*, 53–62.

Beaumont, A. (1959). "Diseases of Farm Crops." Collingridge, London.

Black, W. (1957). Incidence of physiological races of *Phytophthora infestans* in various countries. *Rep. Scott. Pl. Breed. Stn 1957*, 43–49.

Black, W., Mastenbroek, C., Mills, W. R. and Peterson, L. C. (1953). A proposal for an international nomenclature of races of *Phytophthora infestans* and of genes controlling immunity in *Solanum demissum* derivatives. *Euphytica* **2**, 173–178.

Blodgett, E. C. (1945). Water rot of potatoes. *Pl. Dis. Reptr* **29**, 124–126.

Blodgett, E. C. and Ray, W. W. (1945). Leak, caused by *Pythium debaryanum* Hesse produces typical "shell rot" of potato in Idaho. *Am. Potato J.* **22**, 250–253.

Boerema, G. H. (1967). The *Phoma* organisms causing gangrene of potatoes. *Neth. J. Pl. Path.* **73**, 190–192.

Boerema, G. H. (1969). The use of the term *forma specialis* for *Phoma*-like fungi. *Trans. Br. mycol. Soc.* **52**, 509–513.

Boerema, G. H. and Howeler, L. H. (1967). *Phoma exigua* Desm. and its varieties. *Persoonia* **5**, 15–28.

Boyd, A. E. W. (1952a). Dry rot disease of the potato. IV. Laboratory methods used in assessing variations in tuber susceptibility. *Ann. appl. Biol.* **39**, 322–329.

Boyd, A. E. W. (1952b). Dry rot disease of the potato. V. Seasonal and local variations in tuber susceptibility. *Ann. appl. Biol.* **39**, 330–338.

Boyd, A. E. W. (1952c). Dry rot disease of the potato. VI. Varietal differences in tuber susceptibility obtained by injection and riddle abrasion methods. *Ann. appl. Biol.* **39**, 339–350.

Boyd, A. E. W. (1952d). Dry rot disease of the potato. VII. The effect of storage temperature upon subsequent susceptibility of tubers. *Ann. appl. Biol.* **39**, 351–357.

Boyd, A. E. W. (1957). Field experiments on potato skin spot disease caused by *Oospora pustulans* Owen & Wakef. *Ann. appl. Biol.* **45**, 284–292.

Boyd, A. E. W. (1960a). Fungicidal dipping and other treatments of seed potatoes in Scotland. *Eur. Potato J.* **3**, 137–154.

Boyd, A. E. W. (1960b). Size of potato tubers and natural infection with blight and pink rot. *Pl. Path.* **9**, 99–101.

Boyd, A. E. W. (1967). The effects of the length of the growth period and of nutrition upon potato tuber susceptibility to dry rot (*Fusarium caeruleum*). *Ann. appl. Biol.* **60**, 231–240.

Boyd, A. E. W. (1972). Potato storage diseases. *Rev. Pl. Path.* **51**, 297–321.

Boyd, A. E. W. (1973a). Potato blight control in East and South-East Scotland 1959–1968. *Ann. appl. Biol.* **74**, 41–58.

Boyd, A. E. W. (1973b). Potato disease research in the Edinburgh School of Agriculture 1950–1973. *A. Rep. Edinb. Sch. Agric. 1973*, 47–63.

Boyd, A. E. W. (1975). Fungicides for potato tubers. *Proc. 8th Br. Insectic. Fungic. Conf.*, 1035–1044.

Boyd, A. E. W. (1977). Control of certain potato tuber diseases by thiabendazole and 2-aminobutane. *Proc. Symp. Probl. Pest Dis Control North. Br., Univ. Dundee*, 42–44.

Boyd, A. E. W. (1981). Control of potato storage diseases. *Proc. Crop Prot. Conf. North. Br., Univ. Dundee 1981*, 111–118.

Boyd, A. E. W. and Lennard, J. H. (1961). Some effects of potato skin spot (*Oospora pustulans*) in Scotland. *Eur. Potato J.* **4**, 361–377.

Boyd, A. E. W. and Lennard, J. H. (1962). Seasonal fluctuation in potato skin spot. *Pl. Path.* **11**, 161–166.

Boyd, A. E. W. and Penna, R. J. (1967). Seed tuber disinfection. *Proc. 4th Insectic. Fungic. Conf.*, 294–302.

Boyd, A. E. W. and Tickle, J. H. (1972). Dry rot of potato tubers caused by *Fusarium sulphureum* Schlecht. *Pl. Path.* **21**, 195.

Boyd, A. E. W., McGee, D. C. and Lennard, J. H. (1968). Plant emergence and skin spot development in relation to variety and levels of seed tuber infection. *Exp. Wk. Edinb. Sch. Agric. 1967*, 43–44.

Brenchley, G. H. and Wilcox, H. J. (1979). "Potato Diseases." H.M.S.O., London.

Brien, R. M. (1940). "Leak", a watery wound-rot of potatoes in New Zealand. *N.Z. J. Sci. Tech., A* **22**, 228–231.

Cairns, H. and Muskett, A. E. (1933). Pink rot of the potato. *Ann. appl. Biol.* **20**, 381–403.

Cairns, H. and Muskett, A. E. (1939). *Phytophthora erythroseptica* Pethybr. in relation to its environment. *Ann. appl. Biol.* **26**, 470–480.

Carnegie, S. F. (1980). Aerial dispersal of the potato gangrene pathogen, *Phoma exigua* var. *foveata*. *Ann. appl. Biol.* **94**, 165–173.

Carnegie, S. F., Adam, J. W., MacDonald, D. M. and Cameron, A. M. (1981). Contamination by *Polyscytalum pustulans* and *Phoma exigua* var. *foveata* of seed stocks derived from stem cuttings in Scotland. *Potato Res.* **24**, 389–397.

Chamberlain, E. E. (1935). Fungi present in the stem end of potato tubers. *N.Z. J. Sci. Tech.* **16**, 242–246.

Clinton, G. P. (1908). Report of the botanist: IV Artificial cultures of *Phytophthora*, with special reference to oospores. *A. Rep. Conn. agric. Exp. Stn.* **32**, 891–907.

Cooke, L. R. (1981). Resistance to metalaxyl in *Phytophthora infestans* in Northern Ireland. *Proc. Br. Crop Prot. Conf. 1981, Pests Dis.*, 641–649.

Copeland, R. B. and Logan, C. (1976). The effect of interval from haulm destruction to harvest on the gangrene potential of tubers in storage. *Potato Res.* **19**, 203–213.

Copeland, R. B. and Logan, C. (1980). Curing tuber damage at the time of grading. *Ann. appl. Biol.* **96**, 374–377.

Copeland, R. B., Logan, C. and Little, G. (1980). Studies on the effect of chemical treatments on potato disease control in progeny tubers. *Rec. agric. Res. Minist. Agric. North. Ire.* **28**, 49–53.

Cotton, A. D. (1921). Fungus diseases of plants in England and Wales, 1920–1921. *Misc. Publ. Minist. Agric. Fish.*, No. 33, 26–28.

Cotton, A. D. (1922). Potato pink rot: a disease new to England. *J. Minist. Agric.* **28**, 1126–1130.

Croke, F. M. (1980). Studies on the aetiology and life cycle of the potato gangrene pathogen. Thesis, Queen's University of Belfast.

Cruikshank, I. A. M. (1953). A note on the use of autoclaved pea seed as a culture medium. *N.Z. J. Sci. Technol, B* **34**, 248.

Cunliffe, C., Londsdale, D. and Epton H. A. S. (1977). Transmission of *Phytophthora erythroseptica* on stored potatoes. *Trans. Br. mycol. Soc.* **69**, 27–30.

Cutter, E. J. (1978). Structure and development of the potato plant. *In* "The Potato Crop. The Scientific Basis for Improvement" (P. M. Harris, Ed.), 70–152. Chapman and Hall, London.

Davidse, L. C., Looijen, D., Turkensteen, L. J. and Van der Wal, D. (1981). Occurrence of metalaxyl-resistant strains of *Phytophthora infestans* in Dutch potato fields. *Neth. J. Pl. Path.* **87**, 65–68.

Dennis, R. W. C. (1946). Notes on some British fungi ascribed to *Phoma* and related genera. *Trans. Br. mycol. Soc.* **29**, 11–42.

Doke, N. and Tomiyama, K. (1980). Suppression of the hypersensitive response of potato tuber protoplasts to hyphal wall components by water soluble glucans isolated from *Phytophthora infestans. Physiol. Pl. Path.* **16,** 177–186.

Eckert, J. W. (1977). Control of post harvest diseases. *In* "Antifungal Compounds", (M. R. Siegel and H. D. Sisler, Eds), Vol. I, pp. 270–352. Marcel Dekker, New York.

Edie, H. H. (1966). Skin spot (*Oospora pustulans*) disease of potatoes—its possible spread in storage. *Eur. Potato J.* **9,** 161–164.

El-Helaly, A. F., Abo-el-Dahab, M. K. and El-Godrani, M. A. (1971). New diseases of potato tubers in Egypt; leak, grey mould and charcoal rot. *Unit. Arab Repub. J. Phytopath.* **2,** 1–7.

Ellis, M. B. (1976). "More Dematiaceous Hyphomycetes." Cambrian News, Aberystwyth.

Entwistle, A. R. (1972). Study of *Phoma exigua* populations in the field. *Trans. Br. mycol. Soc.* **58,** 217–223.

Faulkner, M. J. and Colhoun, J. (1976). Aerial dispersal of pycnidiospores of *Leptosphaeria nodorum. Phytopath. Z.* **86,** 357–360.

Foister, C. E. (1940). Description of new fungi causing economic diseases in Scotland. *Trans. bot. Soc. Edinb.* **33,** 65–68.

Foister, C. E. and Wilson, A. R. (1950). Dry rot of potatoes. *J. Minist. Agric.* **57,** 229–233.

Foister, C. E., Wilson, A. R. and Boyd, A. E. W. (1945a). Potato dry rot and gangrene as soil-borne diseases. *Nature, Lond.* **155,** 793.

Foister, C. E., Wilson, A. R. and Boyd, A. E. W. (1945b). Control of dry rot of seed potatoes by dusting. *Nature, Lond.* **156,** 394.

Foister, C. E., Wilson, A. R. and Boyd, A. E. W. (1952). Dry rot disease of the potato. I. Effect of commercial handling methods on the incidence of the disease. *Ann. appl. Biol.* **39,** 29–37.

Forsund, E. (1966). Skin spot (*Oospora pustulans*) controlled by thermotherapy. *Proc. 3rd Trienn. Conf. Eur. Ass. Potato Res. 1966,* 221–222.

Fox, R. A. and Dashwood, P. (1969). Mycology. Gangrene of potato. *Rep. Scott. hort. Res. Inst. 1968,* 32–33.

Fox, R. A. and Dashwood, P. (1972). Mycology. Gangrene of potato. *Rep. Scott. hort. Res. Inst. 1971,* 51–52.

Fox, R. A. and Dashwood, P. (1978). Observations on potential alternative hosts of the potato gangrene fungus *Phoma exigua* var. *foveata. 7th Trienn. Conf. E.A.P.R., Warsaw,* 217–218 (Abstract).

Fox, R. A., Dashwood, P. and Wilson, H. M. (1970). Mycology. Gangrene of potato. *Rep. Scott. hort. Res. Inst. 1969,* 30–31.

French, W. M. (1976). Effects of chlorpropham and storage temperature on an abnormal form of skin spot on potato tubers. *Pl. Path.* **25,** 144–146.

Fry, W. E. (1977). Integrated control of potato late blight—effects of polygenic resistance and techniques of timing fungicide applications. *Phytopathology* **67,** 415–420.

Gans, P. T. (1978). Physiological response of potato tubers to damage and to infection by *Phoma exigua* var. *foveata. Ann. appl. Biol.* **89,** 307–309.

Garas, N. A., Doke, N. and Kuc, J. (1979). Suppression of the hypersensitive reaction in potato tubers by mycelial components from *Phytophthora infestans. Physiol. Pl. Path.* **15,** 117–126.

Graham, D. C. and Hardie, J. L. (1971). Prospects for the control of potato blackleg disease by the use of stem cuttings. *Proc. 6th Br. Insectic. Fungic. Conf.,* 219–224.

Graham, D. C., Hamilton, G. A., Quinn, C. E. and Ruthven, A. D. (1973). Use of 2-aminobutane as a fumigant for control of gangrene, skin spot and silver scurf diseases of potato tubers. *Potato Res.* **16,** 109–124.

Graham, D. C., Hamilton, G. A., Quinn, C. E. and Ruthven, A. D. (1981). Summary of results of experiments on the control of the potato tuber fungal diseases, gangrene, dry rot and skin spot with various chemical substances. *Potato Res.* **24,** 147–158.

Greeves, T. M. and Muskett, A. E. (1939). Skin spot (*Oospora pustulans* Owen & Wakef.) of the potato and its control by tuber disinfection. *Ann. appl. Biol.* **26,** 481–496.

Griffin, G. J. (1964). Effect of chilling on phenol metabolism and *Fusarium* infection of cut potato tissue. *Phytopathology* **54,** 1275–1277.

Griffith, R. L. (1969). Potato tuber diseases. *Rep. Rothamsted Exp. Stn. 1968,* 145–146.

Griffith, R. L. (1970). Pathogen, wound type, temperature and gangrene infection. *Rep. Rothamsted Exp. Stn 1969,* 165–166.

Hall, A. M. (1959). The culture of *Phytophthora infestans* in artifical media. *Trans. Br. mycol. Soc.* **42,** 15–26.

Hampson, C. P., Dent, T. J. and Ginger, M. W. (1980). The effect of mechanical damage on potato crop wastage during storage. *Ann. appl. Biol.* **96,** 366–370.

Harris, J. E. and Dennis, C. (1976). Antifungal activity of post-infectional metabolites from potato tubers. *Physiol. Pl. Path.* **9,** 155–165.

Harrison, D. E. (1959). *Phoma* or gangrene of potatoes in Victoria. *J. Dep. Agric. Vict.* **57,** 381.

Hawkins, L. A. (1916). The disease of potatoes known as "leak". *J. agric. Res.* **6,** 627–639.

Henriksen, J. B. (1975). Prevention of gangrene and *Fusarium* dry rot by physical means and with thiabendazole. *Proc. 8th Br. Insectic. Fungic. Conf.* **2,** 603–608.

Henriksen, J. B. (1976). Infection by gangrene as influenced by temperature and humidity. *Proc. 6th Trienn. Conf. Eur. Ass. Potato Res. 1975,* 186–187.

Hesen, J. C. (1981). Storage and quality of potatoes. *In* "Survey Papers", *8th Trienn. Conf. Eur. Ass. Potato Res. 1981,* 123–132.

Hide, G. A. (1978). Incidence of pathogenic fungi on Scottish potato seed stocks derived from stem cuttings. *Potato Res.* **21,** 277–289.

Hide, G. A. (1981). Fungus diseases of potato seed tubers planted in England and Wales, 1963–76. *Ann. appl. Biol.* **98,** 377–393.

Hide, G. A., Hirst, J. M. and Salt, G. A. (1968). Methods of measuring the prevalence of pathogenic fungi on potato tubers. *Ann. appl. Biol.* **62,** 309–318.

Hide, G. A., Hirst, J. M. and Griffith, R. L. (1969). Control of potato tuber diseases with systemic fungicides. *Proc. 5th Br. Insectic. Fungic. Conf.,* 310–314.

Hims, M. J. and Fletcher, J. T. (1980). Damage to potato tubers caused by fumigation with 2-aminobutane. *Potato Res.* **23,** 365–369.

Hirst, J. M. and Hide, G. A. (1967). Attempts to produce pathogen-free stocks. *Rep. Rothamsted Exp. Stn 1966,* Part 1, 129.

Hirst, J. M. and Salt, G. A. (1959). *Oospora pustulans* Owen & Wakefield as a parasite of potato root systems. *Trans. Br. mycol. Soc.* **42,** 59–66.

Hirst, J. M., Salt, G. A. and Hide, G. A. (1963). Potato diseases. *Rep. Rothamsted Exp. Stn 1962,* 120–121.

Hirst, J. M., Hide, G. A., Griffith, R. L. and Stedman, O. J. (1970). Improving the health of seed potatoes. *Jl. R. agric. Soc. Engl.* **131,** 87–106.

Holden, J. H. W. (1977). Potato breeding at Pentlandfield. *Scott. Pl. Breed. Stn Rep. April 1976 to March 1977*, 66–87.

Hooker, W. J. (1967). Tuber diseases. *In* "Potato Processing" (W. F. Talburt and O. Smith, Eds), 121–153. Avi Publishing, Connecticut.

Humphreys-Jones, D. R. (1969). Rubbery rot (*Oospora lactis* (Fres) Sacc) of potatoes. *Pl. Path.* **18**, 186–187.

Ives, J. V. (1955). An abnormal form of skin spot on potatoes. *Pl. Path.* **4**, 17–21.

Jeffries, C. J. (1978). Transmission studies on the potato pathogens *Fusarium solani* var. *coeruleum* and *Fusarium sulphureum*. Ph.D. Thesis, University of Edinburgh.

Jeffries, G. J. and Boyd, A. E. W. (1977). The control of tuber diseases of potatoes other than blight. Experimental work. Edinburgh School of Agriculture (1976). pp. 116–118.

Keay, M. (1953). Media for the culture of *Phytophthora infestans*. *Pl. Path.* **2**, 103.

Kerkenaar, A. (1981). On the antifungal mode of action of metalaxyl, an inhibitor of nucleic acid synthesis in *Pythium splendens*. *Pest. Biochem. Physiol.* **16**, 1–13.

Khan, A. A. (1967). A study of some factors affecting infection and development of potato gangrene. *Rec. agric. Res. Ministr. Agric. North. Ire.* **16**, 97–101.

Khan, A. A. and Logan, C. (1968). A preliminary study of the sources of potato gangrene infection. *Eur. Potato J.* **11**, 77–87.

Kharkova, A. P. (1961). On the biology of the causal agent of oosporosis, *Oospora pustulans* Owen & Wakef. *Bot. Zh. S.S.S.R.* **46**, 399–407.

Kranz, J. (1958). Studies of *Phoma* rot of potato tubers with particular reference to the host–parasite relationship. *Phytopath. Z.* **33**, 153–196.

Lacey, J. (1966). The distribution of healthy and blighted tubers in potato ridges. *Eur. Potato J.* **9**, 86–98.

Lacey, J. (1967). The role of water in the spread of *Phytophthora infestans* in the potato crop. *Ann. appl. Biol.* **59**, 245–255.

Lapwood, D. H. (1964). Potato diseases. *Rep. Rothamsted Exp. Stn 1963*, 111–112.

Lapwood, D. H. (1965). Potato diseases. *Rep. Rothamsted Exp. Stn 1964*, 136–137.

Lapwood, D. H. (1966). Potato diseases. *Rep. Rothamsted Exp. Stn 1965*, 129–130.

Lapwood, D. H. (1977). Factors affecting the field infection of potato tubers of different cultivars by blight (*Phytophthora infestans*). *Ann. appl. Biol.* **85**, 23–42.

Leach, S. S. (1971). Postharvest treatments for the control of *Fusarium* dry rot development in potatoes. *Pl. Dis. Reptr* **55**, 723–726.

Lennard, J. H. (1967). The development of skin spot in relation to lifting and storage factors. *Proc. 4th Br. Insectic. Fungic. Conf. 1967*, 269–275.

Lennard, J. H. (1980). Factors affecting the development of potato pink rot (*Phytophthora erythroseptica*). *Pl. Path.* **29**, 80–86.

Logan, C. (1964). Bacterial hard rot of potato. *Eur. Potato J.* **7**, 45–56.

Logan, C. (1967a). Potato gangrene. *Seed Potato* **7**, 78–80.

Logan, C. (1967b). Potato stem infection by *Phoma solanicola* Prill et Delacr. f. *foveata* (Foister) Malcolmson. *Pl. Path.* **16**, 64–67.

Logan, C. (1967c). The control of potato gangrene by tuber disinfection. *Rec. agric. Res. Ministr. Agric. North. Ire.* **16**, 25–28.

Logan, C. (1969). Reduce potato gangrene. *Agric. North. Ire.* **44**, 1–3.

Logan, C. (1970a). The effect of the diseased parent tuber on the incidence of potato gangrene. *Proc. 4th Trienn. Conf. Eur. Ass. Potato Res. 1969*, 189.

Logan, C. (1970b). The effect of potato haulm treatment on the incidence of potato gangrene in storage. *Pl. Path.* **19**, 95–98.

Logan, C. (1974). The effect of soil- and tuber-borne inoculum on the incidence of potato gangrene. *Ann. appl. Biol.* **78**, 251–259.

8. POTATOES

Logan, C. (1976). The spread of *Phoma exigua* within the potato crop. *Ann. appl. Biol.* **82**, 169–174.

Logan, C. and Copeland, R. B. (1975). Potato gangrene—its control during storage by mist application of thiabendazole. *Proc. 8th Br. Insectic. Fungic. Conf.* **2**, 589–595.

Logan, C. and Khan, A. A. (1969). Comparative studies of *Phoma* spp. associated with potato gangrene in Northern Ireland. *Trans. Br. mycol. Soc.* **52**, 9–17.

Logan, C. and O'Neill, R. (1970). Production of an antibiotic by *Phoma exigua*. *Trans. Br. mycol. Soc.* **55**, 67–75.

Logan, C. and Woodward, J. R. (1971). Pathogenicity differences within *Phoma exigua* var. *foveata*. *Rec. agric. Res. Ministr. Agric. North. Ire.* **19**, 27–31.

Logan, C., Copeland, R. B. and Little, G. (1975). Potato gangrene control by ultra low volume sprays of thiabendazole. *Ann. appl. Biol.* **80**, 199–204.

Maas, P. W. Th. (1965). The identity of the footrot fungus of flax. *Neth. J. Pl. Path.* **71**, 113–121.

Mackenzie, M. (1968). Factors influencing the infection of potato tubers by *Phoma solanicola* f. *foveata*. *Eur. Potato J.* **11**, 193 (Abstract).

McGee, D. C., Morton, A. P. and Boyd, A. E. W. (1972). Reactions of potato varieties to skin spot (*Oospora pustulans*) infection and transmission in different soils. *Potato Res.* **15**, 302–316.

McKay, R. (1955). "Potato Diseases." At the Sign of the Three Candles, Dublin.

McKee, R. K. (1952). Dry rot disease of the potato. II. Fungi causing dry rot of seed potatoes in Britain. *Ann. appl. Biol.* **39**, 38–43.

McKee, R. K. (1954). Dry rot disease of the potato. VIII. A study of the pathogenicity of *Fusarium caeruleum* (Lib) Sacc. and *Fusarium avenaceum* (Fr) Sacc. *Ann. appl. Biol.* **41**, 417–434.

McKee, R. K. (1961). Observations on the toxicity of solanine and related alkaloids to fungi. *TagBer. dt. Akad. LandwWiss., Berl.* **27**, 277–289.

Malcolmson, J. F. (1958a). A consideration of the species of *Phoma* which parasitize potatoes. *Trans. Br. mycol. Soc.* **41**, 413–418.

Malcolmson, J. F. (1958b). Some factors affecting the occurrence and development in potatoes of gangrene caused by *Phoma solanicola* Prill. et Delacr. *Ann. appl. Biol.* **46**, 639–650.

Malcolmson, J. F. (1969). Races of *Phytophthora infestans* occurring in Great Britain. *Trans. Br. mycol. Soc.* **53**, 417–423.

Malcolmson, J. F. (1981). Mechanisms of field resistance to potato blight and variability of the pathogen. *A. Rep. Scott. Pl. Breed. Stn 1980–81*, 110–111.

Malcolmson, J. F. and Gray, E. (1968a). The incidence of gangrene of potatoes caused by *Phoma exigua* in relation to handling and storage. *Ann. appl. Biol.* **62**, 89–101.

Malcolmson, J. F. and Gray, E. (1968b). Factors affecting the occurrence of gangrene (*Phoma exigua*) in potatoes. *Ann. appl. Biol.* **62**, 77–87.

Meijers, C. P. (1975). Experiences with thiabendazole in control of potato storage diseases in Holland. *Proc. 8th Br. Insectic. Fungic. Conf.* **2**, 597–602.

Melhus, I. E. (1914). A *Phoma* rot of stored potatoes. *Phytopathology* **4**, 41.

Melhus, I. E., Rosenbaum, J. and Schultz, E. S. (1916). Studies on *Spongospora subterranea* and *Phoma tuberosa* on the Irish potato. *J. agric. Res.* **7**, 213–254.

Meredith, D. S. (1975). Control of fungal diseases of seed potatoes with thiabendazole. *Proc. 8th Br. Insectic. Fungic. Conf.* **2**, 581–594.

Milburn, T. and Bessey, E. A. (1915). "Fungoid Diseases of Farm and Garden Crops." Longmans, Green, London.

Moore, F. J. (1945). A comparison of *Fusarium avenaceum* and *Fusarium caeruleum* as causes of wastage in stored potatoes. *Ann. appl. Biol.* **32**, 304–309.

Moore, W. C. (1948). Report on the occurrence of fungus, bacterial and other diseases of crops in England and Wales for the years 1943–46. *Bull. Ministr. Agric. Fish., Lond.*, No. 139.

Moore, W. C. (1959). "British Parastic Fungi." Cambridge University Press, Cambridge.

Muller, K. O. and Borger, H. (1941). Experimentelle Untersuchungen uber die *Phytophthora*-Resistenz der Kartoffel. *Arb. biol. Abt. (Anst.-Reichsanst.), Berl.* **23**, 189–231.

Murphy, P. A. (1921). The sources of infection of potato tubers with the blight fungus, *Phytophthora infestans. Scient. Proc. R. Dubl. Soc., N.S.* **16**, 353–368.

Nagdy, G. A. and Boyd, A. E. W. (1965). Susceptibility of potato varieties to skin spot (*Oospora pustulans*) in relation to the structure of the skin and eye. *Eur. Potato J.* **4**, 200–214.

Nikolaeva, V. V. (1970). Zabolevanie kartofelya forozom. *Bull. vses. nauchno-issled Inst. Zasch. Rast* **2**, 50–51.

O'Brien, M. J. and Rich, A. E. (1976) "Potato Diseases", Agriculture Handbook No. 474. United States Department of Agriculture, Washington, D.C.

Owen, M. N. (1919). The skin spot disease of potato tubers (*Oospora pustulans*). *Kew Bull.*, 289–301.

Paulson, G. A. and Schoeneweiss, D. F. (1971). Epidemiology of stem blight of *Vinca minor* incited by *Phoma exigua* var. *exigua. Phytopathology* **61**, 959–963.

Pethybridge, G. H. (1913). On the rotting of potatoes by a new species of *Phytophthora. Scient. Proc. Roy. Dublin Soc., N.S.* **13**, 529–565.

Pethybridge, G. H. (1915). Investigations on potato diseases. *J. Dep. Agric. Ire.* **15**, 524–542.

Pethybridge, G. H. (1926). Report on the occurrence of fungus, bacterial and allied diseases of crops in England and Wales for the years 1922–24. *Misc. Public. M.A.F.F., Lond.*, No. 52.

Pethybridge, G. H. and Bowers, E. H. (1908). Dry rot of the potato tuber. *Econ. Proc. R. Dubl. Soc.* **1**, 547–558.

Pethybridge, G. H. and Lafferty, H. A. (1917). Further observations on the cause of dry rot of the potato tuber in the British Isles. *Scient. Proc. R. Dubl. Soc.* **15**, 193–222.

Pethybridge, G. H. and Smith, A. (1930). A watery wound rot of the potato tuber. *J. Minist. Agric.* **37**, 335–340.

Pethybridge, G. J., Smith, A. and Moore, W. C. (1934). Report on the occurrence of fungus, bacterial and other diseases of crops in England and Wales, for the years 1928–32. *Bull. Minist. Agric. Fish., Lond.*, No. 79.

Pett, B. and Hahn, E. (1974). On the occurrence of watery wound rot on potatoes. *NachrBl. PflSchutz D.D.R.* **28**, 84.

Pietkiewicz, J. B. and Jellis, G. J. (1975). Laboratory testing for the resistance of potato tubers to gangrene. *Phytopath. Z.* **83**, 289–295.

Prillieux, E. E. and Delacroix, G. (1890). Sur une malade de la pomme de terre produite par le *Phoma solanicola* nov. sp. *Bull. Soc. mycol. Fr.* **6**, 178–181.

Radtke, W. (1969). Defence reaction of potato tissue against *F. coeruleum* with special consideration of the osmotic pressure. *Phytopath. Z.* **64**, 143–147.

Salt, G. A. (1964). The incidence of *Oospora pustulans* on potato plants in different soils. *Pl. Path.* **13**, 155–158.

Scheepens, P. C. and Fehrmann, H. (1978). Cultivation of *Phytophthora infestans* on defined nutrient media. *Phytopath. Z.* **93**, 126–136.

Schoene, K. (1967). Studies on the susceptibility of potato tubers to *Fusarium caeruleum*, the causal agent of white rot. *Phytopath. Z.* **60**, 201–236.

Small, T. (1944a). Dry rot of potato. The soil as a source of infection. *Nature, Lond.* **153**, 436.

Small, T. (1944b). Dry rot of potato (*Fusarium caeruleum* (Lib) Sacc.). Investigation on the sources and time of infection. *Ann. appl. Biol.* **31**, 290–295.

Small, T. (1946). Further studies on the effect of disinfecting and bruising seed potatoes on the incidence of dry rot (*Fusarium caeruleum* (Lib.) Sacc.) *Ann. appl. Biol.* **33**, 211–219.

Smith, W. L. and Wilson, J. B. (1978). "Market Diseases of Potato," Agricultural Handbook No. 479. United States Department of Agriculture, Washington, D.C.

Todd, J. M. and Adam, J. W. (1967). Potato gangrene: some interconnected sources and factors. *Proc. 4th Br. Insectic. Fungic. Conf.*, 276–284.

Tomiyama, K. (1967). Further observations on the time requirement for hypersensitive cell death of potatoes infected by *Phytophthora infestans* and its relation to metabolic activity. *Phytopath. Z.* **58**, 367–378.

Twiss, P. T. C. and Jones, M. P. (1965). A survey of wastage in bulk stored maincrop potatoes in Great Britain. *Eur. Potato J.* **8**, 154–172.

Van Haeringen, G. H. (1938). Some practical observations on *Phytophthora erythroseptica*. *Tijdschr. Plziekt.* **44**, 247–256.

Varns, J. L., Currier, W. W. and Kuc, J. (1971). Specificity of rishitin and phytuberin accumulation by potato. *Phytopathology* **61**, 968–971.

Walker, R. R. and Wade, G. C. (1976). Epidemiology of potato gangrene in Tasmania. *Aust. J. Bot.* **24**, 37–47.

Walker, R. R. and Wade, G. C. (1978). Resistance of potato tubers (*Solanum tuberosum*) to *Phoma exigua* var. *exigua* and *Phoma exigua* var. *foveata*. *Aust. J. Bot.* **26**, 239–251.

Wastie, R. L. and Stewart, H. E. (1977). Potato tuber disease resistance. *Rep. Scott. Pl. Breed. Stn 1976*, 51–52.

White, N. H. (1946). Host parasite relations in pink rot of potato. *J. Aust. Inst. agric. Sci.* **11**, 195–197.

9

Bacterial Spoilage

BARBARA M. LUND

I. Introduction

Bacteria can be the cause of a substantial proportion of the spoilage of vegetables and salad crops. Some fruits, for example, cucumbers, peppers and tomatoes, are affected but microbial spoilage of the majority of other fruits is usually caused by fungi rather than by bacteria. This is probably because the low pH of most fruits (usually <4.5; von Schelhorn, 1951) inhibits growth of the majority of bacteria capable of degrading plant tissues. The higher pH values within vegetables (4.5–7.0) allow growth of spoilage bacteria.

There is relatively little quantitative information regarding the amount of wastage resulting from bacterial spoilage. Some of the more recent, accurate estimates are those reported for selected fruit and vegetables sampled in the New York and Chicago areas at wholesale, retail and consumer levels (Harvey, 1978). Bacterial soft rot was responsible for a major proportion of the microbial spoilage of lettuce (Ceponis, 1970; Ceponis *et al.*, 1970; Beraha and Kwolek, 1975), potatoes (Ceponis and Butterfield, 1973), bell peppers (Ceponis and Butterfield, 1974a), cucumbers (Ceponis and Butterfield, 1974b) and tomatoes (Ceponis and Butterfield, 1979). An average of 8.8% of Florida bell peppers was spoiled on arrival at the Rotterdam market between January and April 1979; bacterial soft rot was responsible for 86% of this decay and for a high proportion of the increase in decay during simulated wholesale and retail periods (McDonald and de Wildt, 1980). In celery grown in Australia, up to 75% of loads transported during the period February to April were destroyed by bacterial soft rot (Wimalajeewa, 1976). However, in surveys of cargoes of fruit and vegetables (including root crops such as beetroot, carrots, ginger, onions and potatoes, and fruits such as apples, bananas, citrus, dates, grapes, melons, peaches, pears, peppers and tomatoes) imported into the U.K. between 1974 and 1977, bacterial disorders were rare in commodities other than potatoes, which suffered severe losses due to bacterial soft rot (Lowings, 1977).

Bacteria are unable to penetrate the intact surface of plant tissue, but can enter through natural openings such as hydathodes, stomata and lenticels (Billing, 1982). Bacteria from the sepals of young tomato fruit were reported to enter the fruit through connective tissue at the stem end (Samish and Etinger-Tulczynska, 1963). Invasion of cucumber fruit by "*Pseudomonas lachrymans*" has been shown to occur through fruit stomata (Wiles and Walker, 1951) or from wounds at the junction of stems and peduncles (Pohronezny *et al.*, 1978). In general, damage caused by agents as varied as fungi, nematodes, animals, birds, rain, hail or the passage of machinery can allow entry of spoilage bacteria.

Table 9.1 *Bacteria that cause soft rot of vegetables and fruit*

Bacterium	Temperatures for Growth (°C)			Produce Affected
	Min.	Opt.	Max.	
E. carotovora subsp. atroseptica	3	27	35[a]	Most vegetables, particularly potatoes, some fruit
E. carotovora subsp carotovora	6	28–30	37–42[a]	Most vegetables and some fruit
E. chrysanthemi	6	34–37	>45[a]	Pineapple
P. marginalis	>0.2[b]	25–30	>41[c]	Many vegetables
P. viridiflava	—	—	—	Beans
P. cichorii[h]	–	c.30	>41[c]	Chicory, endive, cabbage, lettuce
P. cepacia	>4	30–35	40–41[c]	Onion
P. gladioli pv. allicola	>4	30–35	40–41[d]	Onion
B. polymyxa	5–10	—	35–40[3]	Potato, pepper
B. subtilis	5–20	—	45–55[e]	Potato, tomato
Clostridium puniceum	7	—	39[f]	Potato
Low-temperature clostridia	0–4	9–22	17–30[g]	Potato

[a] Pérombelon and Kelman (1980).
[b] Brocklehurst and Lund (1981).
[c] Doudoroff and Palleroni (1974).
[d] Doudoroff and Palleroni (1974) under "*P. marginata*".
[e] Gibson and Gordon (1974).
[f] Lund *et al.* (1981).
[g] Brocklehurst and Lund (1982).
[h] The lesions caused by *P. cichorii* are probably not true soft rots.

Bacteria can cause the following types of symptom that affect the market quality of produce:

(1) *Soft rots.* These are caused by relatively few types of bacteria (Table 9.1). The plant tissue is macerated by bacterial enzymes and decay can

progress very rapidly in favourable conditions of temperature and humidity.

(2) *Other defects.* Many bacteria cause field diseases other than soft rots that affect the market quality of vegetables and fruits and can result in wastage. The disease may be evident as an infection of the vascular system or may result in distinct lesions on leaves, fruit, tubers or roots. Much of the affected produce may be rejected during inspection and packing, but inconspicuous lesions may progress during transit and storage and allow soft-rot bacteria to enter and cause more extensive spoilage.

Because the majority of progressive, post-harvest bacterial spoilage takes the form of soft rots, the greater part of the chapter will be concerned with this type of spoilage.

Unless indicated otherwise, the bacterial names used in this chapter are those that are included in the Approved Lists of Bacterial Names, 1980 (Skerman *et al.*, 1980). For names that have been published validly in the *International Journal of Systematic Bacteriology* since that date the reference is given. Where names are used that are included in the International Society for Plant Pathology (ISPP) list 1980 (Dye *et al.*, 1980), "ISPP List 1980" is added in parentheses. Other names are given in quotation marks and are now without standing in bacterial nomenclature.

II. Spoilage by Soft-Rot *Erwinia* spp.

A. Characteristics of Soft-Rot *Erwinia* spp.

1. *Taxonomy*

The soft-rot *Erwinia* spp. are Gram-negative, non-sporing, motile bacteria capable of growth in both aerobic and anaerobic conditions. Their properties have been summarized by Lelliott (1974) and discussed in detail by Graham (1964, 1972), Dye (1969), Dickey (1979) and Pérombelon and Kelman (1980). Within the *Erwinia carotovora* group, five species or subspecies are recognized that cause soft rots: *E. carotovora* subsp. *atroseptica (Eca), E. carotovora* subsp. *carotovora (Ecc), E. chrysanthemi (Echr.), E. cypripedii* and *E. rhapontici.* The last two species cause, respectively, brown rot of cypripedium orchids and crown rot of rhubarb; as there is little information about these species they will not be considered further. An additional subspecies, "*E. carotovora* subsp. *betavasculorum*", that causes soft-rot and vascular necrosis of sugar beet has recently been described (Thomson *et al.*, 1981).

2. Distribution and Vegetables Attacked

According to Chupp and Sherf (1960), the *E. carotovora* group of bacteria is probably the major single cause of microbial spoilage of vegetables. In a survey of crop losses due to prokaryotic plant pathogens in the U.S.A., the losses due to *E. carotovora* in 1976 were estimated as costing $28 million (Kennedy and Alcorn, 1980).

Ecc causes soft rots of a wide variety of vegetables and vegetable-fruits both in the field (Chupp and Sherf, 1960; Ogilvie, 1969) and after harvest (Ramsey *et al.*, 1959; Ramsey and Smith, 1961; Smith *et al.*, 1966; McColloch *et al.*, 1968; Smith and Wilson, 1978). *Eca* has been found mainly in association with potatoes; in temperate regions it is the predominant cause of blackleg in the field and is responsible for a major part of the post-harvest bacterial spoilage of potatoes (Pérombelon and Kelman, 1980). Although *Eca* is capable of attacking most types of vegetable, it is apparently less frequent as the cause of spoilage of vegetables other than potato than is *Ecc*; however, it has been found as the cause of a stem rot of tomato (Barzic *et al.*, 1976) and has been isolated with *Ecc* from rots of cauliflower (Lund and Wyatt, unpublished) and cabbage (Keller and Knösel, 1980). *Erwinia* spp. have caused soft rot and vascular necrosis of sugar beet in the U.S.A. (Ruppel *et al.*, 1975; Stanghcllini *et al.*, 1977; Thomson *et al.*, 1977) and Californian isolates have been characterized as "*E. carotovora* subsp. *vasculorum*" (Thomson *et al.*, 1981).

Echr. attacks a wide range of tropical and sub-tropical crops (Dickey, 1979) and greenhouse plants such as saintpaulia, carnation, philodendron and dieffenbachia grown in temperate regions. Some strains attack maize (*Zea mays*) (Dickey, 1981). Other strains attack pineapple fruit (Lim and Lowings, 1979); much of the decay occurs in the field but it can also develop after harvest (Thompson, 1937). Both *Ecc* and *Echr.* were reported to cause core rot found in carrots at harvest in Texas (Towner and Beraha, 1976).

B. Characteristics of Spoilage

1. Symptoms

The symptoms of bacterial soft rot are similar on most harvested vegetables. Initially, a small area of the tissue becomes water-soaked and soft. Maceration of the parenchyma occurs, giving rise to an area of rotting that can spread rapidly in suitable conditions, causing complete collapse of the tissue. Growth of secondary, non-rotting bacteria occurs, contributing to the offensive nature of the decayed tissue. On cauliflower curd, infection often starts on regions where florets have become bruised, and on vegetables such as cauliflower, cabbage and lettuce rotting may spread from the cut stem if this has remained moist. Infection of potatoes often starts from lenticels, but can also occur through wounds, via the stolon, or following infection with other pathogens (Lund, 1979; Pérombelon and Kelman, 1980). The extension of soft-rot lesions depends on the presence of moist conditions; in the case of potatoes it depends on the presence of moisture and factors that lead to anaerobiosis in the tubers. If the plant tissue is exposed to a dry atmosphere, lesions that started as soft rots can become limited, and the tissue may collapse, giving rise to crater-like lesions such as those characteristic of "crater rot" and blotch of celery (Ark, 1945) and "hard rot" of potato (Logan, 1964). Rotting of cauliflower florets by *Ecc* resulted in increased formation of ethylene by the tissue (Lund and Mapson, 1970); such an effect could lead to an increase in the rate of senescence of adjacent produce.

2. *Mechanism*

The ability of *Eca*, *Ecc* and *Echr.* to macerate plant tissue results mainly from their ability to degrade the pectic components of plant cell walls (Lund, 1979; Rombouts and Pilnik, 1980; Lund, 1982a). Pectinesterase and several types of lyase and hydrolase enzyme may be formed; the majority of workers have concluded that the most important enzymes involved are probably extracellular, endo-pectate lyases. Whereas a wild-type strain of *Echr.* that formed both a pectate lyase and a pectate hydrolase was able to macerate plant tissue, a mutant strain lacking pectate lyase, but which formed an uncharacterized pectate hydrolase, was unable to cause maceration (Chatterjee and Starr, 1977). There is evidence that an exo-poly-α-D-galacturonosidase formed by a strain of *Echr.* could be implicated in regulation of the production of extracellular endo-pectate lyase (Collmer *et al.*, 1982), but the significance in tissue breakdown of the endo-polygalacturonases formed by *Erwinia* spp. is not clear.

C. Epidemiology

1. *Methodology*

Soft-rot bacteria usually form a high proportion of the micro-organisms present at the leading edge of an actively spreading rot. Soft-rot erwinias can readily be isolated by sampling tissue from this region and streaking onto media that are not necessarily very selective, but on which the ability of these bacteria to degrade pectate can readily be detected. Many media with varying degrees of selectivity have been devised for this purpose, including those listed by Rombouts (1972) and by Cuppels and Kelman (1974). The media can be divided into the following three types:

(1) *Single-layer media solidified with agar*, in which degradation of pectate or pectin is detected by flooding the plates with a solution of a quaternary ammonium compound (e.g. the media of Jayasanker and Graham, 1970; Hankin *et al.*, 1971; Burr and Schroth, 1977).

(2) *Single-layer media solidified with pectates or pectin*; degradation of the polymer by bacteria results in the formation of a depression around the colony (e.g. the medium of Cuppels and Kelman, 1974).

(3) *Double-layer media in which the lower layer is solidified with agar and the upper layer with pectate or pectin*. Colonies of bacteria that degrade pectate or pectin ("pectolytic bacteria") are surrounded by a characteristic depression of the pectate gel (e.g. the media of Paton, 1959; Stewart, 1962; Logan, 1963; Pérombelon, 1971a). To obtain the best results these media must be well dried before use, so that no free moisture is present.

Stewart's medium, which has been widely used for detection of soft-rot erwinias, consists of MacConkey agar with an overlayer of pectate gel. Pérombelon (1971a) reported that the bile salts in MacConkey agar could have a toxic effect on *Eca* and *Ecc* and that substitution with 20 mg litre^{-1} of a suitable anionic surface-active agent resulted in an increase of up to 50% in counts. Cuppels and Kelman's (1974) crystal violet pectate (CVP) medium was reported (O'Neill and Logan, 1975) to be more selective than those of Stewart (1962), Logan (1963) and Pérombelon (1971a), particularly when the CVP medium was modified by the incorporation of manganous sulphate, 0.8 g litre^{-1}. Production of foam during preparation of the medium could be overcome by the addition of an anti-foam agent (O'Neill and Logan, 1975), but the medium has to be poured into Petri dishes immediately after autoclaving, since it solidifies rapidly and cannot be re-melted. Low numbers of soft-rot erwinias in plant tissue or in soil in the presence of a high background population of other bacteria can be counted by a quantal method that is reported to be more sensitive than a method based on a count of numerous colonies (Pérombelon, 1971b).

In order to detect low numbers of soft-rot erwinias in soil or in other situations where little or no enrichment has occurred, enrichment has been induced in plant tissue such as potato or carrot (Graham, 1958; Kikumoto and Sakamoto, 1967; Togashi, 1972). Alternatively, enrichment may be made by anaerobic incubation in a pectate–basal salts solution (Meneley and Stanghellini, 1976) or by incubation in the presence of asparagine (Butler, 1980). Following enrichment, the soft-rot bacteria can be detected by plating on a pectate medium or by serology. The method of Meneley and

Stanghellini (1976) was reported to be more sensitive than the use of potato tissue and to avoid the possible problem of latent infection of the potato tissue (De Boer *et al.*, 1979). Meneley and Stanghellini (1976) reported that, using their technique, soft-rot *Erwinia* spp. could reliably be recovered from soils that were artificially infected with more than 10 bacteria g^{-1} dry weight of soil.

2. *Sources of Inoculum, Survival and Dissemination*

The review of Pérombelon and Kelman (1980) should be consulted for detailed information on this topic.

There is little data on the incidence of soft-rot erwinias on the surface of undamaged, fresh vegetables other than potatoes. Rombouts (1972) reported that of 53 representative isolates of pectolytic bacteria from fresh, sound endive, leek, spinach, green cabbage and chicory, none were *Eca*, *Ecc* or *Echr*. On fresh cauliflower curd the numbers of *E. carotovora* present were too low to detect by direct plating of dilutions, but the presence of these bacteria was revealed by the high numbers found on curds that subsequently showed spoilage (Lund, 1981).

Eca and *Ecc* are commonly present on potatoes in the U.K. Extensive contamination of progeny tubers can occur after decay of the mother tuber, which is probably caused to a large extent by soft-rot erwinias. The bacteria survive in lenticels and to a lesser extent on the surface of the tuber during storage for 6–7 months (Pérombelon, 1974; Pérombelon and Kelman, 1980).

In some circumstances the presence of soft-rot bacteria including *Erwinia* spp. has been reported in the internal tissue of sound vegetables. In field-grown cucumbers, 56% contained soft-rot bacteria including *Erwinia* spp. (Meneley and Stanghellini, 1975). The number of bacteria present per gram of tissue was not determined, and soft rot only occurred after incubation of cucumber at high temperature (37°C) or inoculation with *Pythium aphanidermum*. The percentage of field-grown cucumbers infected varied from 0 to 100% in different locations, and of over 100 greenhouse-grown fruits collected over a period of 1 year most were free of internal microorganisms and none contained soft-rot bacteria. Low numbers of soft-rot erwinias have also been found in stems of healthy Chinese cabbage (Kikumoto and Ômatsuzawa, 1981).

The subject of the survival of soft-rot erwinias in soil and the importance of the soil as a source of inoculum has been controversial, and conclusions have depended on the sensitivity of the detection methods used. Several studies have shown that following introduction into non-sterile soil the numbers of *Eca* and *Ecc* declined rapidly; in general, the soft-rot erwinias appear to survive poorly in soil and detection requires sensitive enrichment procedures. These bacteria can, however, survive over winter in contaminated plant residues that remain in the field after harvest, and in association with volunteer plants of affected crops. They also survive in the rhizosphere of crops such as those of cruciferous plants (Kikumoto and Sakamoto, 1969a, b; Togashi, 1972; Voronkevitch *et al.*, 1972; Kikumoto, 1974; Mew *et al.*, 1976), of lettuce, carrot and certain weeds (Burr and Schroth, 1977) and of sugar beet, wheat, maize (*Zea mays*) and lupin (de Mendonça and Stanghellini, 1979). Soft-rot erwinias could first be detected in the soil of the rhizosphere and phyllosphere of Chinese cabbage about 40 days after planting seed. About 10 days later the bacteria could be detected in soil collected from almost all the individual cabbages tested, the number ranging from 10^6 to 10^7 bacteria g^{-1}; the erwinias could not be detected in non-rhizosphere soil, indicating that they were present at less than 10^2 bacteria g^{-1}. At about 60 days after planting seed, soft-rot erwinias were found within the healthy stems of almost all heads of Chinese cabbage tested, the number present ranging from 10^3 to 10^7 bacteria per stem (Kikumoto and Ômatsuzawa, 1981). The infection did not necessarily result in symptoms of disease; whether it did so was dependent on the resistance of the host plant and on environmental conditions of temperature and humidity.

Whereas soft-rot erwinias do not usually survive in or on true seed, they can be introduced on planting material for vegetatively produced crops (Pérombelon and Kelman, 1980). This has been demonstrated particularly in the case of seed potatoes. Surveys in Scotland, Germany and the U.S.A. have shown that a high proportion of tubers from commercial seed stocks was contaminated with *Eca* and *Ecc*, and *Echr.* has also been detected on tubers grown in Japan and Peru. The erwinias were present on the surface of the tuber, in lenticels and in tissue damaged by cracks and bruises; a high proportion of those present in lenticels survived storage of tubers over winter. In addition to soil and planting material, other sources of soft-rot erwinias are diseased plants, refuse dumps of plant debris and potato cull piles.

Dissemination of these bacteria can occur in soil water, resulting in transfer from decaying mother tubers to daughter tubers on the same plant and on adjacent plants. Aerial transfer of the bacteria between crops or from refuse dumps to crops can result from carriage by insects and possibly by birds. Aerosols containing the bacteria can be generated by the impact of rain on diseased stems or by pulverizing potato stems before harvest, as is done in Scotland; it has been estimated that viable bacteria can thus be blown for a distance of several hundred metres. Soft-rot bacteria can also be spread on contaminated farm implements such as knives used for cutting seed potatoes, graders, tractors and harvesting machinery.

Survival of soft-rot erwinias in association with certain crops may result in the presence of a relatively high inoculum that could increase the risk of spoilage of a subsequent crop. Using the enrichment technique of Meneley and Stanghellini (1976), *Ecc* could be detected, at the time of planting peppers, in field soil at sites previously planted with cabbage or potatoes but not at sites previously planted with soybeans (Coplin, 1980). Fruit from plants on the former sites, when stored at a high humidity, showed a higher incidence of soft rot than fruits from plants on the latter sites. It was suggested that rotation of peppers with crops such as soybean, rather than with cabbage or potatoes, could help to minimize soft rot caused by *Erwinia* spp.

III. Spoilage by *Pseudomonas marginalis*

A. Characteristics of *P. marginalis*

1. *Taxonomy*

Pseudomonas spp. are Gram-negative, non-sporing, motile rods that require aerobic conditions for growth, except that strains capable of denitrification are able to grow anaerobically in media containing nitrate. Many soft-rotting strains of oxidase-positive, fluorescent pseudomonads have been identified as *P. marginalis* or as Group IVb pseudomonads (Lelliot *et al.*, 1966). Apart from their ability to degrade pectate, *P. marginalis* and similar pseudomonads resemble *P. fluorescens* biotype II (Doudoroff and Palleroni, 1974) and other biotypes (Cuppels and Kelman, 1980; Brocklehurst and Lund, 1981). In the present state of knowledge regarding the taxonomy of these bacteria, it has been suggested that the name *P. marginalis* should be retained as a means of identifying all soft-rot-causing, oxidase-positive, fluorescent pseudomonads (Cuppels and Kelman, 1980) and this practice will be followed in the present chapter.

2. *Distribution and Vegetables Attacked*

Strains of *P. marginalis* are capable of causing field disease and post-harvest spoilage of many types of vegetable. They have been shown to cause field disease of lettuce (Brown, 1918; Paine and Branfoot, 1924; Burkholder, 1954; Ramsey *et al.*, 1959; Berger, 1967; Köhn, 1973; Pieczarka and Lorbeer, 1975), forced rhubarb (Clark and Graham, 1962), cucumber (Ohta *et al.*, 1976), leeks (Garibaldi and Tamietti, 1977), fennel (Surico and

Iacobellis, 1978a) and celery (Surico and Iacobellis, 1978b). They have also been reported as a cause of post-harvest spoilage of celery (Harrison and Barlow, 1904), chicory (Friedman, 1951b), lettuce (Dowson, 1941; Ceponis, 1970; Beraha and Kwolek, 1975), Chinese chard (Burton, 1971) and cabbage (Bobbin and Geeson, 1977), and have been associated with pink-rib of lettuce (Hall *et al.*, 1971).

There have been occasional reports of storage rots of potatoes caused by *P. marginalis* (Dowson, 1941; Garrard, 1945). Strains of *P. fluorescens* were reported to be consistently associated with red xylem and pink-eye diseases of potatoes in the U.S.A., suggesting a causal relationship (Folsom and Friedman, 1959), but this has not been conclusively demonstrated (Cuppels and Kelman, 1980). In favourable conditions, some of the isolated strains caused rotting of potatoes, and formation of pectic enzymes by some strains has been demonstrated (Huether and McIntyre, 1969). Red xylem is characterized by the presence of a sunken lesion at the stolon scar, a cavity extending from the scar to the centre of the tuber, and a reddish discolouration of the vascular ring; potatoes affected by pink-eye show pink discolouration of the tissue which remains firm and corky, and black regions and cavities may occur (Folsom and Friedman, 1959; Smith and Wilson, 1978).

Pseudomonas marginalis pv. *pastinaceae* (ISPP List 1980) was isolated by Burkholder (1960) as the cause of rotting of parsnip roots that had over-wintered in the field and were harvested in the spring, and named "*P. pastinaceae*". This organism was indistinguishable from *P. marginalis* in the tests used by Lelliott *et al.* (1966). Burkholder, however, reported that two different isolates of *P. marginalis*, unlike "*P. pastinaceae*", caused no infection when inoculated into parsnips.

B. Characteristics of Spoilage

1. *Symptoms*

The soft-rot symptoms caused by *P. marginalis* are, in general, similar to those caused by *E. carotovora*, but in the most favourable conditions soft-rot pseudonomads are probably less virulent than the erwinias. There are some reports that defects other than soft rots have been caused by *P. marginalis*. Vascular browning and russet-spotting have sometimes been observed on head lettuce infected with *P. marginalis*, and these symptoms have been reproduced by treatment of lettuce with sterile filtrates of *P. marginalis* cultures (Ceponis and Friedman, 1959). An unusual, pectate-degrading strain of *P. fluorescens* biotype 1 that formed a bluish-black, diffusible pigment was reported to be responsible for vascular blackening in witloof chicory which was not accompanied by soft rot (Sellwood *et al.*, 1981).

2. *Mechanism*

The ability of *P. marginalis* to macerate plant tissue results, like that of *E. carotovora*, from its ability to form pectic enzymes. Several types of enzyme have been reported (references cited in Rombouts *et al.*, 1978) and the range of enzymes formed may differ between strains. During growth of two strains in culture medium both formed pectinesterase and pectate lyase, but only one formed detectable amounts of polygalacturonase (Nasuno and Starr, 1966). The pectate lyase enzymes have received the most study: a strain of *P. fluorescens* was reported to form an endo-pectate lyase when grown in a culture medium or on peeled potatoes, and the purified enzyme macerated potato tissue (Rombouts *et al.*, 1978). No detectable pectinesterase or polygalacturonase was formed.

C. **Epidemiology**

1. *Methodology*

The same type of solid media as those used for isolation of *E. carotovora* can also be used for *P. marginalis*. Stewart's (1962) medium is not useful: although colonies are formed, depressions due to degradation of pectate are not obvious. On Paton's (1959) medium several types of pectate-degrading colony are formed by different strains of *P. marginalis*: these colonies are quite distinct from those formed by *E. carotovora*. A comparison of colonies of one group of strains of *P. marginalis* with those of *E. carotovora* on CVP medium was shown by Cuppels and Kelman (1980).

2. *Sources of Inoculum, Survival and Dissemination*

Fluorescent pseudomonads form a major proportion of the bacteria present on the surface of leafy vegetables, at least in moist, temperate regions, and *P. marginalis* and related strains have been reported to form the majority of the actively pectolytic bacteria present. Rombouts (1972) reported that the numbers of pectolytic bacteria detected on fresh spinach and on the inner, sound tissue of fresh cabbage were about $10^4 \, g^{-1}$, whereas those on leek and endive were about $10^5 \, g^{-1}$. On five fresh vegetables, endive, leek, spinach, green cabbage and chicory, the majority of pectolytic bacteria were *Pseudomonas* spp., 75% of which were fluorescent, oxidase-positive strains. On the outer leaves of freshly harvested, winter-white cabbage, up to 1.2×10^4 pectolytic bacteria cm^{-2} were detected, 80% of which were fluorescent pseudomonads resembling *P. marginalis* (Geeson, 1979). Up to 10^5 pectolytic bacteria were present per gram of freshly harvested, sound cauliflower curd, the majority being *P. marginalis* (Lund, 1981). In their study of bacteria from the internal tissues of healthy, surface-sterilized, field-grown cucumbers Meneley and Stanghellini (1972) reported that the most frequently occurring soft-rotting bacteria were *Pseudomonas* spp., and it is likely that many of these were *P. marginalis*. Pectolytic pseudomonads have also been found on the surface of potato tubers, where their numbers ranged from 2×10^3 to $2 \times 10^6 cm^{-2}$ of tuber surface (Cuppels and Kelman, 1980). There is evidence that, like other Gram-negative bacteria, *P. marginalis* is often more numerous in the rhizosphere of plants than in plant-free soil (Cuppels and Kelman, 1980).

As a result of isolations using immature potato tubers as an enrichment medium, Graham (1958) reported that most isolations from soil in Scotland yielded soft-rotting, fluorescent pseudomonads rather than *E. carotovora*; these pseudomonads, but not added *Eca*, survived over winter in soil. *Pseudomonas marginalis*, but not *E. carotovora*, could be isolated by direct plating from field soils collected in early spring in Wisconsin, U.S.A., where it appeared to have survived over winter (Cuppels and Kelman, 1980). In soil sampled from Wisconsin corn fields immediately after the ground had thawed, the level of *P. marginalis* was approximately 6.8×10^3 to $8.5 \times 10^3 g^{-1}$ dry wt of soil, and under unusually wet conditions in the muck soil of a carrot field the numbers reached $10^9 g^{-1}$ dry wt of soil (Cuppels and Kelman, 1980).

It is clear from these observations that *P. marginalis* is regularly present in relatively high numbers in soil and on plant surfaces. Although some measures can be taken to minimize the numbers present, for example by avoidance of a build-up of decaying plant debris, it must be assumed that relatively high numbers of these bacteria will always be present on vegetables and will be a potential cause of soft rot.

IV. Rots Caused by Other Species of *Pseudomonas*

A. *Pseudomonas viridiflava*

This organism is a fluorescent pseudomonad that can be distinguished from *P. marginalis* by the fact that it gives a negative reaction in tests for oxidase and for arginine dihydrolase activity, and from *P. syringae* by the fact that it causes soft rot of potato (Lelliott *et al.*, 1966; Billing, 1970; Doudoroff and Palleroni, 1974).

Pseudomonas viridiflava was first described as the cause of reddish-brown necrotic lesions on a market sample of green bean pods (Burkholder, 1930), and was reported to invade bean plants, producing necrotic lesions without water-soaking. Billing (1970) reported that this bacterium was a common member of the epiphytic flora of dwarf beans, and had been isolated from a variety of plants. It has also been associated with field outbreaks of disease in several plants, including soft rot of cauliflower leaves, internal stem rot of tomato and wet rot of pea plants, and has been shown to cause soft rot of cabbage and lettuce leaves, as well as necrotic lesions on bean pods (Wilkie *et al.*, 1973).

Part of the effect of *P. viridiflava* on plants is probably due to its ability to form pectic enzymes. It caused pitting of pectate gels at pH 6.9–7.1 and pH 8.3–8.5 but not at pH 4.9–5.1 (Hildebrand, 1971). Pectic and cellulolytic enzymes were formed in culture, and after inoculation into *Phaseolus vulgaris* and *Vicia faba* the diseased plants showed increased levels of poly-galacturonase, and pectate and pectin lyases, compared with healthy plants (Cabezas de Herrera and Jurado, 1975).

Pseudomonas viridiflava forms craters on the medium of Paton (1959), which can be used to aid isolation; biochemical tests are then necessary to distinguish the organism from *P. marginalis*. Burkholder (1930), Lelliott *et al.* (1966) and Billing (1970) commented on the ease with which some cultures lost their virulence, but this was not the experience of Wilkie *et al.* (1973). It is possible that this bacterium is occasionally the cause of soft rot of vegetables, and it may have been reported as *P. syringae* or *P. marginalis*.

B. *Pseudomonas cichorii*

Pseudomonas cichorii is a fluorescent pseudomonad that can be distinguished from remaining fluorescent species by the fact that it gives a positive reaction in the oxidase test, a negative reaction for "arginine dihydrolase", fails to cause soft rot of potato slices and causes a hypersensitive reaction in tobacco (Lelliott *et al.*, 1966; Doudoroff and Palleroni, 1974).

It has been isolated as the cause of field disease in chicory and endive (Swingle, 1925; Kotte, 1930), cabbage (Wehlburg, 1963), celery (Thayer and Wehlburg, 1965; Wilkie and Dye, 1974), tomato (Wilkie and Dye, 1974), lettuce (Burkholder, 1954; Grogan *et al.*, 1977), cauliflower (Coleno *et al.*, 1972) and of other plants (Wilkie and Dye, 1974) and of post-harvest spoilage of cabbage (Smith and Ramsey, 1956) and lettuce (Ramsey *et al.*, 1959; Ceponis, 1970; Ceponis *et al.*, 1970).

On cabbage, two main types of lesion have been described (Wehlburg, 1963). The first type occurred as grey-brown to dark-brown, slightly sunken, round or oval spots from 1 to 5 mm in diameter ("plain spots"). The second type of lesion was characterized by the formation of concentric rings of light-coloured zones alternating with brown lines; this type of lesion was described by Smith and Ramsey (1956) as "bacterial zonate spot". According to Wehlburg (1963), the plain type of lesions was formed when unwounded cabbage leaves were sprayed with a suspension of the bacterium, whereas the zonate spot lesion developed on heads that were inoculated in wounds made with a needle or scalpel. The tissue was reported to remain firm unless soft-rot bacteria also became involved. Burkholder (1954) referred to the symptom produced on lettuce by *P. cichorii* as a rot and stated that the organism rotted slices of potato tubers, a report that was not confirmed by Lelliott *et al.* (1966). According to Burkholder, only a few isolates liquefied a pectate gel. Ramsey *et al.* (1959), Ceponis (1970) and Ceponis *et al.* (1970) referred to isolation of *P. cichorii* from lettuce affected by bacterial soft rot.

Pseudomonas cichorii was shown to cause "varnish spot", a destructive disease of mature head lettuce that occcurred sporadically in California (Grogan *et al.*, 1977). This disease was characterized by the formation of dark brown, shiny, firm necrotic spots on the blades and petioles of leaves underneath the second or third outermost leaves. Symptoms were only evident after heads were cut open, preventing selective harvest of non-infected heads. The disease was only observed in sprinkler-irrigated fields.

There appears to be no information on the mechanism of formation of lesions by this bacterium and it is doubtful whether it causes a true soft rot. Five strains studied by Hildebrand (1971) failed to show breakdown of pectate at pH 4.9–5.1, pH 6.9–7.1 or pH 8.3–8.5.

C. *Pseudomonas cepacia* (Palleroni and Holmes, 1981)

This pseudomonad does not form fluorescent pigments but produces a variety of non-fluorescent pigments, often yellow or greenish, that may diffuse into the medium or remain bound to the cells (Palleroni and Holmes, 1981). The use of ultraviolet light at a wavelength of 254 nm is necessary to distinguish the fluorescent from non-fluorescent pigments. *Pseudomonas cepacia* is an obligate aerobe and is nutritionally the most versatile pseudomonad at present known, being capable of using a wide variety of organic compounds as sole sources of carbon and energy.

Burkholder (1950) described *P. cepacia* as the cause of "sour skin", a rot observed on onions at harvest and during early storage that affected certain of the outer, fleshy scale leaves, although not necessarily the outermost (Smith *et al.*, 1966; Bazzi, 1979). Infected scales were slimy and yellow and had a sour, vinegar-like odour. According to Burkholder (1950), necrotic tissue yielded a high number of bacteria, most of which were saprophytes, and the pathogen formed a very low proportion of the bacteria isolated. It was suggested that infection occurred when onion tops were cut at harvest, and that the increased use of mechanical toppers might contribute to the incidence of the disease (Burkholder, 1950). Kawamoto and Lorbeer (1974) concluded that *P. cepacia* was not very invasive in onion, infection requiring free moisture and high temperature. They attributed the frequency of rots of onion caused by *P. cepacia* in New York State to the occurrence of rainstorms and flooding, which resulted in wounding and localized congestion of onion leaves.

Five strains of *P. cepacia* studied by Hildebrand (1971) caused pitting of pectate gels at pH 4.9–5.1 and 6.9–7.1 but not at pH 8.3–8.5. Inoculation of *P. cepacia* onto onion slices resulted in maceration and a decrease in the pH of the juice from 5.5 to about 4 (Ulrich, 1975). Pectinesterase and polygalacturonase were present in the macerate, but pectate lyase was seldom detected. In contrast, when the bacterium was grown in a culture medium at pH 7.2 with pectate as substrate very high levels of pectate lyase activity were detected. The purified polygalacturonase showed maximum activity at pH 4.4–4.6, probably attacked the substrate in a random manner, and caused maceration of onion tissue.

Pseudomonas cepacia is thought to be widely distributed in soil but there appears to be little quantitative information on its incidence. This organism is also of clinical importance: it is regarded as a low-grade pathogen that is potentially dangerous because of its ability to survive and possibly grow in some of the disinfectants used in hospitals (Snell *et al.*, 1972). There is evidence that strains of *P. cepacia* isolated from plants can, however, be distinguished from those of clinical origin by differences in production of bacteriocins, maceration of onion slices, hydrolysis of pectate agar at pH 4.9 and the size of resident plasmids (Gonzalez and Vidaver, 1979).

D. *Pseudomonas gladioli* pv. *alliicola* (ISPP List, 1980)

"*Phytomonas alliicola*" was isolated by Burkholder (1942) as the cause of rot of onion bulbs. Subsequent workers (Ballard *et al.*, 1970; Hildebrand *et al.*, 1973) have reported that using nutritional and biochemical tests and DNA–DNA homology "*Pseudomonas alliicola*" was indistinguishable from "*P. marginata*" and *P. gladioli*, two species first described as the cause of rots of leaves and of corms, respectively, of *Gladiolus* spp. *"Pseudo-monas alliicola"* and *"P. marginata"* have been reported to cause similar symptoms when inoculated into onion bulbs (Taylor and Holden, 1977). The close relationship between these bacteria is reflected in the currently accepted name. *Pseudomonas gladioli* pv. *alliicola* does not form a fluore-scent pigment, but forms a diffusible, non-fluorescent, pale yellow to yellowish-green pigment, and rarely a reddish pigment (Hildebrand *et al.*, 1973).

According to Burkholder (1942), rots caused by *"Phytomonas alliicola"* were observed infrequently in onions. In the early stages of the disease the bulb could appear sound on the outside, but on cutting it open some of the inner scales appeared water-soaked and soft. In advanced stages the entire bulb could be affected. The organism also caused soft rot, affecting the fleshy scale leaves of onions imported into the U.K. (Roberts, 1973) and of onions stored in the U.K. (Taylor, 1975). After inoculation into onions the bacterium caused rots within 4 weeks in onions stored at 25°C, but not in those stored at 15°C (J. D. Taylor, personal communication). Rotting is likely, therefore, to be facilitated by high-temperature drying of onions but arrested during subsequent low-temperature storage (Taylor *et al.*, 1980).

Burkholder (1942) noted that even at the advancing margin of a rot the plant tissue was filled with bacteria. Hildebrand (1971) reported that 7/7 cultures named *"P. allicola"* and 6/7 named *"P. marginata"* caused con-siderable degradation of pectate gels at pH 4.9–5.1 and pH 6.9–7.1, but not at pH 8.3–8.5. There is little further information relating to the mechanism of symptom formation by this organism or to its survival and dissemination (see Ch. 4).

E. *Pseudomonas aeruginosa*

Pseudomonas aeruginosa typically forms fluorescent pigment(s) but can be distinguished from the remaining species of fluorescent pseudomonads by formation of the blue pigment pyocyanin, possession of a single polar flagellum and the ability to grow at 41°C (Doudoroff and Palleroni, 1974). This bacterium has played a particularly important role as an opportunist human pathogen. The properties of the bacterium that lead to this situation are: resistance to antibiotics and disinfectants and consequent selection in the hospital environment, ability to grow in water with very low levels of nutrients, and production of enzymes and toxins affecting mammalian tissue (Lowbury, 1975).

There have been a few recent reports that *P. aeruginosa* can occasionally cause post-harvest spoilage of vegetables. An outbreak of spoilage of onion bulbs in Australia in 1974 was attributed to *P. aeruginosa* (Cother *et al.*,

1976). The symptoms took the form of a yellow-brown discolouration of leaf bases, which remained firm until soft rot occurred due to secondary infection. The pathogenic isolates showed greater similarity to *P. aeruginosa* than to *P. fluorescens* and formed a blue-green diffusible pigment that resembled pyocyanin, but different from typical strains of *P. aeruginosa* in frequently possessing more than one polar flagellum. When inoculated into whole onions by stabbing, or onto sterilized transverse sections through leaf bases, the isolates caused discolouration. They failed to cause symptoms when injected into leaves or when infected scissors were used to cut the leaves or roots of bulbs which were then stored at 20–27°C with or without exposure for 48 h to high humidity; it was concluded that specific, extreme conditions were required for infection to take place. The disease has since occurred as an occasional problem in crops treated by overhead irrigation (E. J. Cother, personal communication).

In a shipment of tomatoes from Florida that showed an excessive amount of decay, 17% of the "decay bacteria" isolated were identified as *P. aeruginosa*, 17% were *P. marginalis* and 66% were *Erwinia* spp. (Bartz, 1980). The isolates of *P. aeruginosa* were much less virulent at 25°C than those of *P. marginalis* and *E. carotovora*. After inoculation by vacuum-infiltration and incubation at 25°C, *E. carotovora* and *P. marginalis* caused rots in 24 h on 93% and on 53% of fruit, respectively, whereas on fruit inoculated with *P. aeruginosa*, lesions were visible only after 6 days and on 30% of fruit. The latter organism caused lesions more rapidly at 30–35°C, temperatures that were reached during the period in which the tomatoes were harvested. It was concluded that the lesions originated internally and that fruit became inoculated with bacteria by vacuum-infiltration when tomatoes were immersed in water in dump tanks.

The extent to which *P. aeruginosa* can be regarded as a potential cause of disease or spoilage of plant tissue is controversial. Elrod and Braun (1942) concluded that *P. aeruginosa* was indistinguishable from two isolates of "*P. polycolor*", a bacterium that was isolated as the cause of an important disease of tobacco in the Philippines (Clara, 1930), and reported that *P. aeruginosa* caused pathogenic symptoms when spray-inoculated into water-soaked tobacco leaves and caused soft-rot on sterile slices of potato, cucumber and onion. This report differs from that of Clara (1934), who found that "*P. polycolor*" caused lesions when inoculated into tobacco plants, but not in plants of 17 other types, whereas *P. aeruginosa* failed to cause lesions on tobacco or on any of the other plants. The significance of Elrod and Braun's (1942) report seems doubtful in view of the fact that they considered that *P. aeruginosa* was identical with "*B. marginale*" (*P. marginalis*). These species can be clearly differentiated (Friedman, 1960; Misaghi and Grogan, 1969; Doudoroff and Palleroni, 1974) and the ability of *P. marginalis* to form pectic enzymes and to cause soft rots is well established (see section III), whereas formation of pectic enzymes by *P. aeruginosa* has not been reported. Friedman (1960) considered that there was insufficient evidence to show that *P. aeruginosa* played a role as a plant pathogen. On the basis of a study of nutritional and biochemical properties of plant pathogenic and saprophytic pseudomonads, a high percentage similarity was reported between "*P. polycolor*" and *P. aeruginosa* (Misaghi and Grogan, 1969). Neither bacterium caused soft rot of potato slices. "*Pseudomonas polycolor*" failed to cause pitting of pectate and pectin gels unless media were supplemented with 0.5% glucose or 0.5% succinate (Hildebrand, 1971). Although pitting of pectate media is presumptive evidence of formation of pectic enzymes, it remains to be demonstrated that such enzymes are formed by "*P. polycolor*" or by *P. aeruginosa*.

When leaves of lettuce or bean plants were vacuum-infiltrated with a suspension containing 7×10^6 *P. aeruginosa* ml^{-1} and maintained at 27°C and 80–95% relative humidity, the bacteria multiplied within the tissue and the onset of yellowing and wilting of leaves was more rapid than that in leaves similarly treated but not inoculated (Green *et al.*, 1974). Inoculation of isolates of *P. aeruginosa* from plants and from clinical sources onto wounded leaves of lettuce, onto celery stalks or onto potato slices resulted in rots (Cho *et al.*, 1975), but no evidence was given that Koch's postulates were fulfilled, and it is not clear to what extent the rots involved breakdown of tissue.

Severe conditions, such as the presence of high inocula and water-soaking of plant tissue, appear to be necessary to enable infection of plant tissue by *P. aeruginosa*. Prolonged water-soaking alone can damage plant tissue and has been shown to cause necrosis in tobacco leaves (Valleau *et al.*, 1939). Water-soaking can greatly increase the susceptibility of plant tissue to infection and on water-soaked tomatoes a small amount of necrosis could be induced by spraying with *Escherichia coli* (Johnson, 1937). When bean leaves were maintained in a water-soaked condition, the non-pathogens *Pseudomonas putida* and *Erwinia herbicola* were able to multiply in the tissue (Young, 1974). In these circumstances an increase in yellowing and wilting of leaves, such as that observed by Green *et al.* (1974) in lettuce inoculated with *P. aeruginosa*, may result from the effect of a variety of bacterial metabolites or from the action of proteolytic or other enzymes.

In certain circumstances such as those described by Cother *et al.* (1976) and by Bartz (1980), a high inoculum of *P. aeruginosa* may occur in conjunction with extreme conditions that enable the bacterium to infect plant tissue and cause lesions. The circumstances in which this occurs appear to be exceptional. In considering the importance of *P. aeruginosa* as a cause of spoilage, further studies are required of the strains involved and of the mechanism of their effect on plant tissue.

V. Spoilage by *Bacillus* spp.

Bacillus spp. are rod-shaped, spore-forming bacteria; they may be Gram-positive, Gram-positive only during the early stages of growth or Gram-negative. Many species, including *Bacillus subtilis*, require aerobic conditions for good growth, but some, including *B. polymyxa*, grow well in both aerobic and anaerobic conditions (Gibson and Gordon, 1974).

There are a few reports that species of *Bacillus* are able to cause spoilage of vegetables. Brierley (1928) isolated "*B. mesentericus*" (probably *B. subtilis*) from a wound rot of stored potato tubers. The organism readily caused rotting of tubers at 20°C and above but invaded tubers very feebly at temperatures lower than this. The *Bacillus* failed to attack carrot or turnip roots, cabbage heads and other plant parts. Brierley commented that "*B. mesentericus* is so common on imperfectly sterilized potato tissue that it is sometimes called the potato bacillus in American and European texts on bacteriology." A *Bacillus* considered to belong to the *B. subtilis* group was isolated as the cause of soft rot in tomatoes and shown to cause extensive rot at 36–40°C (Madhok and Fazel-Ud-Din, 1943). Dowson (1943) reported that *B. polymyxa*, isolated from stored potatoes affected by bacterial soft rot, caused rots of slices of potato, carrot, onion and cucumber at temperatures varying from ambient to 37°C. *Bacillus polymyxa* was readily isolated from all the types of Alberta soil tested (Jackson and Henry, 1946). All the isolates were able to rot potato slices within 3 days at 30°C. Pure cultures caused rots of potato slices at temperatures of 20–45°C, whereas whole tubers were not rotted below 30°C; this was despite the fact that some of the strains of *B. polymyxa* were capable of fermenting wheat mashes at a temperature as low as 15°C. A fruit spot of pepper, which occurred widely in Palestine, was reported to be caused by *B. polymyxa*, which was shown to cause lesions when inoculated onto wounded fruit that was then maintained at 28–30°C in humid conditions (Volcani and Dowson, 1948).

The observation by Jackson and Henry (1946) that *B. polymyxa* could readily be isolated from soil has been confirmed by other workers. During isolation of soft-rot bacteria using immature·potato tubers as an enrichment medium at 26°C, Graham (1958) reported that isolation of *B. polymyxa* was fairly common. When enrichment in carrot slices was used to isolate soft-rot bacteria from soil, incubation at 30°C resulted mainly in isolation of *B. polymyxa*, and it was necessary to incubate at 20°C in order to isolate soft-rot erwinias (Togashi, 1972). In this last study it was concluded that soil temperatures were too low for *B. polymyxa* to cause soft rot of Chinese cabbage in the field.

Strains of several species of *Bacillus* form pectic enzymes (Rombouts and Pilnik, 1980); in particular, strains of *B. polymyxa* (Nagel and Vaughn, 1962; Nagel and Wilson, 1970) and *B. subtilis* (Kurowsky and Dunleavy, 1976; Chesson and Codner, 1978) have been shown to form active, extracellular, endo-pectate lyase enzymes. It is therefore not surprising that some strains of *Bacillus* are able to attack plant tissue. The minimum temperature for growth of *B. subtilis* has been reported as between 5 and 20°C, depending on the strain, and that for *B. polymyxa* between 5 and 10°C (Gibson and Gordon, 1974); nevertheless, the relatively few cases in which spoilage due to *Bacillus* spp. has been investigated indicate that these bacteria are only likely to cause decay if produce is exposed to temperatures in the region of 28–30°C. The occurrence of psychrotrophic strains of *Bacillus* able to grow at 1–4°C has, however, been reported in pasteurized milk (Coghill and Juffs, 1979), and the possibility that psychrotrophic strains may in some situations cause spoilage of vegetables should not be ruled out.

VI. Spoilage by *Clostridium* spp.

Clostridium spp. are rod-shaped, spore-forming bacteria that usually require strictly anaerobic conditions for growth, although strains differ in their sensitivity to oxygen. They are generally Gram-positive at least in the early stages of growth (Smith and Hobbs, 1974). Pectate-degrading strains can frequently be isolated from potatoes that show extensive soft rot (Campos *et al.*, 1982; Lund, 1982b). They are usually present with *E. carotovora*, but in certain conditions they are able to act as the primary cause of rots. Some strains only show comparable pathogenicity to that of *E. carotovora* at 22°C (Pérombelon *et al.*, 1979), but others do so at lower temperatures (Lund, 1982b). The susceptibility of potatoes to *E. carotovora* results largely from conditions that lead to depletion of oxygen in the tuber, for example, the presence on tubers of a film of water for several hours, exposure to high temperatures or storage in closed containers (Lund, 1979). This depletion of oxygen also provides suitable conditions for clostridia. It seems probable that in suitable conditions pectolytic clostridia can contribute to spoilage of other root vegetables in addition to potatoes.

Several morphological types of *Clostridium* have been observed in, and isolated from, rotting potatoes (Lund, 1979; Pérombelon *et al.*, 1979; Campos *et al.*, 1982), but few have been characterized. Potato tissue attacked by clostridia is usually completely macerated. Some, but probably not all, strains cause the formation of a slimy rot (Campos *et al.*, 1982). Pink-pigmented strains, *C. puniceum* (Lund, *et al.*, 1981) and a group of low-temperature clostridia when grown in potato tissue caused maceration and formed pectate lyase but no detectable polygalacturonase (Lund and Brocklehurst, 1978; Brocklehurst and Lund, 1982).

A double-layer pectate medium containing polymyxin and incubated under anaerobic conditions (Lund, 1972) has been used for isolation of these bacteria. The principal habitat of clostridia is considered to be soil, and stimulation of anaerobic bacteria, in particular of gas-producing clostridia, has been observed in the rhizosphere of plants (Katznelson, 1946). Between 4.4 and 23.5 × 10³ colony-forming units of pectolytic clostridia were detected per gram of dry soil in which carrots or other arable crops had been grown (Perry, 1982); the number in rhizospheres was up to eight times greater than that in soil and it was suggested that this may be due in part to the occurrence of anaerobic microsites around the root.

VII. Defects Other than Soft Rots

Bacteria that cause field diseases other than soft rots and affect the market quality of vegetables and fruits are listed in Table 9.2.

In other cases, e.g. greywall (bacterial necrosis) of tomato (Beraha and Smith, 1964; Segall, 1967; Stall and Hall, 1969) and rind necrosis of cantaloup and watermelon (Thomas, 1976; Hopkins and Elmstrom, 1977), it has been suggested that defects arise from the presence in the plant tissue of endophytic bacteria that usually fail to multiply but may be enabled to do so when the plant tissue has been subjected to stress, e.g. by chilling. Pink disease of pineapple fruit, which results in the development of a dark brown discolouration of the flesh when heated during the canning process, has been attributed to the entry of bacteria into the fruit and formation of 2,5-diketogluconic acid (Cho *et al.*, 1978).

VIII. Factors Affecting the Development of Soft Rots

A. Intrinsic Factors

A feature of soft-rot bacteria is the lack of specificity with regard to the tissue attacked, so that a wide range of types of vegetable and fruit can be affected. Development of spoilage is likely to be affected by the following factors (Lund, 1982a):

Table 9.2 *Bacteria that cause field disease other than soft rot affecting the market quality of vegetables and fruit*

Bacterium[a]	Produce Affected	References
Corynebacterium flaccumfaciens pv. *flaccumfaciens*	Bean	Smith *et al.* (1966)
C. michiganense pv. *michiganense*	Tomato	McColloch *et al.* (1968)
C. michiganense pv. *sepedonicum*	Potato	Smith and Wilson (1978)
Erwinia ananas	Pineapple	Smoot *et al.* (1971)
Pseudomonas solanacearum	Potato	Smith and Wilson (1978)
P. syringae pv. *apii*	Celery	Smith *et al.* (1966)
P. syringae pv. *lachrymans*	Cucumber, honeydew melon	Ramsey and Smith (1961)
P. syringae pv. *maculicola*	Cauliflower	Ramsey and Smith (1961)
P. syringae pv. *phaseolicola*	Bean	Smith *et al.* (1966)
P. syringae pv. *pisi*	Pea	Smith *et al.* (1966)
P. syringae pv. *syringae*	Bean, citrus fruits (particularly lemon)	Smith *et al.* (1966) Smoot *et al.* (1971)
P. syringae pv. *tomato*	Tomato	McColloch *et al.* (1968)
P. tolaasii	Mushroom	Wong and Preece (1979)
"*Streptomyces scabies*"	Potato, beet	Smith and Wilson (1978) Ramsey *et al.* (1959)
Xanthomonas campestris pv. *campestris*	Cabbage, cauliflower	Ramsey and Smith (1961)
X. campestris pv. *phaseoli*	Bean	Smith *et al.* (1966)
X. campestris pv. *pruni*	Peach, nectarine, apricot, plum	Harvey *et al.* (1972)
X. campestris pv. *vesicatoria*	Tomato, pepper, radish	McColloch *et al.* (1968)

[a] The names of pathovars of *Corynebacterium* sp., *P. syringae* and *X. campestris* are those used in the ISPP List 1980 (Dye *et al.*, 1980).

(1) the extent of damage at harvesting
(2) the maturity of the plant
(3) the turgidity of the plant tissue
(4) the structure of the cell wall and its vulnerability to bacterial enzymes
(5) the concentration in the plant of bacterial nutrients and of anti-bacterial compounds
(6) the ability of the plant tissue to form barriers to bacterial infection.

B. Extrinsic Factors

The marked effect of environmental conditions on bacterial spoilage of produce is due in part to an effect on the bacteria and in part to an effect on the interaction between the bacteria and the plant tissue.

1. Temperature

Although there is some information regarding the temperature limits for growth of spoilage bacteria (Table 9.1), there is little information on the effect of temperature on rates of growth. It is noticeable that the bacteria that only appear to be associated with spoilage at relatively high temperatures, *Echr.*, *P. cichorii*, *P. cepacia*, *P. gladioli* pv. *alliicola*, *B. polymyxa* and *B. subtilis,* tend to grow optimally at temperatures higher than 30°C.

Many vegetables can be stored at 0–1°C in order to retard senescence and spoilage; in the case of produce that cannot be stored at such a low temperature, e.g. cucumbers, peppers, tomatoes and potatoes, the risk of spoilage is greater. At low temperatures *P. marginalis* may be a more important cause of spoilage than *E. carotovora*. Decay of witloof chicory transported in refrigerated conditions from Belgium to New York was due to *P. marginalis* rather than to *E. carotovora* (Friedman, 1951b). After inoculation into chicory leaves *E. carotovora* caused more extensive rotting than did *P. marginalis* at 24°C, whereas at temperatures of <2–21°C *P. marginalis* caused the greater damage. The minimum growth temperatures for *Ecc* and *Eca* have been reported as approximately 6 and 3°C, respectively (Burkholder and Smith, 1949), but there appears to be little information on the rate of growth of these bacteria at low temperatures. Strains of *P. marginalis* isolated from rots of celery and cabbage stored at about 1°C showed doubling times at 0.2°C of 15–20 h (Brocklehurst and Lund, 1981), whereas a strain of *Eca* showed a doubling time of 15.4 h at 5.7°C (Lund, 1982a).

2. *Relative Humidity and Free Moisture*

Although a relative humidity (r.h.) of about 95% has been recommended for storage of most vegetables, storage at 98–100% r.h. results in decreased loss of water and wilting, and may not incur an increase in microbial decay provided that the temperature is kept below 4°C (Grierson and Wardowski, 1978; van den Berg and Lentz, 1978). At high r.h. values, small fluctuations in temperature (<1°C) can result in condensation and formation of water films on the surface of stored vegetables, leading to an increased risk of bacterial soft rot. An r.h. of 94–95% is probably low enough to prevent the growth of spoilage bacteria on fruit and vegetable surfaces. Multiplication of spoilage bacteria in such conditions will depend on the extent to which a high humidity occurs locally as a result of:

(1) the humidity of the environment
(2) the local rate of air movement
(3) the presence of cell sap from damaged tissue
(4) the density of packing of produce
(5) the type of packing.

The use of permeable PVC films for wrapping produce such as lettuce and cauliflower during transit and sale can maintain quality by preventing loss of water, but incurs the risks of increased microbial spoilage (Lund, 1981) and perforation of these packs is advisable unless they are maintained at <4°C.

In addition to disseminating spoilage bacteria, washing processes that involve immersion in water have been shown to cause an increase in infection of potatoes through lenticels and wounds (Dewey and Barger, 1948). Increases in the depth and time of immersion and the use of water at a lower temperature than the produce resulted in an increased decay of tomatoes (Segall *et al.*, 1977) and potatoes (A. Kelman and E. A. Maher, personal communication). Immersion of tomatoes in a suspension of *Ecc* cooler than the fruit resulted in uptake of water and infiltration of the bacterium, probably through the stem scar, and an increase in the incidence of bacterial decay of the tomatoes during storage (Bartz and Showalter, 1981). The presence of a film of water on vegetables in general favours development of soft rot, and on potatoes limits the diffusion of oxygen into tubers resulting in susceptibility to soft rot (Lund, 1979).

3. *Gaseous Atmosphere*

The controlled or modified atmospheres used in storage of fruit and vegetables usually contain between 2 and 10% v/v oxygen and carbon dioxide, the precise conditions depending on the commodity (Ryall and Lipton,

1972; Eckert, 1978; Brecht, 1980). The beneficial effect of these atmospheres is probably due mainly to a lowering of the rate of respiration of the produce, rather than to an effect on soft-rot bacteria. Reduction of the oxygen concentration in the atmosphere to less than 0.8% (v/v) was necessary to decrease the rate of growth of a psychrophilic *Pseudomonas* sp. at 5°C, but no growth occurred in the absence of oxygen (Shaw and Nicol, 1959). In the presence of 10% (v/v) carbon dioxide the growth rate was halved; this effect of carbon dioxide was independent of oxygen concentration over the range 1.2-18.7% (v/v). The concentration of oxygen in the atmosphere could be lowered to 2% (v/v) without inhibiting the overall growth rate of a pseudomonas at 22°C (Clark and Burki, 1972); reduction to less than 0.5% (v/v) was necessary in order to increase the generation time during the log phase. In the absence of oxygen, *P. marginalis* is inhibited, except in the case of denitrifying strains in the presence of nitrate, whereas *Ecc* is capable of growing at a rate that is half of that in air (B. M. Lund and K. M. K. Lau, unpublished). According to Wells (1974), carbon dioxide concentrations greater than 10% (v/v) were necessary to cause any significant inhibition of *Eca*, *Ecc* or *P. fluorescens*.

IX. Control of Bacterial Spoilage

It is clear that the following measures can be used to minimize bacterial spoilage of vegetables and fruits.

(1) Reduction in dissemination of spoilage bacteria in the field by agronomic measures, such as the use of disease-free seed and crop rotation (Pérombelon and Kelman, 1980).
(2) Reduction of the numbers of spoilage bacteria in the post-harvest environment by preventing the accumulation of plant debris and by addition of chlorine or chlorine dioxide to water used for washing, hydrocooling and fluming (Eckert, 1977).
(3) Harvest of the crop at the optimum stage of maturity, with the minimum mechanical damage (Eckert, 1978).
(4) Rapid cooling of the produce, e.g. by vacuum cooling or use of an ice-bank cooling system and storage or transport in controlled, optimum conditions of temperature, relative humidity and gaseous atmosphere in order to maintain the quality of the crop and inhibit multiplication of spoilage bacteria (Ryall and Lipton, 1972).

Attempts to control bacterial spoilage of vegetables by the use of chemicals have met with only limited success and the chemicals at present available appear not to be sufficiently effective, inexpensive and safe for

post-harvest application as anti-bacterials. Any possible benefits of applying a chemical in solution may be outweighed by the increased risk of
rotting that results from water remaining on the produce.

Chlorine compounds added to water as chlorine gas or as hypochlorite salt
can be used to rapidly kill bacteria in process-water and prevent dissemination of spoilage bacteria. When chlorine is added to water a proportion
will be reduced by impurities present in the water. The total residual available chlorine will exist in two major forms:(a) free available chlorine and (b)
combined available chlorine. The free available chlorine consists of:

(1) elemental chlorine (Cl_2)
(2) hypochlorous acid (HOCl)
(3) hypochlorite ion (OCl^-).

The combined available chlorine consists of chlorine combined with
ammonia and nitrogenous compounds to form chloramines and N-chloro-
compounds (Dychdala, 1977). The most actively bactericidal of these forms
of chlorine is undissociated hypochlorous acid, $pK_a = 7.46$, and the proportion of the free, available chlorine present in this form depends on the pH
of the solution (Dychdala, 1977; Eckert, 1977). Any report of the anti-bacterial effect of chlorine solutions should, therefore, specify the concentrations of free, available chlorine and of combined, available chlorine, both
of which can be measured by the use of diethyl-p-phenylene diamine (Palin,
1967), and the pH. The total residual available chlorine can be measured by
titration with iodide (Eckert, 1977).

Concentrations of less than $10 \, mg \, litre^{-1}$ free, available chlorine in water
at neutral pH are sufficient to kill vegetative bacteria within a few minutes.
In commercial practice with water containing a large amount of soil and
organic matter, concentrations of $50–100 \, mg \, litre^{-1}$ total available chlorine
are frequently used to control the numbers of bacteria. Despite the lethal
action of available chlorine against bacteria suspended in water, chlorine
treatments are rarely effective in preventing spoilage by bacteria closely
associated with plant tissue (Eckert, 1977). This is probably because hypochlorous acid is reduced by constituents of the host tissue before it in-
activates the spoilage bacteria. Treatment of wash-water with sodium
hypochlorite to give $5 \, mg \, litre^{-1}$ total available chlorine reduced the
number of bacteria present in the water and on spinach leaves, but treatment with up to $100 \, mg \, litre^{-1}$ total available chlorine failed to reduce
decay of the packaged vegetable (Friedman, 1951a). The addition of
sodium hypochlorite ($300 \, mg \, litre^{-1}$ chlorine, pH 8.8) to hot water used to
treat peppers improved the control of soft rot compared with the effect of
hot water alone, but the treatment was completely nullified by subsequent
hydrocooling in water with or without chlorine (Johnson, 1966). Addition

of chlorine, 40–60 mg litre^{-1}, to wash-water sprayed onto radishes, or addition of sodium hypochlorite, giving 10–250 mg litre^{-1} chlorine, to water used for hydrocooling radishes reduced the incidence of black spot caused by "*Xanthomonas vesicatoria*" (Segall and Smoot, 1962). Addition of sodium hypochlorite to maintain 50 mg litre^{-1} chlorine in wash-water reduced the incidence of bacterial soft rot and bacterial necrosis in field-washed tomatoes (Segall, 1968). The incidence of bacterial soft rot on carrots immersed in packing-house waters containing high numbers of *E. carotovora* and total bacteria was higher than that on carrots immersed in water in which the numbers of bacteria were reduced by maintenance of 2–25 mg litre^{-1} chlorine (Segall and Dow, 1973).

Immersion of potatoes in water containing 0.5 g litre^{-1} chlorine in the presence of a detergent reduced lenticel infection of potatoes (Wilson and Johnston, 1967) and treatment with solutions containing 0.6–2.0 g litre^{-1} free available chlorine (pH 9.4–9.7) reduced the incidence of bacterial soft rot in undamaged tubers incubated at 20°C for 5 days (Lund and Wyatt, 1979). Scholey *et al.* (1968) reported that although washing potatoes in water containing sodium hypochlorite (2 g litre^{-1} available chlorine) reduced the incidence of soft rot in tubers stored for 2 days it resulted in increased rotting of tubers stored for 16 days, presumably due to a toxic effect on the surface tissue of the tuber.

Treatment with chlorine dioxide in a stabilized form, equivalent to ClO_2, 2 g litre^{-1}, reduced the incidence of bacterial soft rot in potatoes stored for 5 days at 20°C but was less effective when the treated tubers were stored for 10 days (Wyatt and Lund, 1981). Several other chemicals also reduced the incidence of bacterial soft rot but only partially controlled the spoilage.

The successful use of antibiotics in the treatment of bacterial disease in humans and animals suggested that they might also be effective in the control of bacterial disease in plants and of post-harvest spoilage. The few antibiotics that appear to be potentially useful for this purpose include medically important antibiotics, the use of which for non-medical purposes is to be strongly discouraged. Treatment with streptomycin, oxytetracycline, polymyxin or neomycin reduced spoilage of spinach due to bacterial soft rot (Smith, 1955; Carroll *et al.*, 1957). A combination of oxytetracycline and streptomycin controlled soft rot of lettuce (Cox, 1955) and oxytetracycline controlled radish pit disease (Cox *et al.*, 1957). When a range of vegetables including cucumber, chicory, escarole and iceberg lettuce was treated by a brief immersion in a solution containing 50 parts $\times$ 10^{-6} of antibiotic, oxytetracycline gave good control of bacterial spoilage whereas streptomycin and polymyxin were less effective (Carroll *et al.*, 1957). It is reported that streptomycin is used in Japan against bacterial diseases in vegetables (Egli and Sturm, 1980).

Several workers have investigated the use of antibiotics to prevent bacterial soft rot of potatoes. Decay of freshly cut potato slices by *Eca* was prevented by streptomycin but not by aureomycin, whereas the reverse was true in relation to decay by *P. fluorescens* (*P. marginalis*) (Bonde, 1953). Treatment with streptomycin solutions and with Agrimycin, a formulation containing 15% streptomycin activity and 1.5% oxytetracycline activity (Cates and Van Blaricom, 1961), controlled seed-piece decay by *Eca* without reducing field emergence, but an increase in rots caused by *Fusarium* and *Phoma* was observed in treated seed pieces (Bonde and Malcolmson, 1956). Cates and Van Blaricom (1961) reported that Agrimycin reduced soft rot of potato plugs by *E. carotovora* but failed to control rotting of whole tubers inoculated with an extract of rotting tubers. In the conditions used by Wyatt and Lund (1981), treatment with streptomycin sulphate, 0.5 g litre^{-1}, failed to reduce soft rot of potatoes. In the last two reports pectolytic clostridia as well as *Erwinia* spp. probably contributed to rotting. The use of streptomycin to prevent decay of pre-cut seed potatoes was found to have a detrimental effect on the subsequent stand of plants if treated seed was stored before planting (Duncan and Gallegly, 1963). A product containing zineb and chloramphenicol has been proposed as a seed potato dressing and claimed to be especially active against *E. carotovora* (Egli and Sturm, 1980).

The fact that some control of bacterial soft rot has been achieved with antibiotics indicates the possibility of achieving chemical control. Genetically stable and inheritable resistance to antibiotics can, however, develop in plant pathogenic bacteria, as illustrated by the emergence of resistance to streptomycin in *Erwinia amylovora* (Egli and Sturm, 1980). This resistance could be transmitted to other groups of bacteria including human pathogens. For this reason the use of medically important antibiotics to control post-harvest spoilage of vegetables is not permitted in the U.K.

X. Conclusions

There is a need for more information regarding the amount of wastage caused by bacterial spoilage, particularly in regions with warm temperatures and high rainfall. Such information could indicate the extent of the need for improved practices and for anti-bacterials to aid in control. In developing countries, changes in cultivation procedures, for example, increases in irrigation and in mechanical harvesting, may increase the problem of bacterial spoilage. With increased international transport of produce, spoilage problems typical of particular regions are likely to be experienced more widely.

References

Ark, P. A. (1945). Crater rot and blotch of celery, a new aspect of soft rot caused by *Erwinia carotovora*. *Phytopathology* **35**, 140–142.

Ballard, R. W., Palleroni, N. J., Doudoroff, M. and Stanier, R. Y. (1970). Taxonomy of the aerobic Pseudomonads: *Pseudomonas cepacia, P. marginata, P. alliicola* and *P. caryophylii*. *J. gen. Microbiol.* **60**, 199–214.

Bartz, J. A. (1980). Causes of postharvest losses in a Florida tomato shipment. *Plant Dis.* **64**, 934–937.

Bartz, J. A. and Showalter, R. K. (1981). Infiltration of tomatoes by aqueous bacterial suspensions. *Phytopathology* **71**, 515–518.

Barzic, M. R., Samson, R. and Trigalet, A. (1976). Pourriture bactérienne de la tomate cultivée en serre. *Ann. Phytopathol.* **8**, 237–240.

Bazzi, C. (1979). Identification of *Pseudomonas cepacia* on onion bulbs in Italy. *Phytopath. Z.* **95**, 254–258.

Beraha, L. and Kwolek, W. F. (1975). Prevalence and extent of eight market disorders of Western-grown head lettuce during 1973 and 1974 in the Greater Chicago, Illinois Area. *Pl. Dis. Reptr* **59**, 1001–1004.

Beraha, L. and Smith, M. A. (1964). A bacterial necrosis of tomatoes. *Pl. Dis. Reptr* **48**, 558–561.

Berger, R. D. (1967). Marginal leaf blight of lettuce. *Proc. Fla. St. hort. Soc.* **80**, 134–138.

Billing, E. (1970). *Pseudomonas viridiflava* (Burkholder, 1930; Clara 1934). *J. appl. Bact.* **33**, 492–500.

Billing, E. (1982). Entry and establishment of pathogenic bacteria in plant tissues. *In* "Bacteria and Plants" (M. E. Rhodes-Roberts and F. A. Skinner, Eds), 51–70. Society for Applied Bacteriology Symposium Series, No. 10. Academic Press, London and New York.

Bobbin, P. and Geeson, J. D. (1977). Effects of benzimidazole fungicide drenches on cabbage storage disorders. *In* "Experiments and Development in the Eastern Region 1976", 151–154. Agricultural Development and Advisory Service.

Bonde, R. (1953). Preliminary studies on the control of bacterial decay of the potato with antibiotics. *Am. Potato J.* **30**, 143–147.

Bonde, R. and Malcolmson, J. F. (1956). Studies in the treatment of potato seed pieces with antibiotic substances in relation to bacterial and fungous decay. *Pl. Dis. Reptr* **40**, 615–619.

Brecht, P. E. (1980). Use of controlled atmospheres to retard deterioration of produce. *Fd Technol., Champaign* **34**, (3), 45–50.

Brierley, P. (1928). Pathogenicity of *Bacillus mesentericus, B. aroideae, B. carotovorus* and *phytophthorus* to potato tubers. *Phytopathology* **18**, 819–838.

Brocklehurst, T. F. and Lund, B. M. (1981). Properties of pseudomonads causing spoilage of vegetables stored at low temperature. *J. appl. Bact.* **50**, 259–266.

Brocklehurst, T. F. and Lund, B. M. (1982). Isolation and properties of psychrotrophic and psychrophilic, pectolytic strains of *Clostridium*. *J. appl. Bact.* **53**, 355–361.

Brown, N. A. (1918). Some bacterial diseases of lettuce. *J. agric. Res.* **13**, 367–388.

Burkholder, W. H. (1930). The bacterial diseases of the bean. *Mem. Cornell Univ. agric. Exp. St.,* No. 127.

Burkholder, W. H. (1942). The bacterial plant pathogens: *Phytomonas caryophylli* sp.n., *Phytomonas alliicola* sp.n., and *Phytomonas manihotis* (Arthaud-Berthet et Bondar) Viegas. *Phytopathology* **32**, 141–149.

Burkholder, W. H. (1950). Sour skin, a bacterial rot of onion bulbs. *Phytopathology* **40**, 115–117.

Burkholder, W. H. (1954). Three bacteria pathogenic on head lettuce in New York State. *Phytopathology* **44**, 592–596.

Burkholder, W. H. (1960). A bacterial brown rot of parsnips. *Phytopathology* **50**, 280–282.

Burkholder, W. H. and Smith, W. L., Jr (1949). *Erwinia atroseptica* (Van Hall) Jennison and *Erwinia carotovora* (Jones) Holland. *Phytopathology* **39**, 887–897.

Burr, T. J. and Schroth, M. N. (1977). Occurrence of soft-rot *Erwinia* spp. in soil and plant material. *Phytopathology* **67**, 1382–1387.

Burton, C. L. (1971). Bacterial soft rot and black spot disease of Bok choy (Chinese chard). *Pl. Dis. Reptr* **55**, 1037–1039.

Butler, L. D. (1980). *Erwinia carotovora var. carotovora*, a competitive rhizosphere inhabitant of tomatoes and cucumbers. Ph.D. Thesis, University of Arizona.

Cabezas de Herrera, E. and Jurado, O. G. (1975). Chemical and physical changes produced in vegetable tissues by *Pseudomonas viridiflava*. *Phytopath. Z.* **84**, 360–368.

Campos, E., Maher, E. A. and Kelman, A. (1982). Relationship of pectolytic clostridia and *Erwinia carotovora* strains to decay of potato tubers in storage. *Plant Dis.* **66**, 543–546.

Carroll, V. J., Benedict, R. A. and Wrenshall, C. L. (1957). Delaying vegetable spoilage with antibiotics. *Fd Technol., Champaign* **11**, 490–493.

Cates, F. B. and Van Blaricom, L. O. (1961). The effect of several compounds on post-harvest decay of potatoes. *Am. Potato J.* **38**, 175–181.

Ceponis, M. J. (1970). Diseases of California head lettuce on the New York market during the spring and summer months. *Pl. Dis. Reptr* **54**, 964–966.

Ceponis, M. J. and Butterfield, J. E. (1973). "The Nature and Extent of Retail and Consumer and Losses in Apples, Oranges, Lettuce, Peaches, Strawberries and Potatoes Marketed in Greater New York", Market Res. Rep. No. 996. U.S. Dep. Agric. Washington D.C.

Ceponis, M. J. and Butterfield, J. E. (1974a). Causes of cullage of Florida bell peppers in New York wholesale and retail markets. *Pl. Dis. Reptr* **58**, 367–369.

Ceponis, M. J. and Butterfield, J. E. (1974b). Market losses in Florida cucumber and bell peppers in metropolitan New York. *Pl. Dis. Reptr* **58**, 558–560.

Ceponis, M. J. and Butterfield, J. E. (1979). Losses in fresh tomatoes at the retail and consumer levels in the Greater New York area. *J. Am. Soc. hort. Sci.* **104**, 751–754.

Ceponis, M. J. and Friedman, B. A. (1959). Pectolytic enzymes of *Pseudomonas marginalis* and their effects on lettuce. *Phytopathology* **49**, 141–144.

Ceponis, M. J., Kaufman, J. and Butterfield, J. E. (1970). Relative importance of gray mold rot and bacterial soft rot of Western lettuce on the New York market. *Pl. Dis. Reptr* **54**, 263–265.

Chatterjee, A. K. and Starr, M. P. (1977). Donor strains of the soft rot bacterium *Erwinia chrysanthemi* and conjugational transfer of the pectolytic capacity. *J. Bact.* **132**, 862–869.

Chesson, A. and Codner, R. C. (1978). The maceration of vegetable tissue by a strain of *Bacillus subtilis*. *J. appl. Bact.* **44**, 347–364.

Cho, J. J., Schroth, M. N., Kominos, S. D. and Green, S. K. (1975). Ornamental plants as carriers of *Pseudomonas aeruginosa*. *Phytopathology* **65**, 425–431.

Cho, J. J., Rohrbach, K. G. and Hayward, A. C. (1978). An *Erwinia herbicola* strain causing pink disease of pineapple. *Proc. 4th Int. Conf. Phytopath. Bac. 1978*, 433–440.

Chupp, C. and Sherf, A. F. (1960). "Vegetable Diseases and their Control." The Ronald Press, New York.

Clara, F. M. (1930). A new bacterial leaf disease of tobacco in the Philippines. *Phytopathology* **20**, 691–706.

Clara, F. M. (1934). A comparative study of the green-fluorescent bacterial plant pathogens. *Mem. Cornell Univ. agric. Exp. St*, No. 159, 36 pp.

Clark, D. S. and Burki, T. (1972). Oxygen requirements of strains of *Pseudomonas* and *Achromobacter*. *Can. J. Microbiol.* **18**, 321–326.

Clark, M. R. M. and Graham, D. C. (1962). A disease of forced rhubarb caused by *Pseudomonas marginalis*. *Pl. Path.* **31**, 33.

Coghill, D. and Juffs, H. S. (1979). Incidence of psychrotrophic sporeforming bacteria in pasteurized milk and cream products and the effect of temperature on their growth. *Aust. J. Dairy Technol.* **34**, 150–153.

Coleno, A., Le Normand, M. and Hingand, L. (1972). Sur une affection bactérienne de la pomme de chou-fleur. *C.R. hebd. Séanc. Acad. Agric. Fr.* **57**, 650–652.

Collmer, A., Whalen, C. H., Beer, S. V. and Bateman, D. F. (1982). An exo-poly-α-D-galacturonosidase implicated in the regulation of extracellular pectate lyase production in *Erwinia chrysanthemi*. *J. Bact.* **149**, 626–634.

Coplin, D. L. (1980). *Erwinia carotovora* var. *carotovora* on Bell peppers in Ohio. *Plant Dis.* **64**, 191–194.

Cother, E. J., Darbyshire, B. and Brewer, J. (1976). *Pseudomonas aeruginosa*: cause of internal brown rot of onion. *Phytopathology* **66**, 828–834.

Cox, R. S. (1955). A preliminary report on diseases of lettuce in the Everglades and their control. *Pl. Dis. Reptr* **39**, 421–423.

Cox, R. S., Carroll, V. J. and Benedict, R. A. (1957). Studies on the etiology and control of the radish pit disease. *Phytopathology* **47**, 7 (Abstract).

Cuppels, D. A. and Kelman, A. (1974). Evaluation of selective media for isolation of soft-rot bacteria from soil and plant tissue. *Phytopathology* **64**, 468–475.

Cuppels, D. A. and Kelman, A. (1980). Isolation of pectolytic fluorescent pseudomonads from soil and potatoes. *Phytopathology* **70**, 1110–1115.

De Boer, S. H., Allen, E. and Kelman, A. (1979). Survival of *Erwinia carotovora* in Wisconsin soils. *Am. Potato J.* **56**, 243–252.

de Mendonça, M. and Stanghellini, M. E. (1979). Endemic and soilborne nature of *Erwinia carotovora* var. *atroseptica*, a pathogen of mature sugar beets. *Phytopathology* **69**, 1096–1099.

Dewey, D. H. and Barger, W. R. (1948). The occurrence of bacterial soft rot on potatoes resulting from washing in deep vats. *Proc. Am. Soc. hort. Sci.* **52**, 325–330.

Dickey, R. S. (1979). *Erwinia chrysanthemi*: a comparative study of phenotypic properties of strains from several hosts and other *Erwinia* species. *Phytopathology* **69**, 324–329.

Dickey, R. S. (1981). *Erwinia chrysanthemi*; reactions of eight plant species to strains from several hosts and to strains of other *Erwinia* species. *Phytopathology* **71**, 23–29.

Doudoroff, M. and Palleroni, N. J. (1974). *Pseudomonas*. *In* "Bergey's Manual of Determinative Bacteriology" (R.E. Buchanan and N. E. Gibbons, Eds), 8th edn, 217–243. Williams and Wilkins, Baltimore.

Dowson, W. J. (1941). Soft-rots due to green fluorescent bacteria. *Trans. Br. mycol. Soc.* **25**, 215–216.

Dowson, W. J. (1943). Spore-forming bacteria in potatoes. *Nature, Lond.* **152**, 331.

Duncan, H. E. and Gallegly, M. E. (1963). Field trials for chemical control of seedpiece decay and blackleg of potato. *Am. Potato J.* **40**, 279–284.

Dychdala, G. R. (1977). Chlorine and chlorine compounds. *In* 'Disinfection, Sterilization and Preservation" (S. S. Block, Ed), 2nd edn, 167–195. Lea and Febiger, Philadelphia.

Dye, D. W. (1969). A taxonomic study of the genus *Erwinia* II. The *carotovora* group. *N.Z. Jl Sci.* **12**, 81–97.

Dye, D. W., Bradbury, J. F., Goto, M., Hayward, A. C., Lelliott, R. A. and Schroth, M. N. (1980). International standards for naming pathovars of phytopathogenic bacteria and a list of pathovar names and pathotype strains. *Rev. Pl. Path.* **59**, 153–168.

Eckert, J. W. (1977). Control of postharvest diseases. *In* "Antifungal Compounds" (M. R. Siegel and H. D. Sisler, Eds), Vol. 1, 269–352. Marcel Dekker, New York.

Eckert, J. W. (1978). Pathological diseases of fresh fruits and vegetables. *In* "Postharvest Biology and Biotechnology" (H. O. Hultin and M. Milner, Eds). Food and Nutrition Press, Westport, Connecticut.

Egli, T. and Sturm, E. (1980). Bacterial plant diseases and their control. *In* "Chemie der Pflanzenschutz und Schädlingsbekämpfungsmittel" (R. Wegler, Ed), Band 6, 345–388. Springer-Verlag, Berlin, Heidelberg and New York.

Elrod, R. P. and Braun, A. C. (1942). *Pseudomonas aeruginosa*; its role as a plant pathogen. *J. Bact.* **44**, 633–645.

Folsom, D. and Friedman, B. A. (1959). *Pseudomonas fluorescens* in relation to certain diseases of potato tubers in Maine. *Am. Potato J.* **36**, 90–97.

Friedman, B. A. (1951a). Control of decay in prepackaged spinach. *Phytopathology* **41**, 709–713.

Friedman, B. A. (1951b). *Pseudomonas marginalis* as the cause of soft rot of imported witloof chicory. *Phytopathology* **41**, 880–888.

Friedman, B. A. (1960). Status of synonymy and plant pathogenicity of *Pseudomonas marginalis* and *P. aeruginosa*. *Int. Bull. bact. Nomencl. Taxon.* **10**, 197–204.

Garibaldi, A. and Tamietti, G. (1977). Eziologia ed alcuni aspetti epidemiologici di marciumi batterici del Porro in Liguria. *Phytopathol. mediterr.* **16**, 27–29.

Garrard, E. H. (1945). A storage rot of potatoes caused by a fluorescent organism resembling *Pseudomonas fluorescens* (Flügge) Migula. *Can. J. Res.* **23**, 79–84.

Geeson, J. D. (1979). The fungal and bacterial flora of stored white cabbage. *J. appl. Bact.* **46**, 189–193.

Gibson, T. and Gordon, R. E. (1974). *Bacillus.* *In* "Bergey's Manual of Determinative Bacteriology" (R. E. Buchanan and N. E. Gibbons, Eds), 8th edn, 529–550. Williams and Wilkins, Baltimore.

Gonzalez, C. F. and Vidaver, A. K. (1979). Bacteriocin, plasmid and pectolytic diversity in *Pseudomonas cepacia* of clinical and plant origin. *J. gen Microbiol.* **110**, 161–170.

Graham, D. C. (1958). Occurrence of soft rot bacteria in Scottish soils. *Nature, Lond.* **181**, 61.

Graham, D. C. (1964). Taxonomy of the soft rot coliform bacteria. *A. Rev. Phytopath.* **2**, 13–42.

Graham, D. C. (1972). Identification of soft rot coliform bacteria. *In* "Proceedings of the Third International Conference on Plant Pathogenic Bacteria" (H. P. Maas

Geesteranus, Ed.), 273–279. Centre for Agricultural Publishing and Document-ation, Wageningen.

Green, S. K., Schroth, M. N., Cho., J. J., Kominos, S. D. and Vitanza-Jack, V. B. (1974). Agricultural plants and soil as a reservoir for *Pseudomonas aeruginosa*. *Appl. Microbiol.* **28**, 987–991.

Grierson, W. and Wardowski, W. F. (1978). Relative humidity effects on the post-harvest life of fruits and vegetables. *Hortscience* **13**, 570–574.

Grogan, R. G., Misaghi, I. J., Kimble, K. A., Greathead, A. S., Ririe, D. and Bardin, R. (1977). Varnish spot, destructive disease of lettuce in California caused by *Pseudomonas cichorii*. *Phytopathology* **67**, 957–960.

Hall, C. B., Stall, R. E. and Burdine, H. W. (1971). Association of *Pseudomonas marginalis* with pink rib of lettuce. *Proc. Fla. St. hort. Soc.* **84**, 163–165.

Hankin, L., Zucker, M. and Sands, D. C. (1971). Improved solid medium for the detection and enumeration of pectolytic bacteria. *Appl. Microbiol.* **22**, 205–209.

Harrison, F. C. and Barlow, B. (1904). Some bacterial diseases of plants prevalent in Ontario. *Bull. agric. Coll. Exp. Farm.* **136**, 1–20.

Harvey, J. M. (1978). Reduction of losses in fresh market fruits and vegetables. *A. Rev. Phytopath.* **16**, 321–341.

Harvey, J. M., Smith, W. L., Jr. and Kaufmann, J. (1972). "Market Diseases of Stone Fruits: Cherries, Peaches, Nectarines, Apricots and Plums", Agriculture Handbook No. 414. United States Department of Agriculture, Washington, D.C.

Hildebrand, D. C. (1971). Pectate and pectin gels for differentiation of *Pseudomonas* sp. and other bacterial plant pathogens. *Phytopathology* **61**, 1430–1436.

Hildebrand, D. C., Palleroni, N. J. and Doudoroff, M. (1973). Synonymy of *Pseudomonas gladioli* Severini 1913 and *Pseudomonas marginata* (McCulloch 1921) Stapp 1928. *Int. J. Syst. Bacteriol.* **23**, 433–437.

Hopkins, D. L. and Elmstrom, G. W. (1977). Etiology of watermelon rind necrosis. *Phytopathology* **67**, 961–964.

Huether, J. P. and McIntyre, G. A. (1969). Pectic enzyme production by two strains of *Pseudomonas fluorescens* associated with the pink eye disease of potato tubers. *Am. Potato J.* **46**, 414–423.

Jackson, A. W. and Henry, A. W. (1946). Occurrence of *Bacillus polymyxa* (Praz). Mig. in Alberta soils with special reference to its pathogenicity in potato tubers. *Can.J. Res., Sect. C.* **24**, 39–46.

Jayasanker, N. P. and Graham, P. H. (1970). An agar plate method for screening and enumerating pectinolytic microorganisms. *Can. J. Microbiol.* **16**, 1023.

Johnson, H. B. (1966). "Bacterial Soft Rot in Bell Peppers", Marketing Research Report No. 738. United States Department of Agriculture, Washington, D.C.

Johnson, J. (1937). Relation of water-soaked tissues to infection by *Bacterium angulatum* and *Bact. tabacum* and other organisms. *J. agric. Res.* **55**, 599–618.

Katznelson, J. (1946). The rhizosphere effect of mangels on certain groups of soil microorganisms. *Soil Sci.* **62**, 343–354.

Kawamoto, S. O. and Lorbeer, J. W. (1974). Infection of onion leaves by *Pseudomonas cepacia*. *Phytopathology* **64**, 1440–1445.

Keller, H. and Knösel, D. (1980). Untersuchungen über die bakterielle Strunkfäule des Lagerkohls. *NachrBl. dt. PflSchutz., Stuttg.* **32**, 161–163.

Kennedy, B. W. and Alcorn, S. M. (1980). Estimates of U.S. crop losses due to procaryote plant pathogens. *Plant Dis.* **64**, 674–676.

Kikumoto, T. (1974). Ecological studies of the soft rot bacteria of vegetables. (13) The role of Chinese cabbage culture in seasonal trends in the populations of the soft-rot bacteria in soil. *Bull. Inst. Agric. Res. Tohoku Univ.* **25**, 125–137 (in Japanese).

Kikumoto, T. and Ômatsuzawa, T. (1981). Ecological studies on the soft rot bacteria of vegetables (15) Latent infection of Chinese cabbage by the soft rot bacteria in the field. *Sci. Rep. Res. Insts Tohoku Univ., Ser. D, Agric.* **32**, 27–35.

Kikumoto, T. and Sakamoto, M. (1967). Ecological studies on the soft rot bacteria of vegetables. III Application of immunofluorescent staining for the detection and counting of *Erwinia aroideae* in soil. *Ann. phytopath. Soc. Japan* **33**, 181–186 (in Japanese).

Kikumoto, T. and Sakamoto, M. (1969a). Ecological studies on the soft-rot bacteria of vegetables VI. Influence of the development of various plants on the survival of *Erwinia aroideae* added to soil. *Ann. phytopath. Soc. Japan* **35**, 29–35 (in Japanese).

Kikumoto, T. and Sakamoto, M. (1969b). Ecological studies on the soft-rot bacteria of vegetables. VII. The preferential stimulation of the soft-rot bacteria in the rhizosphere of crop plants and weeds. *Ann. phytopath. Soc. Japan* **35**, 36–40 (in Japanese).

Köhn, S. (1973). *Pseudomonas marginalis* (Brown) Stevens als Erreger einer Bakteriose an Kopfsalat in Deutschland. *Phytopath. Z.* **78**, 187–191.

Kotte, W. (1930). Eine bakterielle Blattfaule der Winter-Endive (*Cichorium endivia* L.). *Phytopath. Z.* **1**, 605–613.

Kurowsky, W. M. and Dunleavy, J. A. (1976). Cellular and environmental factors affecting the synthesis of polygalacturonate lyase by *Bacillus subtilis. Eur. J. appl. Microbiol. 32,* 103–113.

Lelliott, R. A. (1974). *Erwinia. In* "Bergey's Manual of Determinative Bacteriology" (R. E. Buchanan and N. E. Gibbons, Eds) 8th edn, 332–338. Williams and Wilkins, Baltimore.

Lelliott, R. A., Billing, E. and Hayward, A. C. (1966). A determinative scheme for the fluorescent, plant pathogenic pseudomonads. *J. appl. Bact.* **29**, 470–489.

Lim, W. H. and Lowings, P. H. (1979). Pineapple fruit collapse in peninsular Malaysia: symptoms and varietal susceptibility. *Pl. Dis. Reptr* **63**, 170–174.

Logan, C. (1963). A selective medium for the isolation of soft rot coliforms from soil. *Nature, Lond.* **199**, 623.

Logan, C. (1964). Bacterial hard rot of potato. *Eur. Potato J.* **7**, 45–56.

Lowbury, E. J. L. (1975). Ecological importance of *Pseudomonas aeruginosa*: medical aspects. *In* "Genetics and Biochemistry of *Pseudomonas*" (P. H. Clarke and M. H. Richmond, Eds), 37–65. John Wiley, London, New York, Sydney and Toronto.

Lowings, P. H. (1977). Losses in international transportation. *Proc. 1977 Br. Crop Prot. Conf., Pests Dis.* **3**, 747–753.

Lund, B. M. (1972). Isolation of pectolytic clostridia from potatoes. *J. appl. Bact.* **35**, 609–614.

Lund, B. M. (1979). Bacterial soft-rot of potatoes. *In* "Plant Pathogens" (D. W. Lovelock, Ed), Society for Applied Bacteriology Technical Series No. 12, 19–49. Academic Press, London, New York and San Francisco.

Lund, B. M. (1981). The effect of bacteria on post-harvest quality of vegetables. *In* "Quality in Stored and Processed Vegetables and Fruit" (P. W. Goodenough and R. K. Atkin, Eds), 7th Long Ashton Symposium, 287–300. Academic Press, London and New York.

Lund, B. M. (1982a). The effect of bacteria on post-harvest quality of vegetables and fruits, with particular reference to spoilage. *In* "Bacteria and Plants" (M. E.

Rhodes-Roberts and F. A. Skinner, Eds), 133–153, Society for Applied Bacteriology Symposium Series No. 10. Academic Press, London and New York.

Lund, B. M. (1982b). Clostridia and plant disease: new pathogens? *In* "Phytopathogenic Prokaryotes" (M. S. Mount and G. H. Lacy, Eds), Vol. 1, 263–283. Academic Press, New York.

Lund, B. M. and Brocklehurst, T. F. (1978). Pectic enzymes of pigmented strains of *Clostridium. J. gen. Microbiol.* **104,** 59–66.

Lund, B. M. and Mapson, L. W. (1970). Stimulation by *Erwinia carotovora* of the synthesis of ethylene in cauliflower tissue. *Biochem. J.* **119,** 251–263.

Lund, B. M. and Wyatt, G. M. (1979). A method of testing the effect of antibacterial compounds on bacterial soft rot of potatoes, and results for preparations of dichlorophen and sodium hypochlorite. *Potato Res.* **22,** 191–202.

Lund, B. M., Brocklehurst, T. F. and Wyatt, G. M. (1981). *Clostridium puniceum* sp. nov., a pink-pigmented, pectolytic bacterium. *J. gen. Microbiol.* **122,** 17–26.

McColloch, L. P., Cook, H. T. and Wright, W. R. (1968). "Market Diseases of Tomatoes, Peppers and Eggplants", Agriculture Handbook No. 28. United States Department of Agriculture, Washington, D.C.

McDonald, R. E. and de Wildt, P. P. Q. (1980). Cause and extent of cullage of Florida Bell Peppers in the Rotterdam Terminal Market. *Plant Dis.* **64,** 771–772.

Madhok, M. R. and Fazal-Ud-Din. (1943). Bacterial soft rot of tomatoes caused by a spore forming organism. *Indian J. agric. Sci.* **13,** 129–133.

Meneley, J. C. and Stanghellini, M. E. (1972). Occurrence and significance of soft-rotting bacteria in healthy vegetables. *Phytopathology* **62,** 779 (Abstract).

Meneley, J. C. and Stanghellini, M. E. (1975). Establishment of an inactive population of *Erwinia carotovora* in healthy cucumber fruit. *Phytopathology* **65,** 670–673.

Meneley, J. C. and Stanghellini, M. E. (1976). Isolation of soft-rot *Erwinia* spp. from agricultural soils using an enrichment technique. *Phytopathology* **66,** 367–370.

Mew, T. W., Ho, W. C. and Chu, L. (1976). Infectivity and survival of soft-rot bacteria in Chinese cabbage. *Phytopathology* **66,** 1325–1327.

Misaghi, I. and Grogan, R. G. (1969). Nutritional and biochemical comparisons of plant-pathogenic and saprophytic fluorescent pseudomonads. *Phytopathology* **59,** 1436–1450.

Nagel, C. W. and Vaughn, R. H. (1962). Comparison of growth and pectolytic enzyme production by *Bacillus polymyxa. J. Bact.* **83,** 1–5.

Nagel, C. W. and Wilson, T. M. (1970). Pectic acid lyases of *Bacillus polymyxa. Appl. Microbiol.* **20,** 374–383.

Nasuno, S. and Starr, M. P. (1966). Pectic enzymes of *Pseudomonas marginalis. Phytopathology* **56,** 1414–1415,

Ogilvie, L. (1969). "Diseases of Vegetables", Ministry of Agriculture, Fisheries and Food Bulletin 123. H.M.S.O., London.

Ohta, K., Morita, H., Mori, K. and Goto, M. (1976). Marginal blight of cucumber caused by a strain of *Pseudomonas marginalis* (Brown) Stevens. *Ann. phytopath. Soc. Japan ,* 197–203.

O'Neill, R. and Logan, C. (1975). A comparison of various selective isolation media for their efficiency in the diagnosis and enumeration of soft rot coliform bacteria. *J. appl. Bact.* **39,** 139–146.

Paine, S. G. and Branfoot, J. M. (1924). Studies in bacteriosis XI. A bacterial disease of lettuce. *Ann. appl. Biol.* **11,** 312–317.

Palin, A. T. (1967). Methods for the determination, in water, of free and combined available chlorine, chlorine dioxide and chlorite, bromine, iodine and ozone using diethyl-*p*-phenylene diamine (DPD). *J. Instn Wat. Engrs* **21**, 537–547.

Palleroni, N. J. and Holmes, B. (1981). *Pseudomonas cepacia* sp. nov., nom. rev. *Int. J. Syst. Bacteriol.* **31**, 479–481.

Paton, A. M. (1959). An improved method for preparing pectate gels. *Nature, Lond.* **183**, 1812–1813.

Pérombelon, M. C. M. (1971a). A semi-selective medium for estimating population densities of pectolytic *Erwinia* spp. in soil and in plant material. *Potato Res.* **14**, 158–160.

Pérombelon, M. C. M. (1971b). A quantal method for determining numbers of *Erwinia carotovora* var. *carotovora* and *E. carotovora* var. *atroseptica* in soils and plant material. *J. appl. Bact.* **34**, 793–799.

Pérombelon, M. C. M. (1974). The role of the seed tuber in the contamination by *Erwinia carotovora* of potato crops in Scotland. *Potato Res.* **17**, 187–199.

Pérombelon, M. C. M. and Kelman, A. (1980). Ecology of the soft rot Erwinias. *A. Rev. Phytopath.* **18**, 361–387.

Pérombelon, M. C. M., Gullings-Handley, J. and Kelman, A. (1979). Population dynamics of *Erwinia carotovora* and pectolytic *Clostridium* spp. in relation to decay of potatoes. *Phytopathology* **69**, 167–173.

Perry, D. A. (1982). Pectolytic *Clostridium* spp. in soils and rhizospheres of carrot and other arable crops in east Scotland. *J. appl. Bact.* **52**, 403–408.

Pieczarka, D. J. and Lorbeer, J. W. (1975). Micro-organisms associated with bottom rot of lettuce grown in organic soil in New York State. *Phytopathology* **65**, 16–21.

Pohronezny, K., Larsen, P. O. and Leben, C. (1978). Observations on cucumber fruit invasion by *Pseudomonas lachrymans*. *Pl. Dis. Reptr.* **62**, 306–309.

Ramsey, G. B. and Smith, M. A. (1961). "Market Diseases of Cabbage, Cauliflower, Turnips, Cucumbers, Melons and Related Crops", Agriculture Handbook No. 184. United States Department of Agriculture, Washington D.C.

Ramsey, G. B., Friedman, B. A. and Smith, M. A. (1959). "Market Diseases of Beets, Chicory, Endive, Escarole, Globe Artichokes, Lettuce, Rhubarb, Spinach and Sweet Potatoes", Agriculture Handbook No. 155. United States Department of Agriculture, Washington, D.C.

Roberts, P. (1973). A soft rot of imported onions caused by *Pseudomonas alliicola* (Burk) Starr & Burkh. *Pl. Path.* **22**, 98.

Rombouts, F. M. (1972). Occurrence and properties of bacterial pectate lyases. Ph.D. Thesis, Agricultural University, Wageningen, The Netherlands.

Rombouts, F. M. and Pilnik, W. (1980). Pectic enzymes. *In* "Microbial Enzymes and Bioconversions" (A. H. Rose, Ed.), "Economic Microbiology", Vol. 5, 227–282. Academic Press, London and New York.

Rombouts, F. M., Spaansen, C. H., Visser, J. and Pilnik, W. (1978). Purification and some characteristics of pectate lyase from *Pseudomonas fluorescens* GK-5. *J. Fd Biochem.* **2**, 1–22.

Ruppel, E. G., Harrison, M. D. and Nielson, A. K. (1975). Occurrence and cause of bacterial vascular necrosis and soft rot of sugarbeet in Washington. *Pl. Dis. Reptr* **59**, 837–840.

Ryall, A. L. and Lipton, W. J. (1972). "Handling, Transportation and Storage of Fruits and Vegetables." AVI Publishing, Westport, Connecticut.

Samish, S. and Etinger-Tulczynska, R. (1963). Distribution of bacteria within the tissue of healthy tomatoes. *Appl. Microbiol.* **11**, 7–10.

Scholey, J., Marshall, C. and Whitbread, R. (1968). A pathological problem associated with pre-packaging of potato tubers. *Pl. Path.* **17**, 135–139.

Segall, R. H. (1967). Bacterial soft rot, bacterial necrosis and *Alternaria* rot of tomatoes as influenced by field washing and post-harvest chilling. *Pl. Dis. Reptr* **51**, 151–152.

Segall, R. H. (1968). Reducing postharvest decay of tomatoes by adding a chlorine source and the surfactant Santomerse F85 to water in field washers. *Proc. Fla. St. hort. Soc.* **81**, 212–214.

Segall, R. H. and Dow, A. T. (1973). Effects of bacterial contamination and refrigerated storage on bacterial soft rot of potatoes. *Pl. Dis. Reptr.* **57**, 896–899.

Segall, R. H. and Smoot, J. J. (1962). Bacterial black spot of radish. *Phytopathology* **52**, 970–973.

Segall, R. H., Henry, F. E. and Dow, A. T. (1977). Effect of dump-tank water temperature on the incidence of bacterial soft rot of tomatoes. *Proc. Fla. St. hort. Soc.* **90**, 204–205.

Sellwood, J. E., Ewart, J. M. and Buckler, E. (1981). Vascular blackening of chicory caused by a pectolytic isolate of *Pseudomonas fluorescens*. *Pl. Path.* **30**, 179–180.

Shaw, M. K. and Nicol, D. J. (1969). Effect of the gaseous environment on the growth on meat of some food poisoning and food spoilage organisms. *Proc. 15th Eur. Meet. Meat Technol. Univ. Helsinki, Finland,* 226–232.

Skerman, V. B. D., McGowan, V. and Sneath, P. H. A. (1980). Approved lists of bacterial names. *Int. J. Syst. Bacteriol.* **30**, 225–420.

Smith, L. DS., and Hobbs, G. (1974). *Clostridium. In* "Bergey's Manual of Determinative Bacteriology" (R. E. Buchanan and N. E. Gibbons, Eds), 8th edn, 551–572. Williams and Wilkins, Baltimore.

Smith, M. A. and Ramsey, G. B. (1956). Bacterial zonate spot of cabbage. *Phytopathology* **46**, 210–213.

Smith, M. A., McColloch, L. P. and Friedman, B. A. (1966). "Market Diseases of Asparagus, Onions, Beans, Peas, Carrots, Celery and Related Vegetables", Agriculture Handbook No. 303. United States Department of Agriculture, Washington, D.C.

Smith, W. L., Jr (1955). Streptomycin sulphate for the reduction of bacterial soft rot of packaged spinach. *Phytopathology* **45**, 88–90.

Smith, W. L. and Wilson, J. B. (1978). "Market Diseases of Potatoes", Agriculture Handbook No. 479. United States Department of Agriculture, Washington, D.C.

Smoot, J. J., Houck, L. G. and Johnson, H. B. (1971). "Market Diseases of Citrus and Other Subtropical Fruits", Agriculture Handbook No. 398. United States Department of Agriculture, Washington, D.C.

Snell, J. J. S., Hill, L. R., Lapage, S. P. and Curtis, M. A. (1972). Identification of *Pseudomonas cepacia* Burkholder and its synonymy with *Pseudomonas kingii* Jonssen. *Int. J. Syst. Bacteriol.* **22**, 127–138.

Stall, R. E. and Hall, C. B. (1969). Association of bacteria with graywall of tomato. *Phytopathology* **59**, 1650–1653.

Stanghellini, M. E., Sands, D. C., Kronland, W. C. and Mendonca, M. M. (1977). Serological and physiological differentiation among isolates of *Erwinia carotovora* from potato and sugar beet. *Phytopathology* **67**, 1178–1182.

Stewart, D. J. (1962). A selective-diagnostic medium for the isolation of pectinolytic organisms in the Enterobacteriaceae. *Nature, Lond.* **195**, 1023.

Surico, G. and Iacobellis, N. S. (1978a). Ceppi pectolitici di *Pseudomonas fluorescens* Migula biotipo C agenti di un marciume molle del Finocchio (*Foeniculum vulgare* Mill.). *Phytopathol. mediterr.* **17**, 65–68.

Surico, G. and Iacobellis, N. S. (1978b). 'Un marciume batterico del Sedano (*Apium graveolens* L.) causato da *Pseudomonas marginalis* (Brown) Stevens. *Phytopathol. mediterr.* **17**, 69–71.

Swingle, D. B. (1925). Center rot of "French Endive" or wilt of chicory (*Cichorium intybus* L). *Phytopathology* **15**, 730 (Abstract).

Taylor, J. D. (1975). Bacterial rots of stored onion. *Rep. natn. Veg. Res. St. 1974*, 116–117.

Taylor, J. D. and Holden, C. M. (1977). Bacterial rots of stored onions. *Rep. natn. Veg. Res. Stn 1976*, 104–105.

Taylor, J. D., Dudley, C. L. and Littlejohn, I. H. (1980). Bacterial rots of stored onions. *Rep. natn. Veg. Res. Stn 1979*, 76–77.

Thayer, P. L. and Wehlburg, C. (1965). *Pseudomonas cichorii*, the cause of bacterial blight of celery in the Everglades. *Phytopathology* **55**, 554–557.

Thomas, C. E. (1976). Bacterial rind necrosis of cantaloup. *Pl. Dis. Reptr* **60**, 38–40.

Thomson, S. V., Schroth, M. N., Hills, F. J., Whitney, E. D. and Hildebrand, D. C. (1977). Bacterial vascular necrosis and rot of sugar beet. General description and etiology. *Phytopathology* **67**, 1183–1189.

Thomson, S. V., Hildebrand, D. C. and Schroth, M. N. (1981). Identification and nutritional differentiation of the *Erwinia* sugar beet pathogen from members of *Erwinia carotovora* and *Erwinia chrysanthemi*. *Phytopathology* **71**, 1037–1042.

Thompson, A. (1937). Pineapple fruit rots in Malaya. *Malay. agric. J.* **30**, 407–420.

Togashi, J. (1972). Studies on the outbreak of the soft-rot disease of Chinese cabbage by *Erwinia aroideae* (Towns) Holl. *Sci. Rep. Res. Insts Tohoku Univ., Ser. D, Agric.* **23**, 17–52.

Towner, D. B. and Beraha, L. (1976). Core-rot: a bacterial disease of carrots. *Pl. Dis. Reptr* **60**, 357–359.

Ulrich, J. M. (1975). Pectic enzymes of *Pseudomonas cepacia* and penetration of polygalacturonase into cells. *Physiol. Plant Path.* **5**, 37–44.

Valleau, W. D., Diachun, S. and Johnson, E. M. (1939). Injury to tobacco leaves by water soaking. *Phytopathology* **29**, 884–890.

van den Berg, L. and Lentz, C. P. (1978). High humidity storage of vegetables and fruit. *Hortscience* **13**, 565–569.

Volcani, Z. and Dowson, W. J. (1948). A plant disease caused by a spore-forming bacterium under natural conditions. *Nature, Lond.* **161**, 980.

von Schelhorn, M. (1951). Control of microorganisms causing spoilage in fruit and vegetable products. *Adv. Fd Res.* **3**, 429–482.

Voronkevitch, I. V., Matveeva, E. A. and Odintsova, M. A. (1972). The rhizosphere as the habitat of phytopathogenic bacteria. II. Significance of the rhizosphere in the life cycle of causal agents of soft rots. *Biol. Nauki* **15**, 103–109 (in Russian). Cited in *Rev. Pl. Path.* **52**, 897 (1973).

Wehlburg, C. (1963). A bacterial spot of cabbage caused by *Pseudomonas cichorii*. *Proc. Fla. St. hort. Soc.* **76**, 119–122.

Wells, J. M. (1974). Growth of *Erwinia carotovora*, *E. atroseptica* and *Pseudomonas fluorescens* in low oxygen and high carbon dioxide atmospheres. *Phytopathology* **64**, 1012–1015.

Wiles, A. B. and Walker, J. C. (1951). The relation of *Pseudomonas lachrymans* to cucumber fruits and seeds. *Phytopathology* **41**, 1059–1064.

Wilkie, J. P. and Dye, D. W. (1974). *Pseudomonas cichorii* causing tomato and celery diseases in New Zealand. *N.Z. J agric. Res.* **17**, 123–130.

Wilkie, J. P., Dye, D. W. and Watson, D. R. W. (1973). Further hosts of *Pseudomonas viridiflava*. *N.Z. J agric. Res.* **16**, 315–323.

Wilson, J. B. and Johnston, E. F. (1967). Reducing the incidence of bacterial lenticel infection in fall-washed Maine potatoes. *Am. Potato J.* **44,** 342 (Abstract).

Wimalajeewa, D. L. S. (1976). Studies on bacterial soft rot of celery in Victoria. *Aust. J. Exp. Agric. Anim. Husb.* **16,** 915–920.

Wong, W. C. and Preece, T. F. (1979). Identification of *Pseudomonas tolaasi:* the white line in agar and the mushroom tissue block rapid pitting tests. *J. appl. Bact.* **47,** 401–407.

Wyatt, G. M. and Lund, B. M. (1981). The effect of antibacterial products on bacterial soft rot of potatoes. *Potato Res.* **24,** 315–329.

Young, J. M. (1974). Effect of water on bacterial multiplication in plant tissue. *N.Z. Jl agric. Res.* **17,** 115–119.

Index

FOOD SCIENCE AND TECHNOLOGY
A SERIES OF MONOGRAPHS

FOOD SCIENCE AND TECHNOLOGY

A SERIES OF MONOGRAPHS

Malcolm C. Bourne, FOOD TEXTURE AND VISCOSITY: CONCEPT AND MEASUREMENT. 1982.

R. Macrae (ed.), HPLC IN FOOD ANALYSIS. 1982.

Héctor A. Iglesias and Jorge Chirife, HANDBOOK OF FOOD ISOTHERMS: WATER SORPTION PARAMETERS FOR FOOD AND FOOD COMPONENTS. 1982.

John A. Troller, SANITATION IN FOOD PROCESSING. 1983.

Colin Dennis (ed.), POST-HARVEST PATHOLOGY OF FRUITS AND VEGETABLES. 1983.

In preparation

Joe M. Regenstein and Carrie E. Regenstein, FOOD PROTEIN CHEMISTRY. 1984.